Thinking with Objects

Thinking with Objects

The Transformation of Mechanics in the Seventeenth Century

DOMENICO BERTOLONI MELI

The Johns Hopkins University Press

Baltimore

The Johns Hopkins University Press
2715 North Charles Street
Baltimore, Maryland 21218-4363
www.press.jhu.edu

Library of Congress Cataloging-in-Publication Data
Bertoloni Meli, Domenico.
Thinking with objects : the transformation of mechanics in the seventeenth century /
Domenico Bertoloni Meli.
p. cm.
Includes bibliographical references and index.
ISBN 0-8018-8426-8 (hardcover : alk. paper) — ISBN 0-8018-8427-6 (pbk. : alk. paper)
1. Mechanics—History—17th century. 2. Motion—History—17th century. 3. Physics—
History—17th century. I. Title.
QC125.2.M45 2006
531.09'032—dc22   2006002642

A catalog record for this book is available from the British Library.

Illustration credits: 1.1, 1.2, 1.5, by courtesy of the Biblioteca Oliveriana, Pesaro; 3.9, 6.1, 6.2, 8.3, 8.4, by courtesy of the Lilly Library, Indiana University, Bloomington; 4.9, 6.7, 6.8, 6.9, 8.11, by courtesy of the Burndy Library, Dibner Institute.

To Alexander and Sofia

CONTENTS

strange justificate

A few years ago, while I was immersed in my studies on mechanistic medicine, I received a request by Dan Garber for a chapter on motion and mechanics in the seventeenth century. Although I had often toyed with the idea of working on that topic, Dan's request forced me to order and develop my thoughts and served as a catalyst for my project. I am indebted to Dan for his initial stimulus and for his encouragement and support over the ensuing years. I am most grateful to Niccolò Guicciardini, who was kind enough to read an earlier version of the entire manuscript and to offer criticisms and suggestions that saved me from many inaccuracies and omissions.

The bulk of this work was written while I was a fellow at the Dibner Institute, which provided a most congenial environment; the riches of the Burndy Library proved invaluable to my research. I am very grateful to Jürgen Renn for his invitation to spend several months at the Max-Planck-Institut in Berlin. I benefited from discussions at the reading groups on mechanics both at the Dibner Institute and the Max-Planck. I would like to thank all participants and especially Moti Feingold, Andrew Janiak, Al Martinez, George Smith, Jim Voelkel, Jochen Büttner, Peter Damerow, Peter McLaughlin, and Matthias Schemmel.

A number of colleagues and friends over the years offered help and advice on a number of topics. I wish to thank Roger Ariew, Antonio Becchi, Fabio Bevilacqua, Jordi Cat, Raine Daston, Michael Dickson, Michael Friedman, Franco Giudice, Noretta Koertge, Nancy Nersessian, Bill Newman, Mike Mahoney, Mark Schiefsky, Alan Shapiro, Jonathan Sheehan, and Edith Sylla. I also thank for his comments an anonymous reader for Cambridge University Press.

I am most grateful to Indiana University for research leave, sabbatical leave, and an Arts and Humanities Fellowship. A portion of my research leave was funded by grant NSF/SES-0115154, which is gratefully acknowledged. I wish to express my gratitude for their kind help to the staffs of the Burndy Library at the

Dibner Institute; the Rare Books Room at Cambridge University Library; the Houghton Library at Harvard University; the library at the Max-Planck-Institut für Wissenschaftsgeschichte in Berlin; and the Lilly Library and the Interlibrary Loan department at the Wells Library, both at Indiana University.

I claim sole responsibility for all remaining errors and omissions.

Thinking with Objects

Introduction

1. Beyond Inertia: From Laws to Objects

Over the last decades the historiography of motion and mechanics in the seventeenth century has been dominated by the notion of laws of nature, especially the law now known as the law of inertia. In the introduction to *Galileo Studies* (originally published in 1939), Alexandre Koyré credited historians such as Ernst Cassirer, Kurd Lasswitz, Ernst Mach, and Emil Wohlwill with having identified in the concept of inertia the key notion "unknown to the ancients" but characteristic, in his words, of "classical physics" or "classical science." While partly agreeing with their tenet, Koyré argued that it was not sufficient to take notice of the existence of the law of inertia, but rather its emergence had to be explained. In his view two closely linked events were at the basis of this phenomenon: on the one hand, the geometrization of space and the dissolution of the ancient cosmos; on the other, the replacement of the immediately sensible space of pre-Galilean physics with the abstract space of Euclidean geometry. It was this replacement that allowed the formulation of the law of inertia. In his 1943 monograph on Mersenne, Robert Lenoble could claim as a matter of fact, "Modern mechanics was born with the principle of inertia."[1]

More recently, Richard Westfall in *Force in Newton's Physics: The Science of Dynamics in the Seventeenth Century* (1971) chose as his research topic the history of Newton's second law of motion, a law often expressed somewhat anachronistically as $F = ma$. The historiographic link with Koyré is especially clear and strong if one bears in mind that the law of inertia is Newton's first law. By quoting the text of the first three laws on page 1 of his book, Westfall could present it as a direct and natural continuation of a lineage going through Koyré. Westfall introduced his work as a history of conceptual developments steering away from social issues on the one hand and technical mathematical questions on the other. In subsequent years Westfall departed from those historiographic principles, devoting celebrated essays to patronage and the relationships between Galileo and hydraulic engineers, for example.[2] These works, however, did not lead to a new, comprehensive picture of mechanics in the seventeenth century, replacing accounts such as those by René Dugas, *La mécanique au XVIIe siècle* (1954) and István Szabò, *Geschichte der mechanischen Prinzipien* (1977). Although Dugas focused on key figures while Szabò arranged his work thematically, they shared the concern of providing a useful chronicle of who-discovered-what-when, rather than documenting changing methods of investigation, styles of presentation, audiences, and reading practices.[3] Other works that have appeared more recently have raised a number of themes and shed light on many chapters of mechanics and its transformations, but they have not provided an overall picture of the field. Thus Alan Gabbey's conclusion in 1990 that Westfall's *Force in Newton's Physics* is the "best overall modern survey" still holds true.[4]

The line of research centered on laws of nature followed by Koyré, the early Westfall, and others has produced important and lasting results based on a careful intellectual analysis of many primary texts, of key notions, and of terminology. Laws, however, especially one so abstract and idealized as the law of inertia, fail to capture the complexity of the field. It is one thing to know Newton's laws of motion and universal gravity; it is quite another to calculate the motion of the moon, the speed of efflux of water from a hole at the bottom of a vessel, the resistance of a beam, or air resistance to motion. Unbalanced attention to laws fails to convey the scholars' practices and their engagement with the features and peculiarities of the objects of mechanics. Even while working in their studies, scholars did not rely only on paper and ink but often used objects constructed to be tested anywhere. I am reluctant to call them "instruments," since that term conveys the idea of something elaborate, whereas I am dealing largely with simple and mundane tools. Galileo's pendulum, inclined planes, beams, rolling and projected balls, Benedetto Castelli's barrel and taps, Evangelista Torricelli's pierced cisterns, Marin Mersenne's vibrating strings, pendulums, and rolling spheres, Descartes' slings, Marcus Marci's billiard balls, Huygens's cycloidal and colliding pendulums, Hooke's springs, Newton's pendulums and balls of wool, glass, cork, and steel—these are just some notable examples of the engagement of scholars with the material world around them. Close attention not just to abstract laws and principles but also to the material world of objects promises to provide a more comprehensive picture of mechanics. In many instances mathematicians displayed an intimate knowledge of the behavior of bodies, and this knowledge was an integral and important part of their work. Consider these two quotations, which convey an almost tactile sense of that knowledge:

> I use an exquisitely round bronze ball, no larger than a nut; this is rolled on a metal mirror held not vertically but rather tilted, so that the ball in motion runs over it and

presses it lightly. In moving, it leaves a parabolic line, very thin, and smoothly traced . . . To describe parabolas in this way, the ball must be somewhat warmed and moistened by manipulating it in the hand, so that the traces it will leave shall be more apparent on the mirror.

I have tested this as follows with tightly wound balls of wool strongly compressed. First, releasing the pendulums and measuring their reflection, I found the quantity of their elastic force; then from this force I determined what the reflections would be in other cases of their collision, and the experiments which were made agreed with the computations. The balls always rebounded from each other with a relative velocity that was to the relative velocity of their colliding as 5 to 9, more or less. Steel balls rebounded with nearly the same velocity and cork balls with a slightly smaller velocity, while with glass balls the proportion was roughly 15 to 16.

These two passages come from Galileo's *Discorsi* and Newton's *Principia*. From Guidobaldo dal Monte and Galileo to Hooke and Newton we detect a constant attention being paid to such mundane issues as sweaty bronze spheres and the behavior of strongly compressed balls of wool.[5]

Focus on objects may help to remedy a problem that has affected several accounts of the history of mechanics. Many histories have been rather insensitive to mapping its actual historical contents and shifting contours. Entire areas that did not fit the historiographic tradition based on the law of inertia have been ignored. One would be hard-pressed to figure out from that tradition which was the first science in Galileo's *Discourses and Mathematical Demonstrations about Two New Sciences,* or *Discorsi*. Incidentally, it was in the discussion of that first science of the resistance of materials that Galileo talked of parabolic beams and described the way of drawing parabolas quoted above. The treatment of elastic bodies and the science of fluids are further glaring examples of areas left at the margin and sometimes outside certain historical accounts.[6]

Several scholars have focused on objects, either studying their properties and peculiarities or replicating some of the experiments performed in the past.[7] My work aims at extending those investigations and using them to rethink the history of mechanics. Several accounts of the so-called Scientific Revolution focus on instruments and their practitioners. Most of them emphasize the new instruments that emerged in the seventeenth century, such as the telescope, the microscope, the barometer, the thermometer, and the air pump. At the time these were called "philosophical" instruments, in contrast to "mathematical" instruments such as surveying equipment, quadrants, sectors, and other measuring tools. Jim Bennett has argued for a more prominent role in our histories for mathematical instruments and their users—the mathematical practitioners—both for the patronage niches they established for mathematics and especially for the fertile milieu they opened for experimental philosophy. The objects at the center of this essay relate to and extend Bennett's thesis, especially his emphasis on "operative knowledge" or "knowing by doing," notions that can be found already in the seventeenth century in Mersenne and Gassendi. Bennett mentions the construction of a rotating celestial globe before the publication of Copernicus's *De revolutionibus*, for example, but his reasoning can easily be extended to inclined planes, pendulums, springs, and a host of devices used by seventeenth-century scholars.[8]

Despite their apparent simplicity and availability, objects were not always easy to handle, either conceptually or practically. They required training—for example, in dropping balls at the same time or in counting pendular oscillations over several hours—and ingenuity. At times they required intellectual tools developed only much later, but often it was not necessary to have a complete understanding, in the form of a fully developed mathematical theory of how the objects worked, to start dealing with them in a variety of ways. Mathematicians learned to deal with rolling spheres in a comprehensive fashion only in the eighteenth century, for example, but dal Monte and Galileo could learn important lessons from rolling spheres on inclined planes in the late sixteenth century. Mersenne had trouble comparing the motion of a sphere rolling down an inclined plane with one falling vertically, but comparing two spheres rolling down planes of different inclinations proved more manageable.[9] Likewise, pendular oscillations were not rigorously isochronous until Huygens showed that they had to be constrained by cycloidal cheeks. As long as oscillations were small, however, the unconstrained pendulum provided an accurate measure of time, at least on land. Despite problems and anomalies, objects were used at many levels even with a rather primitive theory and understanding of how they worked.

The objects I have selected do not figure among the "philosophical" instruments, although many of those I discuss became more and more relevant to natural and experimental philosophy. They are related to the mathematical disciplines, but they are not strictly speaking "mathematical" instruments in the sense of being measuring tools with practical and utilitarian applications originally used by the mathematical practitioners. For example, dal Monte's machines were all used to raise weights, weigh, or break them, but the same does not hold true for all objects I will be talking about. Some of them stem directly or indirectly from the classical mechanical tradition, such as the lever, the inclined plane, the wedge, floating bodies, the hydrostatic press, and the beam. The string was part of the classical tradition of harmonics. Some objects, such as beams, strings, and springs or other elastic tools, were relatively common in any type of construction, in musical instruments, and in war machines, such as Roman catapults, yet they received a more sophisticated mathematical treatment in the seventeenth century. Most of the objects I describe had practical applications, where practice can be interpreted both in general terms and in the realm of accurate experimentation. For example, the beam was of central importance to engineering whereas the pendulum attracted Galileo's attention as a time-measuring device both in general and for quantitative experimentation, but also for its remarkable mathematical properties.

Most of the seventeenth-century objects I discuss have not survived, possibly because they were too common. If some of them, such as inclined planes for slowing down free fall, appear in a history-of-science museum alongside telescopes and astrolabes, they are recent replicas. Thus museum holdings are not representative of the tools used by seventeenth-century scholars, especially those unadorned objects used for research rather than display. Generally, with the exception of pendulum clocks and spring-regulated watches, the objects I discuss required minimal skill or financial resources to be fabricated or procured: they were not the big science of the time. The Minim Friar Mersenne, not noted for his large purse, was able to perform many experiments in his rooms at the Convent de l'Annonciade on falling bodies, bodies rolling down inclined planes, loaded beams, strings and bows, pendulums, and pierced cisterns.[10] This situation is in noticeable contrast to that of the most prominent "philosophical" instruments, namely the telescope, the microscope, and the air pump, which required skillful technicians and comparatively deep pockets. Boyle employed a number of people constructing and operating air pumps, and from the middle of the century the best telescopes—with the exception of Huygens's—and microscopes were fabricated by specialized technicians, such as Felice Fontana, Eustachio Divini, and Giuseppe Campani. Such differences highlight the existence of a variety of forms of social and intellectual interactions affecting the way experiments were performed.

It would be erroneous to overemphasize the material aspect at the risk of transforming mathematicians into carpenters or joiners. Often the intimate knowledge they developed of the material world around them was aimed at abstracting from the peculiarities and vagaries of matter. Thus we have to consider both the material objects and their abstract counterparts: perfectly smooth inclined planes, perfectly elastic collisions, perfect beams, and so on. The objects I discuss go hand in hand with the abstract geometrical diagrams representing the tangible objects of the world and also with principles and laws directly anchored to their behavior. For example, Galileo conceived inclined planes with progressively smaller and smaller slopes to introduce the notion that horizontal motion requires no force. The pendulum was associated with the principle that a falling body acquires enough power to raise it back to its original height. On a ship, bodies falling in a room under the deck or from the mast behaved as if the ship were standing still, thus instantiating relativity of motion. Far from being a drawback, this polyvalence in the notion of objects allows us to grasp the dual nature of seventeenth-century studies of motion and mechanics. It is the combination of practical and theoretical work that I wish to capture with the phrase "thinking with objects."

A noticeable feature of accounts based on laws and principles is that over the centuries historians have provided widely different pictures based on their own views of mechanics. This would be the topic for an extensive and worthwhile study, but for our purposes here it may suffice to say that at times when minima or conservation principles were in vogue, historians focused on finding traces of those principles in a rather anachronistic fashion.[11] We may feel uncomfortable reading about the roots of the principle of virtual displacements or of conservation of energy stretching to times when the notions of infinitesimal virtual displacements, let alone energy, were ill defined or not present. Moreover, the same mathematical expressions were read and interpreted in different ways at different times, so their occurrence cannot be taken in isolation from those time-dependent interpretations. I believe that an account anchored to objects allows for a study of laws and principles less prone to anachronism.

2. Motion and Mechanics

Writing on mechanics and the mathematical science of motion in the seventeenth century poses problems to the historian. From the end of the sixteenth century to the beginning of the eighteenth, mechanics underwent such remarkable transformations that if a scholar at the beginning of that period, for example one of the leading Italian experts in the field such as dal Monte, could have hypothetically been shown a work written a century later, by Johann Bernoulli for instance, he would have had serious difficulty in grasping its contents. During that period mechanics did not simply grow in complexity with the addition of new and more general theorems; it expanded its domain to areas previously considered to be beyond the conceivable reach of mathematical description.

Around 1600 the study of motion, in the Aristotelian sense of local motion,[12] was at the margins of the science of mechanics, whereas by 1700 it was at the center. In the course of the seventeenth century several domains of knowledge became integral parts of mechanics. These domains include areas traditionally belonging to other disciplines, such as astronomy, optics, harmonics or music, and natural philosophy, as well as new areas of inquiry. Celestial motions were associated with mechanics and machines in connection with planetary orbs, whose models were at times actually built by medieval and Renaissance craftsmen. With the

dissolution of solid celestial spheres, mechanics entered celestial domains in a new
fashion with the analogy between the motion of celestial bodies and that of pendulums and projectiles and then with the notion of central forces. Likewise, the
study of vibrations, previously part of harmonics, and the propagation of light, associated with optics, established links with mechanics. New areas too, such as the
science of the resistance of materials, the motion of waters and more generally
fluids, collision, the motion of bodies in resisting media, and the motion of rigid
bodies, became major components of mechanics. By about 1670, mathematicians
such as John Wallis, Ignace Gaston Pardies, and Claude Dechales listed collision,
the motion of liquids, the motion of waves, and the vibrations of pendulums,
cords, and springs as part of mechanics.[13] Thus practitioners at the end of the sixteenth century and at the beginning of the eighteenth would have given quite
different definitions of "mechanics," and the historian has to take their changing
perspectives into account. The present work aims at identifying the intellectual
and technical factors that made these transformations possible. I shall argue that
not only mathematics and experiment, but also the search for new principles inspired by the Archimedean tradition and the revival by dal Monte of a systematic
foundational project associated with Pappus of Alexandria, were important factors
in this process.

   Mechanics was a mathematical discipline whose academic rise was associated
with a variety of factors, from its growing importance to the art of war in the age
of artillery, especially for the construction of suitable fortifications such as the bastion, to the appearance in print in the sixteenth century of a number of classical
treatises on mechanics. Besides the *Quaestiones mechanicae*, then attributed to
Aristotle and now considered an early product of his school, they included Latin
translations of works by Archimedes and Pappus, as well as Vitruvius's treatise on
architecture. These works exerted an enormous impact on mechanics and its status, in that they provided both a classical pedigree and a considerable body of
knowledge, including refined and sophisticated theorems.[14]

   Mechanics was linked to natural philosophy or *physica*, to mathematics, and to
the practical arts, such as navigation and architecture. Whereas *physica* was a
sprawling discipline providing a causal study of the entire natural world, the
causal nature of mathematical demonstrations was a debated issue. The links
among disciplines were quite complex, and mechanics was said in technical par-
lance to be a subordinate science. According to Aristotle, a subordinate science depends on the superior science for its principles and proofs. The principles of mechanics were derived from geometry. Other examples of subordinate sciences
whose principles were derived from mathematics were harmonics and optics, depending on arithmetic and geometry, respectively. In mechanics, harmonics, and

optics the subordinate science treats knowledge of the fact, whereas the higher science provides demonstrative knowledge; in addition, the subordinate science is perceptual, whereas the higher one abstracts from perception in that it is mathematical. At times Aristotle seemed to refer to triplets of subordinate sciences, such as arithmetic, mathematical harmonics, and acoustical harmonics, or geometry, optics, and the science of the rainbow. One could usefully consider a similar triplet with geometry, theoretical or rational mechanics, and practical mechanics. The terms "rational" and "applied" or "practical" mechanics occur already in antiquity, according to Pappus. Newton mentioned Pappus's distinction and claimed that the former proceeds by rigorous demonstration, whereas the latter is a subject of the manual arts.[15]

These preliminary reflections point to the need to pay close attention to the terminology used by mathematicians in connection with the location of a discipline on the map of knowledge and the profound implications of some key terms. Here I wish to discuss the widespread usage of the terms "kinematics" and "dynamics" as representing some key constituents of mechanics. The former usually means the study of motion simply as a mathematical description without regard to causes or forces. The latter usually means the study of motion taking its causes or forces into account. Despite their Greek roots, the term "kinematics" was coined in the nineteenth century by André Marie Ampère, and "dynamics" was coined by Leibniz in the late seventeenth century with a meaning quite different from what we can find in a modern textbook. Thus the use of these terms in the treatment of motion in the seventeenth century is fraught with danger. We are so used to thinking of kinematics and dynamics as part of mechanics that we may tend to think that the study of motion always belonged to mechanics, which is not the case. Moreover, we are so used to thinking of kinematics and dynamics as mathematical disciplines that we may assume the study of motion always to have been couched in mathematical language, which is also not the case. These difficulties do not exhaust the list of problems but may be sufficient at this stage to alert the reader to the dangers of anachronistic terminology. Although in some specific and limited cases it might be safe to use terms like "kinematics" and "dynamics," I prefer to dispense with the former altogether and use the latter only as it was understood around 1700, namely in a Leibnizian fashion.[16]

During the sixteenth century, mechanics entered the university curriculum and was taught by the mathematics professor. However, motion was not generally part of mechanics and was taught by the philosophy professor relying on a variety of Aristotelian texts, such as *Physica* and *De coelo,* and their commentaries. Although some elementary forms of quantitative reasoning may have entered the discussions, overall the treatment was qualitative and philosophical. Moreover, the status of the discipline changed dramatically over the long century covered by this essay. At the beginning, mechanics, practical but also rational, had quite a low status in the universities and, though to a lesser extent, in society. Treatises devoted to mechanics often started with defenses of its dignity and utility, as if to justify the author's involvement with the "vile mechanical arts." By the eighteenth century, however, mechanics had merged with large portions of natural philosophy to become the archetype of the power of the human mind, and its intellectual status had thus risen dramatically in society as well as in the universities.

3. The Role of Mathematics

Several intellectual and practical themes intersect these developments. Both mechanics and the science of motion, insofar as it was linked to mechanics, were mathematical disciplines. Mathematics itself underwent profound transformations during our period, such as the invention of the method of indivisibles, the rise of algebra and analytic geometry, the discovery of new curves such as the cycloid, and the emergence of infinite series and of calculus. It is impossible to talk about motion and mechanics without referring to the mathematical tools used and at times created to deal with them. Reflections on problems such as those related to centers of gravity, the behavior of bodies at the very beginning of motion, and the definition of instantaneous or mean velocity occurred in conjunction with notions of infinitesimals, the thorny issue of the composition of the continuum, the transition from proportions to equations, quadratures, and integration. Studies on motion and mechanics mirrored and contributed to shaping the mathematical horizon of the time when they were formulated. Therefore my reflection on terminology can be extended to the issue of mathematical concepts, practices, and notation. I will refrain as much as possible from using modern formulations and try instead to follow the conceptual framework of the seventeenth century.

The older tradition of the history of mathematics focused on results rather than methods and concepts, while historians of mechanics have often left mathematics at the margin of their investigations. Over the last few decades, however, historians of mathematics have produced innovative and historically sensitive works that have changed our understanding of the discipline and its methods. Only comparatively recently has the practice of translating seventeenth-century works into modern notation become unacceptable, for example. We are therefore in a much better position than previous historians in having this new and sophisticated literature at our disposal.[17] While keeping my focus on mechanics, I have relied on this literature and shall refer to it below. Despite the growth of algebra, throughout most of the seventeenth century the geometrical diagram was a key tool of investigation. Mathematicians from Galileo to Newton worked and reasoned with the help of geometrical figures. This practice has enabled me to present a layered text where some technical material is discussed in the captions of the figures; scholars interested in that material can study it in somewhat greater detail, while others with different interests can read a text largely unencumbered by mathematics.

It is not surprising that the development of more powerful mathematical tools enabled the growth of mechanics and its extension to new domains. For example, it seems reasonable that the invention of calculus allowed the extension of mechanics to areas undreamed of in the sixteenth century. In several instances, however, mechanics developed using elementary or Greek mathematics, and this calls for a reflection on other themes. If new areas of knowledge came under the compass of mathematical description relying largely or entirely on traditional mathematical techniques, one has to look for an explanation different from that of more powerful mathematical tools. Intellectual, social, and technological factors interacted in a complex fashion and affected in different ways the emergence of a mathematical science of motion, of resistance of materials, and of waters.

A related concern touches on the conceptual and practical ways to produce a plausible account of natural phenomena. This means both dealing with the problem of which phenomena, and to what extent, can conceivably be described in a mathematical fashion, and the practical problem of producing such descriptions. It is one thing to state a priori, following a tradition loosely derived from Plato, that the motion of a cannonball or the flow of water in a river can be described mathematically, but it is quite another actually to produce a meaningful quantitative description of such complex phenomena. One issue that was often debated by seventeenth-century scholars was that of mathematical versus physical rigor. Descriptions of the physical world cannot be generally expected to provide absolute precision, but what is an acceptable level of precision, or of error? To be more specific, can we assume that gravity acts along lines parallel among themselves on the weights attached to a balance, or can we neglect air resistance in the motion of a projectile? In order to find out how and by how much the various factors affect the results, it is necessary to carry out the calculations. But the reason some factors were ignored in the first place was precisely that they made calculations at best unwieldy and often impossible to solve.

4. Experience and Experiment

These themes lead to reflections on the role of experience and experiment. Debates on these issues have been common lore among Galileo scholars, but the entire community of practitioners of mechanics dealt with them in one form or another. Given the focus of this work on mathematical disciplines, I deal mainly with quantitative experiments rather than experimental philosophy in general.

The term "experience" usually characterizes observations of the normal course of nature, whereas "experiment" carries the meaning either of a controlled observation or of an intervention violating to some degree nature's normal course.[18] Experiments, such as dropping weights from a high tower or firing a cannonball, were often subjected to quantitative study. The role of experiments varied greatly depending on the area of investigation. In the study of simple machines, for example, such as the balance or the pulley, theory was established from antiquity and experiments served largely the rhetorical purpose of confirming known theorems or convincing the skeptics. In other areas, however, they served not only to test theoretical predictions and calculations but also heuristically to find out how nature behaved and to inspire more abstract speculations. In the study of the fall of bodies along inclined planes, water flowing in a pipe, the impact of inelastic bodies, or the resistance of air and water to motion, for example, there was no established theory, and the outcome of the experiment was not at all obvious or even predictable. In such cases experiments interacted with theory in a more interesting and problematic way.

Without a canon or even shared views on experimental error, however, different practitioners could have seen, and indeed did see, the same experimental data either confirming or refuting a theory. Reporting numerical data of experimental results exposed the author or experimenter to the charge that there was no perfect match with theory. The standard example of numerical data came from astronomy, where from antiquity tables had provided a way of providing numerical predictions and retrodictions, which were often debated and tested. In other disciplines experience and experiment were often performed and reported, though rarely in quantitative ways.[19] Music or harmonics was another area of the mixed mathematics where precise measurements went hand in hand with theoretical reflections related to consonance and dissonance. The development of quantitative experiments in the study of motion and mechanics brought together in new ways the concern for a fit between theoretical predictions and precision measurements found largely in the older sciences of astronomy and music. It has been often pointed out that Galileo's father was both a mathematician and an accomplished musical theorist who studied harmonies experimentally, and several sixteenth-century mathematicians, such as the Venetian Giovanni Battista Benedetti, contributed to the mathematical science of motion and musical theory.[20]

Galileo hardly ever provided numerical experimental results, while Newton often claimed to have achieved a remarkable fit between theory and experiment and was often able to provide explanations for the minute discrepancies between them.[21] Attitudes changed during the decades between their main works, and overall in the seventeenth century we witness deep transformations in the way experiments were conceived, practiced, reported in print, and received. Thus the historian encounters considerable difficulties in assessing from a printed text of the period how experiments were performed and even whether they were performed at all.

In general, separating the mathematical tradition from the experimental may not be helpful. Works on falling and projected bodies, collision, motion in a resisting medium, just to mention a few, belong to both.[22] One aspect in particular deserves careful scrutiny here, namely the emergence of careful quantitative experiments in the study of motion and mechanics. During the seventeenth century this was a new enterprise requiring commonly accepted and reasonable rules. One of the problems at the center of the debates following Galileo's *Discorsi* was the role of experiments in establishing the foundations of a new science. Several scholars argued that no experiment could be so precise as to establish the validity of a certain law as opposed to another predicting undetectably close results. We cannot use experiments with inclined planes in order to find the law of falling bodies, they argued, because the whole enterprise is subject to the vagaries of experimental trials. Therefore, an experiment in isolation from other factors, such as causal explanations or the belief in the simplicity of nature, can never be used to establish a science. We encounter here profound issues about the status and foundations of a new science. I shall focus also on the communication of quantitative experimental results, largely in tables of numerical values. Their emergence is an important but little-studied area at the intersection between the history of experimental philosophy and the history of the book.[23]

5. Practitioners, Sites, and Forms of Communication

The practitioners of mechanics and of the science of motion tended to be university professors of mathematics, courtiers and aristocrats, members of religious orders, or technicians and engineers. Of course, many belonged to more than one category. Niccolò Tartaglia was a teacher of mathematics, and his pupil Benedetti became court mathematician at Parma under the Farnese and then Turin under the Dukes of Savoy. Giuseppe Moletti was at the Gonzaga court at Mantua before becoming professor of mathematics at Padua. Guidobaldo dal Monte and Christiaan Huygens were aristocrats of independent means, but the latter also became a stipendiary member of the French Académie Royale des Sciences. Galileo and Newton started as university professors of mathematics in 1589 and 1669, respectively, but Galileo ended as mathematician and philosopher to the Medici court, and Newton as Warden and then Master of the Mint. Torricelli was mathematician, though not philosopher, to the Grand Duke of Tuscany, Benedetto Castelli was a Cassinese monk who taught mathematics at Pisa University and was later professor of mathematics at Rome, and was also employed by the Pope in the management of waters in the Papal States. His student Giovanni Alfonso Borelli started teaching mathematics at Messina in 1639 and from 1656 to 1667 was both professor of mathematics at Pisa and a member of the Cimento Academy at the Tuscan court. Stefano degli Angeli, a student of Cavalieri at Bologna, was professor of mathematics at Padua and a member of the Jesuat order, like his teacher. Hooke, Wren, and Wallis were members of the Royal Society and professors of mathematics at Gresham College and Oxford. Moreover, Wren and Hooke were also architects who helped rebuild London after the great fire of 1666. Descartes, a man of sufficient means to devote himself to the pursuit of knowledge, ended as court philosopher to Queen Christina in Sweden. Baliani was an aristocrat at the service of the Genoese Republic. Fabri and Mersenne were members of religious orders. Domenico Guglielmini was a professor first at Bologna and later at Padua. Pierre Varignon was a professor at the Collège Mazarin, a priest, and a member of the Académie des Sciences. Despite their different professional lives, these practitioners were able to interact and exchange views and perspectives. They had a common background in the mathematical disciplines, could read and write Latin, and were university trained; thus they generally shared important areas of their education. They were not radically divided between separate groups of experimentalists and theoreticians. Of course, some, like Mariotte (a member of the Académie des Sciences), were primarily experimentalists, while others, like Leibniz (the librarian and councilor to the Duke of Hanover), were mainly philosophers and mathematicians, but these differences did not prevent them from having extensive correspondences and sharing problems and views, as indeed happened between Mariotte and Leibniz. To what extent those debates, as well as the status of the discipline, were affected by their social and intellectual background is a hotly debated issue.[24]

In addition, many of these practitioners relied on lesser known collaborators, technicians, and instrument makers who were rarely mentioned in the published works. Guidobaldo dal Monte, Robert Hooke, and Christiaan Huygens relied on instrument makers renowned for their skills. In some cases technicians and engineers played an important role in the development of mechanics, as Galileo suggested in the opening of the *Discorsi*, where the Venice Arsenal is referred to as a venue for philosophical speculations and the *proti*, or senior technicians, working there are considered worthy of respect for their knowledge. Galileo and Castelli also interacted with hydraulic engineers in their works on water management. According to one personage in a dialogue on mechanics by Moletti, however, some engineers were so ignorant as to be "senza lettere," a position later qualified by the other interlocutor in the dialogue, who states that the "principal foundations" of engineering are derived from books.[25]

It is helpful to inquire where mechanics was practiced and studied. The first and most obvious space was the scholar's study, where we would find him calculating and experimenting. Mechanics entered the mathematics curriculum at Padua University in the sixteenth century with the printing of the Aristotelian *Quaestiones mechanicae* and with surveying and fortification, which were becoming crucial to the art of war in the age of guns. All these aspects were covered by Galileo at Padua, but his predecessors, Pietro Catena and Giuseppe Moletti, also taught the *Quaestiones*. Petrus Ramus too lectured on it at Paris.[26] Devices for the amusement of the court, following the example of Hero of Alexandria's *Spiritalia*, were probably less frequent and important in the seventeenth century than in the sixteenth.[27] Courts, however, supported practical mechanical pursuits related to the art of war and water management; thus theoretical and practical mechanicians could be found around a bastion, at the arsenal, and at the confluence of two rivers.

Although the seventeenth century saw the emergence of the first scientific journals, such as the *Philosophical Transactions* in London, the *Journal des sçavans* in Paris and Amsterdam, and the *Acta eruditorum* in Leipzig, publication patterns varied considerably from those familiar to modern readers. Original results can be found in manuscripts that had no impact on future developments, such as those of the English mathematician Thomas Harriot. At times, extraordinarily profound practitioners, such as Isaac Beeckman in the Netherlands, were highly influential largely through personal contacts and oral communication, with Descartes, Mersenne, and Gassendi among others. Probably the most common form of communication was through letters, which had an amphibious status between private communication and public report. A letter to Mersenne, for example, often belonged to the latter category.

Printed books too need to be seen in a historical context of style of composition, authorship, publishing, circulation, and reading. Teamwork and collaboration are important issues at many levels. At times, major theoretical works relied on data provided by experimenters and practitioners. Huygens's *Horologium oscillatorium*, for example, relied partly on data provided by sailors; Newton's *Principia* relied on astronomical observations by John Flamsteed and others. Moreover, at times a book was seen through the press by someone other than the author, such as Edmond Halley for the first edition of Newton's *Principia*. The results of nonauthorial interventions by Roger Cotes for the second edition were such as to transform the work substantially. Teamwork and collaboration are important issues at many levels, and several works relied on data provided by experimenters and practitioners as well as colleagues and friends. From Galileo, Vincenzo Viviani, and Evangelista Torricelli, to Newton, John Flamsteed, and Roger Cotes, patterns of collaboration in book production and research are a little-studied but important aspect of the history of mechanics.

Other questions, such as who were the intended and actual readers of Aristotle's *Quaestiones mechanicae,* dal Monte's *Mechanicorum liber,* Galileo's *Discorsi,* and Huygens's *Horologium oscillatorium,* are important in assessing the intellectual and social status of motion and mechanics at the intersection between natural philosophy and the practical arts. Their literary style too was important in this respect. *Mechanicorum liber* and *Horologium oscillatorium,* for example, were composed in a classical mathematical style starting from definitions and axioms and then proceeding with propositions or theorems. This, however, was not the only alternative. *Quaestiones mechanicae* was indeed structured as a series of informal questions and answers on a series of problems having largely to do with simple machines and their remarkable effects. Galileo's *Discorsi* was a hybrid wherein an Italian dialogue served both to deal with a miscellaneous series of themes during the first day and with the science of the resistance of materials on day 2, and also as a framework for a rigorous treatment in Latin of the science of motion on days 3 and 4. These different styles and languages were significant with respect to the formulation and organization of the subject matter and the intended audience of a text.[28]

6. Structure and Organization of the Present Work

This work is arranged thematically and follows broadly chronological developments. The themes discussed above will be interwoven with my narrative. Although in some cases I refer to private manuscripts, I privilege printed texts, letters, or other sources potentially available to contemporary spectators. It would be impossible to present an exhaustive picture of seventeenth-century mechanics; this work, however, attempts to provide a comprehensive analysis of mechanics as it was conceived at the time, and I hope it will stimulate other works in the area and serve as a useful model for them.

The choice of the endpoints of a historical narrative is crucially important. Some accounts have taken Galileo as their starting point, but the drawback of this approach is the lack of historical perspective about his project and practices. It is often stated that dal Monte's *Mechanicorum liber* (1577) was the major systematic treatise on mechanics of the Renaissance, yet its significance has not been fully appreciated. I shall argue that the project associated with Pappus and revived by dal Monte of explaining all simple machines in terms of the lever inspired Galileo and, in looser and broader fashion, became a key component of seventeenth-century mechanics and is therefore a useful tool to explore this field. In choosing dal Monte and some issues from the late sixteenth century, I intend to provide a source and background against which to assess several features of Galileo's work, such as his views about the foundations of mechanics, his reliance on instruments and experiments, his use of mathematics, and his attitude toward the discrepancies between theory and observation.

If dal Monte occupies a nodal position at one end of my account, then Varignon emerges as a natural choice at the other end, for two related reasons. In *Project d'une nouvelle mechanique* (1687), Varignon explicitly rejected dal Monte's approach of seeing simple machines as dependent on the lever and sought instead a more general and abstract principle. In the same year, Newton published *Principia mathematica,* a work that has been considered a watershed in the history of mechanics. Yet Newton's work was couched in a mathematical style that readers found cumbersome and obscure. Varignon embarked on the task of rendering its main results in the language of Leibnizian calculus. Both the search for new principles and the algebrization of mechanics in terms of the new algorithm set the stage for the development of eighteenth-century analytical mechanics, a field laid open by Leibniz, Varignon, and the brothers Jakob and Johann Bernoulli. Taking into account this broad spectrum of practitioners raises issues relating to mathematical styles employed, attitude toward experiments, and philosophical backgrounds, all crucial aspects for understanding mechanics at the outset of the eighteenth century and its later development.

My work consists of two parts with five chapters each. The first two chapters set the scene by charting some sixteenth-century developments, such as the rediscovery of classical sources including Archimedes, Pappus, and the author of *Quaestiones mechanicae,* and the way they inspired dal Monte, Stevin, Benedetti, the young Galileo, and others to systematize mechanics and explore new domains. In chapter 3 we move to the seventeenth century to investigate the works by Galileo and his disciples Castelli and Cavalieri on the sciences of motion, of waters, and of the resistance of materials. Chapter 4 highlights the radically different readings of Galileo in the aftermath of his main publications. Finally, chapter 5 examines the rise of the mechanical philosophy and Descartes, and compares his and Marci's rules of impact.

After a brief intermezzo highlighting the generation and institutional transformation separating the two halves of the century, I explore three main areas of research after 1650. Chapter 6 is dedicated to the science of equilibrium and especially of motion of waters, chapter 7 to the way projected and oscillating bodies were used to explore orbital motion, and chapter 8 to the emergence of elasticity in the study of collision, the behavior of springs, and the resistance of materials. The contents of these chapters are grouped in ways suggested by my focus on the objects of investigation. Chapter 9 hinges on Newton's *Principia*, seen in the context of contemporary work by English mathematicians and astronomers, focusing on his remarkable achievements that were made possible by his mathematical and experimental approach to natural philosophy. The last chapter presents a brief picture of mechanics circa 1700, with special emphasis on the new analysis, conservation, and readings of Newton's *Principia*.

The brief conclusion recapitulates the patterns of transformation of mechanics during the long century discussed here, showing how the material and intellectual manipulation of objects and geometric diagrams was at the center of seventeenth-century mechanics. Later in the century mechanics began to be transformed in a way that the search for increasingly abstract principles and the construction and integration of differential equations took center stage.

CHAPTER ONE

Machines in the Field, in the Book, and in the Study

We shall attempt . . . to reduce the screw to the lever and balance
in order finally to gain a full understanding of it.

Guidobaldo dal Monte

1.1 Between Classical Theory and Engineering Practice

In the sixteenth century the major tradition in mechanics was associated with
the science of machines, especially those known as the simple machines: the lever,
balance, pulley, inclined plane, wedge, and screw. Theoretical concerns about simple machines often went hand in hand with practical considerations, as shown
by the contents and publication history of two classics in the field, the pseudo-
Aristotelian *Quaestiones mechanicae* and dal Monte's *Mechanicorum liber* (Pesaro,
1577). In the latter the author stated, "Mechanics can no longer be called mechanics when it is abstracted and separated from machines."[1] Both works emphasized
the theme of utility and attracted the attention of scholars in mathematics as well
as architects and engineers, as witnessed by the translations they went through,
mostly in Italian. *Quaestiones mechanicae* is largely devoted to problems related to
the behavior of bodies, such as balances, levers, oars, rudders, wedges, and the like.
It went through more than a dozen editions and translations with commentary between the end of the fifteenth century and the beginning of the seventeenth. It appeared first in Greek in the Aldine edition of the works of Aristotle (Venice, 1497),
which was often reprinted, and was translated into Latin by Vittore Fausto (Paris,
1517), Niccolò Leonico Tomeo (Venice, 1525; Paris, 1530 and later editions), and
Alessandro Piccolomini (Rome, 1547; Venice, 1565). It was also translated into Italian by the engineers Antonio Guarino (Modena, 1573) and Oreste Vannoccio
Biringucci (Rome, 1582). Several engineers were familiar with *Quaestiones mechanicae,* including Agostino Ramelli, the celebrated author of *Le diverse et artifi-*

ciose machine (Paris, 1588), in which he described a large number of impressive devices. Ramelli was familiar with dal Monte's *Mechanicorum liber* too.[2]

The Marquis Guidobaldo dal Monte, later patron and correspondent of Galileo, was a military man from the Duchy of Urbino and a mathematician enamored of the Greeks. Recently Paolo Galluzzi has characterized Leonardo as the "artist of machines," meaning that Leonardo used his drawings as a tool of representation and investigation in order to understand how machines—including the human body—are constructed and operate. By analogy, dal Monte may well be called the "mathematician of machines," namely the scholar embodying the alliance between the Greek humanist tradition and the world of technicians and engineers who sought to understand the simple machines by means of geometric drawings. Four years after its original edition, *Mechanicorum liber* appeared in Venice as *Le mechaniche*, an Italian translation produced under dal Monte's supervision by Filippo Pigafetta. Pigafetta dedicated the Italian translation to the individual who had commissioned it, Count Giulio Savorgnan, a military man at the service of the Venetian Republic in charge of fortifications. Savorgnan had been responsible for dozens of bastions across the Venetian domains, and his residence at Osoppo housed an armory and a collection of mechanical devices.[3] The Italian translation of dal Monte's work, much like the pseudo-Aristotle's *Quaestiones*, is indicative of its perceived relevance not just to mathematicians but also to practical men. The title page of the Italian translation states that it had been carried out "to the benefit of those who find pleasure in this most noble science, especially war captains, engineers, architects, and all sorts of artificers who intend to perform wonderful and almost supernatural works, by means of machines. And the most difficult terms and passages are explained."[4] Thus the title page highlights Pigafetta and dal Monte's aim of establishing a rigorous mechanical terminology in Italian. The *editio princeps* was in folio, whereas the Italian translation was in quarto, suggesting a different and less wealthy audience. In 1615 the Latin and Italian versions were reprinted following the same criteria as to format. A German translation appeared in 1629.[5]

The theme of wonder at the almost supernatural works of machines appears in *Quaestiones mechanicae* and *Mechanicorum liber* too. The opening of *Quaestiones mechanicae* expresses wonder that "the less masters the greater, and things possessing little weight move heavy weights." Dal Monte claimed that the tools helping smiths, architects, carriers, farmers, and sailors in opposition to the laws of nature ("repugnantibus naturae legibus") belong to mechanics. Indeed, he argued that to operate in rivalry with the laws of nature ("aemulata leges"), or against nature ("adversus naturam"), is a feat worthy of the highest admiration.[6] These passages should be understood not as suggesting that mechanics subverts the course of nature, however, but rather as providing a special perspective and emphasis. Ultimately, from a more general viewpoint, mechanics operates in accordance with nature, as dal Monte stated in his 1588 paraphrase of *On the Equilibrium of Planes*. His was a legitimate interpretation of the Aristotelian tradition and was shared by his contemporaries. The Padua professor of mathematics Moletti, for example, stressed that mechanics imitates nature, as the grinding of grains imitates mastication. Later authors were even more explicit in associating mechanics with the normal operations of nature, since nature itself operates mechanically. Other mechanical traditions, such as Archimedean hydrostatics, focused primarily on the normal operations of nature and lacked the emphasis on machines acting against nature.[7]

While it is generally acknowledged that *Mechanicorum liber* was the main treatise on mechanics of its time, one rarely reads an account of what dal Monte's project actually was, let alone its significance. Dal Monte was the protagonist of the revival of the work on mechanics by Pappus of Alexandria (fourth century C.E.), whose project played a significant role from the late sixteenth century onward. Rather than reading about this story, one often finds echoes of a misleading but influential picture due to Pierre Duhem, who nurtured a profound dislike for dal Monte and everything that seemed to him to pertain to the Renaissance rather than the Middle Ages. Duhem devoted several pages to a demolition of dal Monte, arguing that he possessed a "punctilious" mind and that his ill-conceived love for the Greeks led him to discard valuable results attained by medieval scholars, notably Jordanus Nemorarius (thirteenth century). Before tackling Duhem's views, I will undertake a brief analysis of some problems and debates regarding simple machines appearing in *Quaestiones mechanicae* and especially dal Monte's work. By discussing simple machines, we will gain a sense of some central areas of the science of mechanics at the end of the sixteenth century, and of its practitioners and their locations.

Section 1.2 discusses some key issues in the two classics in mechanics discussed so far. Although it is generally stated that dal Monte's work is rigorously based on statics, he briefly dealt with the force of a falling body hitting an obstacle, known as the force of percussion, and hinted at a relation between height of fall and speed of a body. Sections 1.3 to 1.5 are devoted to specific machines, notably the balance, pulley, and inclined plane. Section 1.3 analyzes the problem of the equilibrium of the balance and the issue of mathematical versus physical rigor. Section 1.4 deals with a debate on pulleys between dal Monte and some engineers. Finally, section 1.5 discusses the inclined plane and presents additional accounts by Tartaglia and Stevin.

1.2 Machines, Equilibrium, and Motion

Quaestiones mechanicae is a collection of questions, often in the form "Why it is the case that," largely devoted to practical problems. A good number of them relate to navigation, such as "4. Why do the rowers in the middle of this ship contribute most to its movement?" Since Padua University was a center of Renaissance Aristotelianism and navigation was much in the mind of the Venetians, it is not surprising that *Quaestiones mechanicae* was particularly popular there, both among the learned and at the Arsenal. It is not always easy to identify a theoretical basis behind all the questions, but if one has to be chosen, the principle of the balance seems the most appropriate. The author of *Quaestiones mechanicae* identified levers in many settings and circumstances, claiming, "The facts about the balance depend upon the circle, and those about the lever upon the balance, while nearly all the other problems of mechanical movement can depend upon the lever."[8] His work was more concerned with wonder at the peculiar properties of the circle than with rigorous foundations, and it does not deal with all the simple machines. Not only is theory treated in a rather discursive way; the presentation is also at times based on a desire to surprise and a love for the paradoxical or what appears to be contrary to nature. A key common element in the author's discussions is the peculiar nature of the circle, which is exemplified in several ways. For example, a circle rotating around its axis has both motion and rest at the same time, and opposite sides of the circle move in opposite directions. A related feature of the *Quaestiones* is that the treatment of the lever relies on reflections on motion rather than equilibrium, because the operations of the lever are conceived in terms of the greater speed of the longer arm compared with that of the shorter.[9] The final problems in the *Quaestiones* concern the motion of projectiles ("32. Why do objects thrown ever stop traveling?") and the behavior of bodies in a whirlpool of water.[10]

The work's format—questions and answers, rather than a rigorous axiomatic structure—possibly invited discussion and criticisms. Both Tartaglia in the seventh book of *Quesiti et inventioni diverse* and Benedetti in *Diversarum speculationum liber* discussed several of the *Quaestiones*.[11] Galileo too found inspiration in several of them. The extensive question 24 on the so-called Aristotle wheel, wherein two concentric circumferences in a rotating circle move at different speeds and generate paradoxical conclusions when they are considered together, attracted Galileo, who discussed it in the first day of his *Discorsi*. The reader of the *Quaestiones* will find many more themes that inspired Galileo. For example, question 19 deals with the force of percussion, a topic dear to Galileo, and provides an important link with the science of motion:

> Why is it that if one puts a large axe on a block of wood and a heavy weight on top of it, it does not cut the wood to any extent; but if one raises the axe and strikes with it, it splits it in half, even if the striker has far less weight than one placed on it and pressing it down? Is it because all work is produced by movement; and a heavy object produces the movement of weight more when it is moving than when it is at rest? When the weight lies on it, it does not produce the movement of the weight, but when it travels it produces both this movement and that of the striker.[12]

The author continues by arguing that the axe acts like a wedge. This attempt to explain one machine in terms of another was to remain an important feature in mechanics.

Let us now consider *Mechanicorum liber*. Dal Monte's allegiance to Archimedes can be easily detected on the title page of his book (fig. 1.1) and is explicitly stated in the dedication. Works by Archimedes had been known throughout the Middle Ages, but the sixteenth century witnessed a remarkable diffusion by means of more and more refined editions. Dal Monte's teacher Federico Commandino of Urbino was the protagonist of this movement, with his careful editions and at times reconstructions of the texts and proofs from defective manuscripts.[13]

Besides knowledge of mathematics and mechanics, the works by Archimedes were highly influential in the formulation and presentation of the science of mechanics, as testified by dal Monte's *Mechanicorum liber*. No longer do we find informal questions and answers, as in the pseudo-Aristotelian *Quaestiones*, but definitions, postulates, and propositions in the style of Euclid's *Elements*. The canonical text by Archimedes was *On the Equilibrium of Planes*, a work, later edited by dal Monte, wherein the mathematician from Syracuse provided a rigorous axiomatic treatment of the lever. Archimedes' study of equilibrium was based on the notion of center of gravity, and in this respect he was followed by Pappus. *Mechanicorum liber* opens with the definitions of "center of gravity" given by Pappus and Commandino. According to the former, "The center of gravity of any body is a certain point within it, from which, if it is imagined to be suspended and carried, it remains stable and maintains the position which it had at the beginning, and is not set to rotating by that motion." According to the latter, "The center of gravity of any solid shape is that point within it around which are disposed on all sides parts of equal moments, so that if a plane be passed through this point cutting the said shape, it will always be divided into parts of equal weight." Noteworthy here is the use of the term "moment" as distinguished from simple

GVIDIVBALDI

È MARCHIONIBVS

MONTIS

MECHANICORVM

LIBER.

PISAVRI

Apud Hieronymum Concordiam.

M. D. LXXVII.

Cum Licentia Superiorum.

Figure 1.1. Title page of dal Monte's *Mechanicorum liber*. The Archimedean lever dominates dal Monte's title page.

weight, meaning weight in conjunction with the distance from the vertical plane through the center of gravity.[14] The study of the centers of gravity of many geometrical figures was a challenging mathematical problem and an important component of mechanics in the sixteenth century and beyond, involving mathematicians from Galileo and Luca Valerio to Wallis and Huygens.

Mechanicorum liber is notable for its systematic attempt to provide a rigorous treatment of all simple machines by unmasking levers. This key project of dal Monte was inspired by Pappus, the author of *Collectiones mathematicae*. The eighth book of this huge tome was devoted to mechanics and consisted of a heterogeneous collection of theorems on mathematics and descriptions of machines. As dal Monte put it in his preface, "No one could, I believe, blame me for following [Pappus] as my leader. I have more readily done so for the reason that Pappus does not depart even a nail's breadth from the principles of Archimedes."[15] The crucial passage in Pappus's *Collectiones* can be found at the end of book 8, where he claimed that, following Philo of Byzantium and Hero of Alexandria, he was seeking to systematize mechanics by identifying in the lever a common denominator for all simple machines. The passage argues that simple machines may have very different shapes or configurations, but they can be reduced to a common φύσις ("in unam reducentur naturam"), or nature. The verb *reducere*, in the sense of going back to some common notions, is of interest here and will occur again in *Mechanicorum liber*.[16] Since Archimedes had established the axiomatic doctrine of the lever, when a machine was shown to be a lever in disguise, its operations were fully understood and the problem of foundations was solved. For example, in discussing the screw, dal Monte stated that his aim was to "reduce" *(reducere)* it "to the lever and balance in order finally to gain a full understanding of it."[17] In the preface, dal Monte explained that he was going to explain the properties of the balance "in order that my whole work might be more easily built up from its foundation to its very top."[18] Despite some shortcomings in dal Monte's analysis, the program he so thoroughly pursued in a geometric and visual fashion left a deep mark on Galileo and represents a useful starting point from which to reexamine seventeenth-century mechanics.

Following Pappus's analysis of simple machines, dal Monte tried to explain "the many properties present in those machines by virtue of the lever."[19] We shall explore below several instances of how dal Monte tried to account for (or *reducere*) the operation of a machine in terms of the lever. In order to grasp his procedure and the meaning in practice of the verb *reducere*, I use the wheel and axle, shown in figure 1.2 both as a real machine in the tradition of Renaissance engineering texts and as a geometrical diagram.[20]

The works by Hero and Philo were not known in the Renaissance, but Pappus

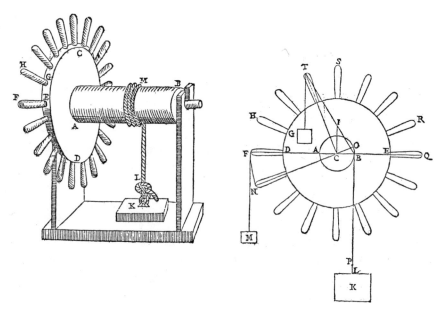

Figure 1.2. Dal Monte's wheel and axle *(Mechanicorum liber).* A rope is wrapped around the cylinder *AB* so that by turning the handles *H, F, N*, etc., weight *K* is raised. The force on the handles is represented by the attached weights *G* or *M.* Here dal Monte identifies the arms of the lever as the radii *CF* and *CB, C* being the fulcrum. By pushing on *F, N*, etc., we can raise the weight *K* with a smaller weight *M*, albeit more slowly than by pulling it directly. The lever as seen by dal Monte has disjointed arms: *CB* is always parallel to the horizon, whereas *CF* may rotate around *C*. Notice that weight *M* is perpendicular to the arm *FC*, whereas in all other positions such as *TC* or *NC* we have to take into account only the component perpendicular to the arm, which changes with the angle of inclination; therefore, as dal Monte states, weight *G* attached in *T* is greater than weight *M.* The illustration reproduced in *Mechanics in Italy,* 318, is inaccurate in that *G* is not drawn larger than *M.*

and dal Monte were by no means the first to rely on the lever to explain other simple machines, as we have seen with *Quaestiones mechanicae.* In the late sixteenth century the Sienese engineer Francesco di Giorgio Martini explained the operation of the wheel in a mill in terms of the "ragione de la lieva," but he also expressed doubts as to its universality, invoking the need to construct actual models. Probably dal Monte had access to the text by Francesco di Giorgio, who had dedicated his work to the Duke of Urbino, Federico da Montefeltro. Dal Monte had access to the Greek manuscript of Pappus's *Collectiones* and to Commandino's translation, and eventually saw it through the press in 1588.[21] Thus dal Monte's project had its roots in the discursive but rather broad tradition of *Quaestiones mechanicae,* and in the rigorous but narrower work by Archimedes on the balance.

Paragraph 19 of the *Quaestiones mechanicae* and Pappus's *Collectiones* also discuss percussion, as when a falling body hits a wedge, breaking a body underneath. The same theme was treated at the end of the section on the wedge in *Mechanicorum liber*, where dal Monte sought to explain the wedge in terms of the lever. But he discussed the force of percussion, too, claiming that it is greater if the falling body is heavier, and if it falls from a greater height, because it acquires more motion and "any heavy object takes on more heaviness when it is in motion than when it is at rest, and the more so the farther it moves." He also discussed percussion by a hammer, claiming that a longer handle generates a greater percussion.[22] This analysis seems to establish a series of proportions involving the force of percussion and the weight, height, and speed of the falling body. Dal Monte was not alone in enunciating this proportionality, since similar results were stated in different contexts by others, such as Tartaglia and Benedetti.[23] Thus although dal Monte adopted a strictly statical approach to the study of the balance and lever, he also mentioned motion as part of mechanics in the case of the force of percussion, trying to establish a link between its effect and the height of fall. In the dedication to the Duke of Urbino, he stressed that mechanics cannot "be considered apart from either geometrical demonstrations or actual motion."[24] Therefore mechanics was linked to the study of motion in more than one way. The Abbot of Guastalla Bernardino Baldi, another student of Commandino and a friend of dal Monte, showed the trajectory of a projectile in the opening of his commentary on *Quaestiones mechanicae*. He extended such a central notion as that of center of gravity to motion, arguing that an impelled body has a violent center of motion, and applied his analysis to the problem of the spinning top. On the other hand, Baldi considered *Quaestiones* 31–33, dealing with motion, as pertaining more to *physica* than to mechanics. Clearly the boundaries of mechanics were rather blurred, and motion in particular was an area on whose location there was no general consensus. Often the issue was how, rather than whether, motion was treated, whether it was natural or violent, or whether it was artificially produced and associated with machines. Natural motion was that of bodies falling downward, if they were heavy, or moving upward, if they were endowed with levity, whereas violent motion was that of projectiles, for example.[25]

1.3 The Balance of dal Monte and the Problem of Rigor

Although the doctrine of the lever had been established since antiquity, surprising concerns about its equilibrium emerge in *Mechanicorum liber* (fig. 1.3) Dal Monte argued correctly that a balance with equal arms and equal weights attached to their extremities, when moved, will return to the equilibrium position if the center of suspension is above the center of the balance or center of gravity,

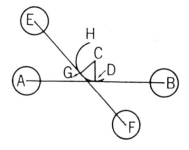

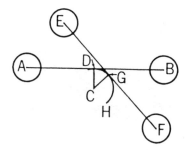

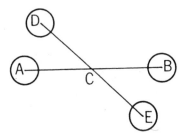

Figure 1.3. Dal Monte's balances *(Mechanicorum liber)*. Here dal Monte describes the equilibrium position of balances in which the center of suspension *C* is above, below, and coincident with the center of gravity.

but will move toward the lower side if the center of suspension is below the center of the balance. A similar issue had been dealt with in the second of the *Quaestiones mechanicae,* but the crucial case when the balance is suspended in its center was not discussed by the follower of Aristotle.[26]

In that case, when the two centers coincide, dal Monte stated that the balance would remain stable in any position into which it is moved. His reasoning was based on the idea that the center of gravity is not changed by rotation and therefore the balance will be in equilibrium in any position. In arguing this proposition in endless detail for over fifty folio pages, dal Monte criticized previous scholars, such as Jordanus, Gerolamo Cardano, and Tartaglia, who had claimed that the balance returns to the horizontal position. Their reasoning was based on the measure of what was then called the "angle of contact," namely the "angle" between a circumference and its tangent. We take the "angle of contact" to be exactly nil, but Tartaglia and even dal Monte thought it to be smaller than any given angle, though not nil.[27] The reasoning based on the "angle of contact" implied that the line of descent of the higher weight was closer to the vertical; therefore the higher body had a stronger tendency to fall than the lower one and the balance returned to the equilibrium position. Similarly, one could argue that a body has a stronger tendency to fall along the vertical than on an inclined plane. This way of conceiving the problem originated in the Middle Ages and can be found in Jordanus, for example. The technical expression was that the higher body is "positionally heavier" than the lower one.[28] In the first supposition in book 2 of his *Nova scientia* of 1537, Tartaglia claimed that all natural movements of heavy bodies were parallel among themselves. He also claimed that he knew this to be, strictly speaking, false, but he argued that the error over a short space was not detectable. Although this supposition sounds reasonable, it becomes problematic if at the same time one takes the "angle of contact" into account, because the angle between the lines of descent of heavy bodies attached to a balance is indeed very small, but it is finite, whereas the angle of contact is not.[29]

Dal Monte disagreed with Tartaglia's analysis, but in typical sixteenth-century fashion he argued that even if one were to adopt it or concede the point, Tartaglia's conclusion still would not follow. The reason would be that the weights of the balance act not along parallel lines, but along lines converging toward the center of the world; therefore taking the angles into account properly would lead to the conclusion that the balance would move downward, not return to the horizontal position. Dal Monte's remark that the lines of descent are not parallel has deceived several commentators, such as Duhem, who have charged dal Monte with being a pedant. In fact, here dal Monte was putting forward this argument as a *concessio*, not because he believed it to be relevant to the outcome. As we have seen, his approach was based on centers of gravity, and from the definitions mentioned above, as well as from proposition 4, the balance is in equilibrium in any position because its center of gravity does not change by rotation. Moreover, in his later discussion dal Monte was ready to concede that the lines of descent could be taken to be parallel among themselves because of the great distance from the center of the world: "We may even concede that . . . the straight descent of the weight must be along a straight line parallel to *FG*" (fig. 1.4)—in other words, that all the lines of de-

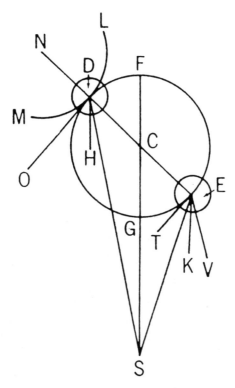

Figure 1.4. Dal Monte's lines of descent *(Mechanicorum liber).* In this complex diagram *DCE* represents the tipping balance with weights in *D* and *E.* Angles *KET* and *HDO* are equal. According to Tartaglia, the body in *D* would be heavier than the body in *E* because the angle *HDG* would be smaller than the angle *KEG.* The difference would come from the angles of contact *TEG* (between the tangent *TE* and the arc *EG*), which has to be added to *KET,* and *ODG* (between the tangent *DO* and the arc *DG*), which has to be subtracted from *HDO.* According to dal Monte's hypothetical reasoning, angle *SEG* would be smaller than *SDG,* where *S* is the center of the Earth; therefore the body in *E* would be heavier. Thus, rather than returning to the horizontal position, the balance would move downward. On the mistaken interpretation by Duhem and his followers, this diagram would be the hallmark of dal Monte's pedantry because of his belief that the lines of descent of the weight of a balance are not parallel among themselves but converge toward *S.*

scent are parallel. Even with this concession, however, the conclusions by Tartaglia and his associates still would not follow. Dal Monte countered their arguments in a variety of ways, arguing, for example, that they had taken the two weights separately, but since they are joined they have to be considered together.[30] Whereas a single weight attached to one arm of a balance would tend to fall obliquely to the center of the world, two weights attached to the arms of a balance would tend to fall parallel to the line between the center of gravity of the balance and the center of the world. The issue was not so much dal Monte's "punctilious" mind but rather the more complex and interesting problem of the relationships between mathematical rigor and the physical world in the late sixteenth century. Duhem's attack on his excessive rigor could be easily leveled against those who had recourse to the angle of contact.[31]

In the Latin version dal Monte argued for a separation between what can be mentally conceived and what can be practically built:

> However, it is to be noticed here that it is difficult in fact to make an actual balance (such as we can imagine mentally) that is supported only at one point and has its arms so exactly equidistant from the center, not only as to length but as to breadth and thickness, that all its parts on both sides weigh precisely the same, because matter does not lend itself easily to such exact measurement. Hence if we consider the center to be in the balance, we need not to have recourse to the senses, for artificial devices cannot be brought to such a degree of perfection.[32]

Despite such a disclaimer, dal Monte must have been quite frustrated at his failure to produce a balance behaving in accordance with his theory. In the Italian translation a note by Pigafetta astonishingly stated that dal Monte had succeeded in having a balance built to prove his point and that Pigafetta had actually seen one in the hands of the bibliophile Gianvincenzo Pinelli at Padua. Thus rather dubious experiential reports entered not the austere *Mechanicorum liber*, but its vernacular version. Dal Monte had dispatched the instrument to his friend Pinelli, whose house and library were the venue for debates among a circle that was to include Galileo. We now know that the entire insert by Pigafetta was provided by dal Monte himself, while it is not clear whether Pigafetta had seen the balance in Pinelli's possession at all. In case Pinelli's balance was malfunctioning, however, dal Monte offered to dispatch one to Pigafetta at his request. Other references testify to the existence of such precision tools proving the indifferent equilibrium of the balance where the center of suspension coincides with the center of gravity. One of them was sent by dal Monte to the Count of Carpegna in 1599. In all probability they are attributable to one of the Urbino workshops, either that of Simone Barocci, brother of the celebrated painter Federico and a former student of Commandino's, or that of the Genga family, adjacent to the Urbino Ducal Palace. Dal Monte relied on the services of these instrument makers specializing in fabricating compasses, squares and drawing devices, astrolabes, and other mathematical instruments.[33] Working to dal Monte's specifications, they also built tools with no practical purpose and whose only task was to confirm his theories. It appears that both dal Monte and his circle of friends and correspondents attached great importance to the behavior of precision instruments. For them, mechanics was a mathematical science that had to function in the real world. These instances highlight the gulf between modern views of an experiment and late-sixteenth-century views of *esperienza*, which in the specific case we have considered was expected to be a mirror image of theory. There is a striking lack of notions such as "order of magnitude" and "experimental error" and of other tools necessary for applying the rigor of mathematics to the imperfections of matter and the vagaries of nature. We will explore similar themes in studying the pulley.

Since dal Monte was not alone in discussing the parallelism of plumb lines, it is worth exploring the opinions of some of his contemporaries. Toward the end of his life Benedetti published *Diversarum speculationum liber* (Turin, 1585), an extensive collection of works and letters on mathematical and physical matters. In a section on mechanics Benedetti first claimed that plumb lines from the arms of a balance are parallel, because the deviation is of insensible magnitude. Later, however, he sided with dal Monte in criticizing Tartaglia and Jordanus for having neglected that plumb lines converge while taking the angle of contact into account.[34]

In the following year Simon Stevin discussed the same issue of the equilibrium of the balance suspended from its center of gravity. Stevin was a near contemporary of dal Monte, providing an interesting Dutch counterpart to the Italian scene. Unfortunately we know very little about his early life and education. Much like his Italian contemporaries, Stevin was as interested in theory as he was in practice and was also a military man, an engineer, and a mathematician. In 1604 he was made quartermaster of the army of the Low Countries and was tutor to Maurice of Nassau, Prince of Orange. His works too exist both in the vernacular and in Latin, but the order of composition is reversed, for Stevin believed that Dutch was superior to other languages and best suited to express his views.[35] He could read Latin, however, and prominent among his sources are Archimedes and the works and editions by Commandino, whom he quoted with respect.[36]

Stevin's main work in mechanics is a collection of texts with separate title pages published in Leyden by Plantijn in 1586, including *The Elements of the Art of Weighing*,[37] *The Practice of Weighing*,[38] and *The Elements of Hydrostatics*.[39] At the end of the last work there are two short additions with continuous pagination, *Preamble of the Practice of Hydrostatics* and *Appendix to the Art of Weighing*.[40] We shall discuss *The Elements of Hydrostatics* and a famous experiment from the *Appendix to the Art of Weighing* in the next chapter. In *The Elements of the Art of Weighing* Stevin discussed the problem of the convergence of the weights of a balance. He argued that, "mathematically speaking," a balance removed from its equilibrium position will tip over, because the lower weight will be heavier, or "heavier in appearance," as he put it, than the higher one. The reason for this is that the lines toward the center of the Earth are not parallel, and the angle between the balance and the line from the higher weight is more acute than the angle between the balance and the line from the lower one. Practically speaking, however, these truths are negligible because they "would be of no advantage for the practice of weighing."[41]

1.4 Pulleys and the Contingency of Matter

Guidobaldo dal Monte was in contact with a group of intellectuals in the Venetian Republic, including Savorgnan, Pigafetta, Giacomo (or Jacomo) Contarini, and Pinelli. Contarini was a patrician from the Veneto, a friend of Savorgnan, a military man in charge of the fortifications of the Venetian Republic, and a collector of history books, art, and mathematical instruments.[42] His 1580 correspondence with dal Monte dealt with pulleys in a fashion that is quite alien to our way of thinking. Partly for this reason, it is helpful to grasp how mechanics was understood at that time. Contarini and Savorgnan performed *esperienze* based on dal Monte's conclusions, finding results different from those in *Mechanicorum liber*. The first step in dal Monte's treatment was to identify a lever in a pulley, namely *DEB* in this case (fig. 1.5).[43]

According to theory, given a certain arrangement of pulleys, a given body on one side was balanced by one weighing one-sixth as much on the other. Contarini and Savorgnan found by trial that the measured proportion was rather one-quarter. From the correspondence it appears that they were using standard ropes and pulleys, rather than precision instruments, but they were still hoping to find confirmation of dal Monte's theoretical predictions. In other words, one could say that they were mixing the left-hand engineering figure in dal Monte's treatise with the geometrical representation on the right. Pigafetta reported having had difficulties with the pulleys during tests performed while he was staying with Savorgnan. This time they appeared to have been finely wrought by a German craftsman and greased with soap. Pigafetta suspected that the resistance of the pulleys, axles, and ropes, and their *gravezza*, might have had something to do with the disagreement between theory and practice.[44]

Dal Monte's replies to Contarini and Pigafetta outline several important points; dal Monte drew a distinction not between theory and practice but rather between different types of practice. The pulleys had to be small, preferably made of brass, with their axles being very thin and made of iron, all so arranged that the wheels would not wobble. Comparing them with scales, dal Monte explained that the pulleys should resemble the small ones used for weighing money rather than the big wooden ones used for weighing meat. Thus precision tools were meant to be a bridge between the real instruments on the left of his figures and their geometri-

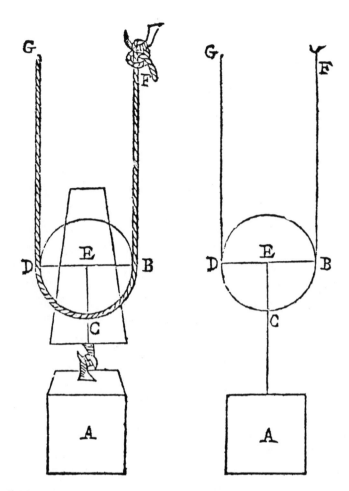

Figure 1.5. Dal Monte's pulley and balance *(Mechanicorum liber)*. Once again, as in figure 1.2, dal Monte depicts both a real machine and a geometrical one. *DEB* is, as it were, a lever with the fulcrum in *B* and the weight appended in *E*. The power is applied in *D* (or *G*). In this case, if we assume the pulley itself to be weightless, the power is half the weight of the body *A* because the arm *BD* is twice the arm *BE*. More elaborate arrangements with more pulleys can reduce the power and resistance further.

cal counterparts on the right. Provided one used the right type of tools, theory will agree with experience because—and this is a crucial point—this is a problem of equilibrium, not of motion. Dal Monte argued that when he talked of a sustaining power *(potentia sustines)* he meant the power required to hold the weight still, not to move it. In the case of equilibrium, he argued that matter does not alter the propositions ascertained by theory, whereas in the case of motion it does. One might add that, in fact, in the case of equilibrium the imperfections of matter and friction make the experiments possible, whereas in the case of motion they introduce a disagreement with theoretical predictions. Whereas in statics one could aim at absolute and total precision, this objective was unattainable in the study of motion, and therefore this far more problematic area was left at the margin of dal Monte's public investigations. Dal Monte claimed in addition that he had performed all the *esperienze* himself and had found them to be invariably in agreement with theory, as in the case of the balance above. His book, however, was not replete with experimental reports. Thus it seems that for late-sixteenth-century mechanics, and for dal Monte especially, there was no unproblematic room for any discrepancy between theory and observation or for some form of error theory. Theoretical predictions, if true, had to be confirmed *in toto*.[45]

This correspondence shows how much mechanics was a science of machines, involving theory and mathematics as well as practice. Rather than considering a simple theory-practice dichotomy, it is more useful to adopt a three-way partition involving mathematical theory, the tools and instruments of everyday experience, and the precision instruments from the Urbino workshops aimed at minimizing or even abolishing the effects of the imperfections of matter. This intermediate level between theory and the rough machines of the practitioners was essential to dal Monte's strategy. On the one hand, the precision tools he had commissioned were material objects suitable for practical demonstrations; on the other hand, they behaved exactly as predicted by his theory. Therefore those tools enabled him to clinch the link between theory and practice.

In a lost letter to Galileo, dal Monte seems to have pointed out a related theme, namely that whenever matter was involved, the mathematician was released from having to deal with the problem, because of the perturbations associated with it. Galileo's reply, however, is extant:

> Please do me the favor of conveying my greetings to Sig. Francesco and tell him that when I have a little leisure I shall write to him of an experiment that has come to my mind for measuring the force of percussion. As to his question, I think that what you say about it is very well put, and that when we commence to deal with matter, because of its contingency, the propositions abstractly considered by the geometrician begin to be altered. Since it is impossible to formulate a certain science of those propositions, being so perturbed, the mathematician is relinquished from having to study them.[46]

It is not clear whether the problem of the contingency of matter was linked to the study of percussion. But whatever its nature, Galileo and dal Monte were in agreement in defending the opinion that mathematicians could not be expected to deal with the contingency of matter; therefore they could not answer Signor Francesco's question. Both motion and the properties of matter, however, were becoming more and more tempting areas of investigation to mathematicians such as dal Monte and Galileo.

1.5 Rival Traditions on the Inclined Plane

The last simple machine I discuss is the inclined plane. Pappus and the medieval scholars produced different solutions to the problem of finding the force required to prevent a body from falling. Tartaglia followed and improved upon the solution of Jordanus, whereas dal Monte in *Mechanicorum liber* referred to Pappus's solution, which must have been available to him in manuscript form. A brief examination of their different approaches reveals important features of theoretical mechanics at the time (figs. 1.6 and 1.7). Since Tartaglia and dal Monte asked slightly different questions about the inclined plane, it seems legitimate to claim that they were dealing with slightly different problems. The latter calculated the force required to prevent a sphere from rolling down, whereas Tartaglia did not have to specify the shape of the body; he was concerned only with its force to descend.

The treatment of the inclined plane in *Mechanicorum liber* appears in the section on the screw, which is conceived of as an inclined plane folded around a cylinder (a further instance of one machine being explained in terms of another). The problem, however, was to go one step further, from the inclined plane to the lever. In the Latin edition dal Monte simply referred to proposition 9 in book 8 of Pappus's *Collectiones*, but in the 1581 Italian translation Pigafetta inserted the relevant portion from the as yet unpublished text of the *Collectiones*.[47] Dal Monte was trying to explain all simple machines by means of the lever and was also eager to promote a style of doing mathematics and mechanics that gave due care to proofs as well as to results. In this case he identified a lever or balance in a sphere on the inclined plane and from there tried to determine the equilibrium conditions. Although his proof is often described as being faulty because it leads to incorrect results, a close analysis shows that dal Monte considered a particular case and drew rigorous conclusions from his assumptions. Pappus assumed from the start that a force is required to move a weight along a horizontal plane instead of trying to find a solution from the limiting case of an inclined plane with no inclination, and dal Monte preferred his solution to Jordanus's. Notice that motion enters this account only nominally, because the force required to generate motion is not quantified.[48] Finally, dal Monte sought to use the balance to explain the inclined plane,

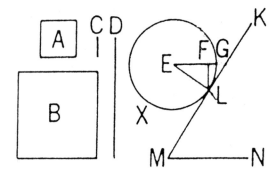

Figure 1.6. Inclined plane according to Pappus *(Collectiones).* According to Pappus, a given force *C* is required to move a weight *A* over a horizontal plane. It is required to find the force needed to move a sphere of weight *A* up on an inclined plane *MK.* The key idea consists in imagining a balance *EG,* parallel to the horizon, where the fulcrum *F* is on the perpendicular from the point of contact *L* of the sphere and the inclined plane. The weight *A* of the sphere acts on its center *E,* and the force *D* is required to keep the weight *B* in equilibrium by pulling down on *G.* This reasoning relies on the assumption that the sphere rolls without sliding. Weight *B* is to weight *A* as *EF* is to *FG,* which can be determined by elementary geometry. Therefore we also have force *D.* This force tends to become infinite as the inclined plane tends toward the vertical. The force required to move the sphere up on the inclined plane is *C* plus *D,* namely the force required to move it along the horizontal plus that required to keep it in equilibrium balancing its weight. Pappus included in his figure two squares representing weights *A* and *B,* and two lines representing forces *C* and *D,* so as to give geometrical representation to all the magnitudes he discusses.

whereas Tartaglia and Jordanus did not try to reduce the problem to the balance and instead juggled with planes of different inclinations. Not surprisingly, their works have been charged with circularity.[49]

The separation between motion and equilibrium is not found only in dealing with the inclined plane but also in the discussions of the lever, pulley, and axle. In these cases dal Monte argued that if a power were to move a weight by means of a machine, the spaces through which the power moves and the weight is moved would be as the moved weight is to the moving power. However, in appropriate corollaries dal Monte modified his conclusion: "The power that sustains is less than the power that moves," and the proportion described above does not hold, because the ratio between the distances covered by the moving power and the moved weight is greater than the ratio between the weight and the power. The reason for this change is motion. Several commentators have drawn the conclusion that in his treatment of simple machines dal Monte introduced a sharp separation between statics and motion that, combined with his unreasonable demands for precision in the convergence of the weights of the balance, made a science of motion

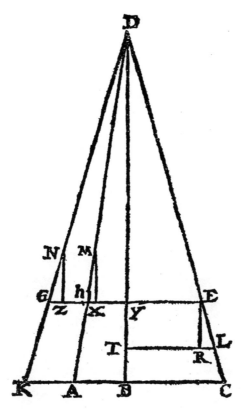

Figure 1.7. Inclined plane according to Tartaglia *(Quesiti).* According to Tartaglia, two weights at *E* and *H* along planes with the same height *DB*, but of different inclinations *DC* and *DA*, have the same powers to descend if the weights are proportional to the lengths of their respective planes. Tartaglia's diagram tries to explain the inclined plane by means of another plane *DK* with a slope equal to that of *DC*. Tartaglia makes no attempt to account for the inclined plane in terms of the lever; rather, he uses the notion that bodies have a greater tendency to descend the closer their descent is to the vertical.

practically impossible.[50] Although dal Monte believed that a force was required to move a body along a horizontal plane and that motion altered the proportions valid in statics, the significance of his claims may have been overemphasized. As we have seen, in some cases he touched on issues pertaining to motion while dealing with the force of percussion, and elsewhere he tried to minimize the imperfections of matter by means of carefully crafted precision instruments. He also performed experiments on motion together with Galileo, as we shall see in the following chapter.

In concluding this discussion, it is instructive to move again from Italy to the

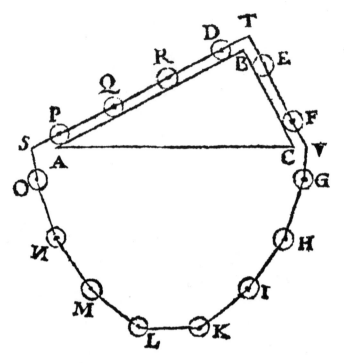

Figure 1.8. Stevin's inclined plane *(Elements of the Art of Weighing)*. In the case studied by Stevin, in the triangle *ABC* the side *AB* is twice *BC,* and the wreath has fourteen spheres overall, four on *AB* and two on *BC.* This conclusion can be easily generalized to the case where one of the planes is vertical. Since the lower part can be removed by symmetry considerations without altering the equilibrium conditions, the two portions at the top must be in equilibrium; hence weights have to be proportional to the length of the inclined planes.

Low Countries, where in 1586 Stevin provided an original solution to the problem of equilibrium on the inclined plane (fig. 1.8). In *The Elements of the Art of Weighing* he argued that a wreath placed around an inclined plane must remain in equilibrium, otherwise it would roll indefinitely in perpetual motion, which is absurd. Stevin's solution agrees with Tartaglia's but is based on the denial of perpetual motion, namely a different principle from dal Monte's.

In his works Stevin refrained from attacking the opinions of other scholars with whom he disagreed. In the *Appendix to the Art of Weighing,* however, he adopted a different style, fearing that his silence might engender misunderstanding.[51] Rather than invidiously challenging many particular errors, Stevin focused on two fundamental errors that were the source of all the others. He challenged the views expressed in *Quaestiones mechanicae,* where conclusions about equilibrium were based on reasoning about motion. This Stevin found objectionable because "the cause of bodies being of equal apparent weight does not reside in the circles described by the extremities of the arms" (obviously referring to the moving arms of a balance, which describe circles).[52] In other words, he rejected the usage of motion as a way to explain the equilibrium of the balance. We shall encounter the other criticism in the following chapter.

CHAPTER TWO

Floating Bodies and a Mathematical Science of Motion

Given two bodies of different material but equal size, to construct
a plane so inclined, that the body which, in a vertical fall, moves
more swiftly than the other will descend on this plane with the
same speed with which the other would fall vertically.

Galileo

2.1 Some Features of Archimedes' *Floating Bodies*

In the previous chapter we considered mechanics as a science of machines and briefly examined some areas of intersection with natural philosophy in the study of motion. We shall now explore another area of mechanics that interacted with natural philosophy in a different fashion and in which machines took a secondary role, namely the tradition of Archimedes' *Floating Bodies*. In addition to Archimedes' treatise, that tradition includes several works on buoyancy dating from antiquity to medieval times. These works appeared in print in the late fifteenth and sixteenth centuries.[1]

Archimedes' work consists of two books quite different in scope. The first deals with the principles of hydrostatics and examines the behavior of bodies in fluids (fig. 2.1). It is here that one finds the celebrated statement that a body in a fluid receives an upward thrust equal to the weight of the volume of the displaced fluid, known as Archimedes' principle. The second book deals with the equilibrium of a sector of a paraboloid floating in water and appears to be closely associated with the problems of equilibrium of a vessel in water and of righting a capsized ship (fig. 2.2). Whereas the first book is rather short and relies on relatively simple mathematics, the second is a veritable mathematical and mechanical tour de force involving such key mechanical notions as equilibrium and centers of gravity, not to mention the practical side related to ship stability. Thus the entire treatise ap-

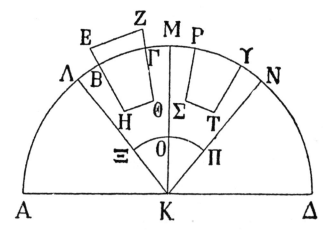

Figure 2.1. An illustration from Archimedes' *Floating Bodies,* book 1. The illustrations of solid bodies EΖΘΗ and ΡΥΤΣ floating in a fluid show the Earth as a whole and suggest cosmological implications about the order of the elements.

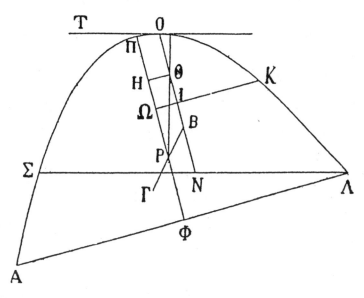

Figure 2.2. A proposition from Archimedes' *Floating Bodies,* book 2. The figure is obtained by rotating a parabolic sector around its axis. The equilibrium conditions depend on the positions of the centers of gravity of the whole body and of the part immersed in the fluid, where ΣΛ is the surface of the fluid.

pears quite plausibly to fall into the tradition of mechanics, or to be a direct extension of it. The first book, however, has a natural-philosophical flavor in that it deals with the natural behavior of bodies, not the operation of machines acting in rivalry with or against nature. It treats some of the same themes discussed in Aristotle's *De caelo* and, to a lesser extent, *Physica*. The propositions in book 1 lead to the problems of book 2, but they also have a cosmological ring to them. For example, proposition 2 states, "The surface of any fluid which is so located that it remains motionless will have the form of a sphere which has the same center as the Earth." The figures in book 1 illustrate equilibrium by showing portions of a sphere obviously representing the Earth, thus reinforcing the impression that the work deals with natural philosophy.[2]

The interactions between mechanics and natural philosophy developed in various ways. Some followers of Aristotle and Archimedes came to disagree about the explanation of buoyancy, as we shall see in chapter 3 in the case of Galileo. Further, in the second half of the sixteenth century several mathematicians tried to formulate a quantitative theory of falling bodies by extending Archimedes' views on buoyancy. Their attempts are a major topic of the present chapter. The following section offers a selective survey of the different traditions of reading and extending *Floating Bodies,* from Tartaglia, Stevin, and Ghetaldi to the very beginning of the seventeenth century. Section 2.3 is devoted to Benedetti, who—possibly in response to an idea suggested by Tartaglia—boldly used the principles of floating bodies as the basis for a mathematical science of falling bodies. Section 2.4 deals with the young Galileo and argues that his early studies on motion, known as *De motu antiquiora,* may reflect the culture of the *circoli* at Pisa University, namely periodic academic disputations involving professors and students. Section 2.5 examines how Jacopo Mazzoni and Stevin read Benedetti's work. I argue that in all probability Galileo either did not read it or, if he did, did not rely much on it.

2.2 Reading *Floating Bodies*

Tartaglia was a crucial figure in the Archimedean and pseudo-Archimedean tradition on floating bodies for more than one reason. He was responsible for the editions of the first book of *On Floating Bodies*, in Latin and in Italian translation (Venice, 1543 and 1551). The latter included a treatise on raising sunken ships, called *La travagliata inventione*. The Italian translation also contains a dialogue between Tartaglia and the English gentleman Richard Wentworth, in which Tartaglia made his interlocutor state that heavy bodies in water "sink to the bottom so much faster, by how much they are [specifically] heavier than water."[3]

Tartaglia repeated the same concept, pointing somewhat incidentally to a link between speed and buoyancy, in *La travagliata inventione*. Similar statements linking differences in specific weight to speed can also be found in later editions of Tartaglia's works, but are not due to Archimedes. Thus Tartaglia did not merely publicize Archimedes' work on floating bodies, he also hinted at an extension of Archimedean doctrines to the study of motion. Tartaglia was not the first to try to extend Archimedean doctrines; Marshall Clagett identified a tradition linking hydrostatics and the speed through a fluid including Leon Battista Alberti, for example. There are some obvious analogies between dal Monte's approach and Tartaglia's strategy of relying on known domains to explore new ones, but whereas dal Monte used geometry and sought rigorous foundations, Tartaglia adopted a looser approach whereby Archimedes served more as a source of inspiration than as bedrock of rigor. Tartaglia was the teacher of Benedetti and probably of Ostilio Ricci, mathematician to the Grand Duke of Tuscany and Galileo's teacher. Both Benedetti and the young Galileo were familiar with Tartaglia's work and attempted to use Archimedean hydrostatics in formulating a mathematical science of motion opposing Aristotle's views.[4]

Other readers of Archimedes focused on different aspects of his work. After Tartaglia's death, his publisher Curtius Troianus made both books of *Archimedis de insidentibus aquae* (Venice, 1565) available from Tartaglia's papers. In the same year Commandino put forward a superior edition in which he paid special attention to the challenging mathematical propositions of book 2. Commandino's edition became known to Stevin and inspired him to write his works on hydrostatics.[5]

Stevin's work consists of a theoretical part, *The Elements of Hydrostatics,* and a practical *Preamble to the Practice of Hydrostatics,* a more complete practical part not having been completed. In a preliminary explanation Stevin addressed two related problems (reminding us of a discussion in chapter 1): whether water surfaces are part of a sphere and whether vertical lines convergence to the center of the world. He argued on practical grounds that water surfaces could be taken to be flat and vertical lines to be parallel to each other. Stevin argued that water keeps its position in water and proved it by claiming that if a portion of the water in a container moved, the part that takes its place would also have to move, for the same reason for which the first one did. But this would amount to a "perpetual motion, which is absurd." Stevin adopted here the same line of reasoning he had employed in dealing with the inclined plane.[6] He proceeded by reformulating Archimedes' principle, stating, "The gravity of any solid body is as much lighter in water than in air as is the gravity of the water having the same volume."[7] Stevin then studied the problem of calculating the weight at the bottom of vessels with different shapes. It is here that he extended Archimedes' doctrine to new areas, studying

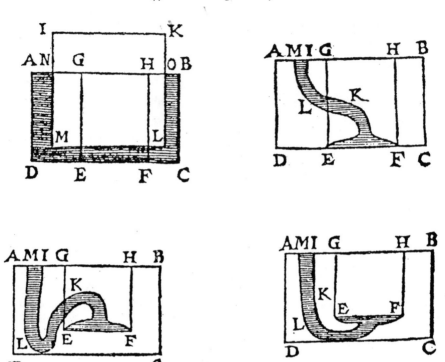

Figure 2.3. Stevin's vessels *(The Elements of Hydrostatics)*. According to Stevin, the weight on the bottom of the vessel *EF* is always equal to the weight of a prism of water whose base is *EF* and height *EG*.

both containers with their bottoms parallel to the horizon and those with their bottoms inclined to the horizon. He examined the first case in theorem 8, stating that the weight on the bottom is given by the gravity of the water whose volume "is equal to that of the prism whose base is that bottom and whose height is the vertical from the plane through the water's upper surface to the base" (fig. 2.3).[8] Stevin illustrated his theorem in several corollaries wherein the water is replaced by a solid body with a peculiar shape and equal specific gravity to the water. Subsequent theorems deal with the weight of water in containers with an inclined base. In the *Preamble of the Practice of Hydrostatics* Stevin provided a number of ingenious examples involving scales and vessels so that "anyone may test and see with his own eyes."[9]

Besides the traditions and issues associated with buoyancy discussed so far, another theme requires our attention here, namely the practical determination of the ratios between weights of equal volumes of different substances. Several scholars provided values of varying accuracy, among them Tartaglia, Giambattista della Porta, and the Jesuit Juan Bautista Villalpando. The most detailed data provided in convenient tables were produced by Marino Ghetaldi, a patrician from Ragusa, now Dubrovnik, and a distinguished mathematician. His work, *Promotus Archimed[e]s*, is an attempt to provide accurate determinations of the weights of equal volumes of several substances, relying on Archimedes' *Floating Bodies* and on a rigorous application of the theory of proportions (fig. 2.4). Despite the potential practical uses of Ghetaldi's tables, his treatise was written in Latin. Ghetaldi used a hydrostatic balance to measure the decrease of weight of bodies of different materials in water. He took great care in the accuracy of his experiments, attaching his bodies to the balance with horse's hairs, whose weight was allegedly comparable to that of an equal volume of water, and working with cylinders rather than spheres, which cannot be wrought with equal precision, then finding the weight of a sphere inscribed in the cylinder using Archimedes' theorem. He printed the unit length of the old Roman foot, but in the *Errata* he specified that the paper had shrunk, so the unit had to be increased by one-fortieth. He relied on special techniques for weighing bodies, such as coating a piece of gold with a thin layer of wax to prevent it from amalgamating when weighed in mercury. He also relied on wax balls attached to lead weights, an idea analogous to that later used by Galileo in a different context in his dispute with the Aristotelians.[10] Ghetaldi's tables could serve different purposes, but they primarily provided a convenient reference tool by visually displaying valuable empirical data.

2.3 Benedetti against the Philosophers

Between 1553 and 1555 the young Venetian mathematician Benedetti published two short treatises, the second in two versions, dealing with hydrostatics and motion. In the first, *Resolutio omnium Euclidis problematum* (1553), he examined Euclidean geometry, but in the preface, addressed to the churchman and diplomat Gabriel de Guzman, he outlined his views on motion, possibly because he feared being anticipated by others. Benedetti may have just read the passage in Tartaglia's 1551 work on the proportionality between the speed of fall of a body in a fluid and the difference of specific weight between the body and the fluid, and therefore wished to put forward his own views on the matter.[11]

Benedetti claimed that if we have two bodies of the same material and shape, but one is four times larger and heavier than the other, they fall in the same amount of time. He defended this statement by considering the larger body to be decomposed into four bodies equal to the smaller one. The four bodies will all fall with the same speed, but if we imagine them joined by a weightless thread, they will still all fall with the same speed and also with the speed of the smaller body,

Altera, ad comparandum interſe duodecim corporum genera, grauitate, & magnitudine, tabella.

	Oleū.	Cera.	Vinū.	Aqua.	Mel.	Stann.	Ferrū.	Aes.	Arg.	Plum.	Ar.Vi.	Aurū.
Aurum.	$4\frac{47}{57}$	$5\frac{5}{209}$	$5\frac{10}{57}$	$5\frac{5}{19}$	$7\frac{12}{19}$	$38\frac{18}{19}$	$42\frac{2}{19}$	$47\frac{7}{19}$	$54\frac{22}{57}$	$60\frac{10}{19}$	$71\frac{3}{7}$	100
Arg. Viuū.	$6\frac{41}{57}$	$7\frac{7}{209}$	$7\frac{14}{57}$	$7\frac{7}{19}$	$10\frac{13}{19}$	$54\frac{10}{19}$	$58\frac{18}{19}$	$66\frac{6}{19}$	$76\frac{8}{57}$	$84\frac{14}{19}$	100	
Plumbum.	$7\frac{67}{69}$	$8\frac{76}{253}$	$8\frac{8}{69}$	$8\frac{16}{23}$	$12\frac{19}{23}$	$64\frac{8}{23}$	$69\frac{13}{23}$	$78\frac{6}{23}$	$89\frac{59}{69}$	100		
Argentum.	$8\frac{17}{31}$	$9\frac{81}{341}$	$9\frac{16}{31}$	$9\frac{21}{31}$	$14\frac{1}{31}$	$71\frac{19}{31}$	$77\frac{13}{31}$	$87\frac{3}{31}$	100			
Aes.	$10\frac{5}{47}$	$10\frac{20}{33}$	$10\frac{25}{27}$	$11\frac{1}{9}$	$16\frac{1}{9}$	$82\frac{2}{3}$	$88\frac{8}{9}$	100				
Ferrum.	$11\frac{11}{14}$	$11\frac{41}{44}$	$12\frac{7}{24}$	$12\frac{1}{2}$	$18\frac{1}{8}$	$92\frac{1}{2}$	100					
Stannum.	$12\frac{41}{111}$	$12\frac{366}{407}$	$13\frac{31}{111}$	$13\frac{19}{37}$	$19\frac{27}{37}$	100						
Mel.	$63\frac{12}{87}$	$65\frac{265}{319}$	$67\frac{71}{87}$	$68\frac{28}{29}$	100							
Aqua.	$91\frac{2}{7}$	$95\frac{5}{11}$	$98\frac{1}{3}$	100								
Vinum.	$93\frac{13}{59}$	$97\frac{47}{649}$	100									
Cera.	$96\frac{2}{63}$	100										
Oleum.	100											

Figure 2.4 Ghetaldi's table (*Promotus Archimedes*). In his treatise Ghetaldi provided several tables of data from his weighing experiments with the hydrostatic balance. The table here shows the relative weights of twelve substances: gold, quicksilver, lead, silver, copper, iron, tin, honey, water, wine, wax, and oil. From the first row, the weights of equal volumes of silver and gold are in the ratio of $54\frac{22}{57}$ to 100. Other tables seemed geared toward goldsmiths and those interested in determining the ratio of silver to gold in an alloy.

even though together they would weigh four times as much. Thus simple weight is not the correct magnitude in the proportion. Benedetti argued that the speeds of bodies falling in a fluid are proportional to the difference between the weight of the bodies and the weight of identical volumes of the fluid.[12]

Benedetti also considered shape in connection with the body's ability to penetrate the fluid, arguing that a body with an acute shape would move faster than one with an obtuse one, whereas bodies with the same shape behave similarly, regardless of their sizes (fig. 2.5). One often thinks in terms of falling bodies, but Benedetti talked of gravity or levity, suggesting both that he was considering extrusion too, and that he was considering levity as resulting from the smaller weight of the body compared with an equal volume of the fluid.[13] Aristotle, by contrast, had taken levity to be an intrinsic property of bodies. Benedetti used the theory of proportions and tried to avoid ratios between inhomogeneous magnitudes, such as weights and volumes, or distances and times. Therefore he did not

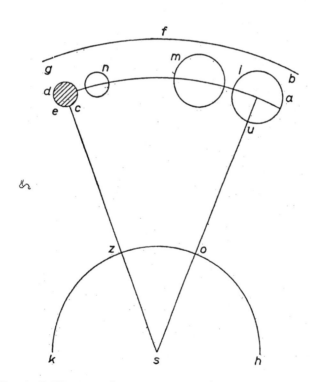

Figure 2.5. Benedetti's falling bodies (adapted from *Resolutio*). Bodies of similar shape *aui* and *ced* float in a medium such as water *bfg, s* being the center of the Earth; *m* is a volume of water equal to *aui; n* is a volume of water equal to *ced.* Setting the density of water as 1, the density of *ced* is 8 and the density of *aui* is 2. The speed of *ced* is as 7 (= 8 − 1), and that of *aui* is as 1 (= 2 − 1). They experience a resistance (buoyancy) as ⅛ and ½ of their density.

compare bodies with different specific gravities, but rather compared the weights of equal volumes of different bodies. Similarly, when he compared two motions, he took the space traversed to be the same; therefore for him *motus* or speed was inversely proportional to time.[14] At times Benedetti talked of "resistance," meaning the buoyancy of the medium rather than the impediment to motion proportional to the surface of the body. In the *Resolutio* Benedetti also stated that violent motions are in fact mixed of violent and natural, and that all motions outside the perpendicular are mixed. During the sixteenth century, discussions about motion were framed according to the Aristotelian dichotomy of natural versus violent. Benedetti's statement was a departure from the Peripatetic doctrine, according to which natural and violent motions were fundamentally different and could not be mixed, a view still held by Tartaglia.[15]

In the *Demonstratio propotionum motuum localium contra Aristotilem et omnes philosophos* (1554), or **Demonstratio*, Benedetti sought to refute several passages by Aristotle, *De coelo* and *Physica*, and also several commentators. Benedetti objected to the statements that weight is proportional to speed and inversely proportional to resistance. In this way, according to Aristotle, motion in a vacuum would encounter no resistance, which would lead to the absurdity of an infinite speed. By contrast, Benedetti claimed that the resistance of the medium had to be subtracted from the speed in an empty space. As in the *Resolutio*, he used "resistance" to mean "buoyancy," but this time he added a few statements about motion in a vacuum. Surprisingly, he did not treat the problem as an extension of motion in a medium with zero density; while one would have expected the speeds of the bodies to be proportional to their densities, Benedetti stated that in a vacuum all bodies would fall with the same speed. The **Demonstratio* was plagiarized by the mathematician Jean Taisnier, who in *Opusculum perpetua memoria dignissimum, de natura magnetis, et eius effectibus* (1562) reproduced Benedetti's work with no reference to its author. Not surprisingly, Benedetti was incensed. Taisnier's edition was subsequently used by Stevin and was also translated into English by the promoter of colonial and navigation projects Richard Eden, thus attaining European circulation.[16]

However, Benedetti was dissatisfied with his own opinions, and approximately one year after the **Demonstratio* he published a revised edition with the same title, or ***Demonstratio* (1555), in which he adopted a slightly less belligerent tone regarding Aristotle. While restating some conclusions he had reached in the previous edition, he also modified his views on the resistance encountered by falling bodies. As we have seen, in the *Resolutio* he had used "resistance" to mean "buoyancy" and had paid attention to shape rather than size, but now he claimed that size too mattered. Benedetti was unable to provide a quantitative measure of the resistance to motion for bodies with different shapes, such as a cube and a sphere, but for bodies of similar shape but different sizes—such as two cubes with different sides or two spheres with different radii—he claimed that resistance is proportional to the surface perpendicular to the direction of motion; hence the meaning of "resistance" changed. In conclusion, the speed of a body falling in a medium is proportional to the difference in density between the body and the medium, and it also depends on the body's shape and surface in the direction of fall.[17]

At the end of the 1550s Benedetti left Venice and moved to the court of Ottavio Farnese at Parma, where he stayed for about eight years. In 1567 he moved to the Savoy court at Turin, where he remained until his death in 1590 and where he published his magnum opus, *Diversarum speculationum mathematicarum et physicarum liber* (1585), a collection of previously published as well as new works on a wide range of topics. Benedetti devoted additional reflections to falling bodies within the framework of the **Demonstratio. He maintained that resistance depends on the surface of the body in the direction of motion and argued that bodies of the same material and shape, but of different sizes, do not fall in a given medium with equal speeds, because their surfaces are different. Benedetti did not deal explicitly with the speed of bodies of different size and material in a void, as he had done in the *Demonstratio. However, he added a discussion on the acceleration of falling bodies. Against Aristotle, he argued that air does not propel a body moved by force or in violent motion, but rather hinders its motion, and that the speed of a falling body increases indefinitely because its *impetus* continually increases. This increase depends not on the closeness of the body to its terminal goal, as claimed by Aristotle, but rather on its distance from the starting point of its fall because the body continually receives new *impetus*. Therefore the body continually accelerates. Although Benedetti's analysis of fall is generally conducted in terms of constant speed, here he introduced the notion that speed in free fall is not constant, but he did not see the need to rethink his entire analysis of the problem. His formulation gives the impression that speed is proportional to distance.[18]

Benedetti faced a major problem in formulating a mathematical science of motion because motion is subjected to so many accidents and perturbations that it seems hardly the subject of a science. A mathematical science requires a regularity that is hard to find in our experiences, except perhaps in celestial motions. Astronomers too, however, had recourse to other disciplines for making sense of observations. For example, the positions of stars near the horizon differ from predicted values and have to be adjusted for atmospheric refraction with optics. (It may be worth mentioning here that Benedetti was a proficient astrologer.) Similar techniques of separating the primary from the secondary phenomenon, or perturbation, and then using a range of mathematical methods and disciplines to deal with them, are at the basis of late-sixteenth-century attempts to formulate a mathematical science of motion.

A related problem concerns the mathematical tools required to deal with perturbations. In the case of motion, the issue was to correct numerical values as with atmospheric refraction and also to deal with the relations among variables such as speed, size, and differences in specific gravity. We tend to think in terms of equations wherein parameters with suitable physical dimensions allow one to add and subtract different variables. This method was not developed until the second half of the seventeenth century, however. Before that time, the theory of proportions was the chief mathematical tool available, and perturbations were awkwardly treated in terms of composed proportions, whereby the proportion between a magnitude a first with b and then with c can be expressed as a proportion of a with b and c simultaneously. We shall see below that Galileo, whose early astronomical interests can be inferred from a now lost commentary on Ptolemy's *Almagest* dealing also with optical phenomena, faced similar problems on a number of occasions.[19]

2.4 Galileo's Early Speculations

Galileo's earliest work on mechanics, *La bilancetta* of 1586, is an elaboration on Archimedes' treatise on floating bodies. Galileo's brief essay circulated in manuscript form and was not published in his lifetime, but it appeared in 1644, soon after his death, thanks to the Sicilian mathematician Giovanni Battista Hodierna. Archimedes' most celebrated discovery was that his ruler's crown was made not of pure gold, as claimed by the artifex, but of an alloy in which gold was mixed with a less noble metal. Since surviving works on the crown problem were not by Archimedes himself and seemed to suggest rather awkward solutions, some scholars, including Galileo, attempted to find better ones involving more appropriate weighing tools and a more refined approach. Galileo elaborated on a method for finding a solution by weighing bodies in water and in air. His main interest was not in providing data about the densities of bodies, which are presented only as examples, but in the procedure and the experimental apparatus. A fine metal wire was tightly wound around one arm of a balance, whereby distances could be measured very accurately by counting the number of coils. Since the wire was very thin, Galileo suggested counting the coils by using the sound produced by a metal pointer sliding across them. Presumably in conjunction with *La bilancetta*, Galileo produced a manuscript table of the differing weights of gold, silver, and several precious stones when weighed first in air and then in water. Moreover, analogous data were tabulated in the geometric and military compass, thus documenting Galileo's early interest in precision measurement.[20]

At about the same time, Galileo was also working on centers of gravity, an area of mechanics established by Archimedes. Galileo's aim was to gain a chair of mathematics at Bologna that had recently become available. His efforts brought him into contact with dal Monte; with the dean of mathematical studies at the Collegio Romano, Christophorus Clavius, whom he met in Rome in 1587; with the Padua professor of mathematics Giuseppe Moletti; and with the Flemish mathematician Michel Coignet. Galileo's results were impressive and extended the Archimedean doctrine to new solids. His treatise was later included at the end of the *Discorsi.* However, both dal Monte and Clavius expressed concern with one of his proofs, claiming that it begged the question. Whereas dal Monte later became convinced of its soundness, Clavius was apparently too busy to devote much attention to the matter. Possibly at about the same time, Galileo obtained sets of lecture notes from the Collegio Romano on logic and natural philosophy that sharpened his background on matters relevant to a range of his interests. William Wallace suggests that the notes were sent by Clavius to Galileo in connection with their dispute over centers of gravity, but evidence for his claim is lacking. Given Clavius's well-attested preference for mathematics over philosophy, I doubt whether he would have had recourse to those extensive lecture notes to settle a disagreement with Galileo about a simple mathematical proof.[21]

The key documents on Galileo's early speculations on motion and mechanics are a series of manuscripts called *De motu antiquiora,* composed before he moved to Padua in 1592.[22] Although those manuscripts were left unpublished at that time and do not appear to have circulated, there are good reasons to discuss them here. Scholars have identified a Pisan context for Galileo's studies and experiments on motion, notably in the works of the philosophy professors Girolamo Borro and Francesco Buonamici. It now seems established that some of Galileo's contemporaries at Pisa, such as Borro, debated and experimented on falling bodies. I propose that the *circoli pisani,* academic functions at which professors debated in front of students, who were allowed to intervene, favored a culture of public debate and experimentation, and indeed may have been a venue for some of Galileo's experiments. Viviani reports that Galileo took part in those functions, and the frequent references in his texts to "my adversaries" and "my opponents" suggest a setting linked to the culture of public disputation. Thus at least some of his views, such as those associated with the leaning-tower experiment, may have been publicly discussed. Galileo kept his manuscripts and referred to some of their contents in his later works on hydrostatics. As late as the *Discorsi* he considered perturbations to falling bodies due to buoyancy. Thus *De motu antiquiora* is a helpful tool for following the debates within Pisa University as well as Galileo's later work on motion.[23]

The main works of *De motu antiquiora*, in the order in which they appear in the manuscript, are a dialogue on motion (D), a ten-chapter essay (E10), a twenty-three-chapter essay (E23), and variants of the first two chapters of E23 (E2). Recently, after several unconvincing attempts, a thorough study of insertions and cancellations has provided evidence in favor of the following sequence: D, E23, E2, E10. I will follow that sequence here, focusing on the first two, the second two being shorter reelaborations.[24]

The dialogue takes place between two personages, Alessandro, representing Galileo, and Domenico, who is readily convinced by Alessandro's reasoning. The subject is motion, especially the motion of heavy and light bodies. Domenico states that since the topic had been dealt with by Girolamo Borro, he would like to focus on only six specific questions. Borro had in all likelihood been one of Galileo's teachers at Pisa, and the reference to the views he expressed in *De motu gravium et levium* (1576) is especially important. Borro reported and discussed several opinions on motion, notably those of the Spanish Muslim philosopher Avempace or Ibn Bājja. Avempace's views share common traits with Galileo's, such as the notion that the *virtus impressa*, on a projected body for example, is self-destroying and the idea that when there is resistance to motion, the speed of a body should be the difference between its unimpeded speed and the resistance, rather than the quotient.[25]

The six questions mentioned at the outset are (1) whether there is rest at the turning point of motion; (2) why, when two bodies of the same size but different materials, such as wood and iron, are dropped from a height, the lighter moves more swiftly than the heavier, if this truly does happen; (3) why natural motion continuously becomes swifter while forced motion continuously slows down; (4) why bodies fall more swiftly in air than in water; (5) why cannonballs travel farther when the cannon is fired at right angles to the horizon (vertically) than when fired parallel to it (horizontally); and (6) why heavier balls shot from a cannon travel more swiftly than lighter ones, even though the latter are easier to move.

Galileo did not treat these questions systematically but devoted considerable space to answering the fourth, first, and third, in that order. In the course of the discussion he argued that the speed of a body falling in a fluid depends on the difference between the specific weight of the body and that of the fluid. In a void, all bodies with the same specific gravity fall with the same speed, and the void is simply treated as a medium with nil specific gravity. A crucial discussion occurs in answering question 3, where in his effort to identify regularities in nature, Galileo took speed to be the crucial variable in free fall and acceleration to be a transient perturbation of the law of falling bodies. He argued that before falling, the body is kept in equilibrium by a force equal and opposite to its weight—for example, a hand holding it. When the body is released, the force exerted by the hand does not vanish instantaneously but disappears over an interval of time during which the body tends to reach its natural speed in the medium. In the interval between the time when the body is released and the time when it reaches its final speed, the body accelerates, or rather, to use Koyré's felicitous expression, is "deretarded." This would be the reason why initially the lighter body is faster than the heavier, as mentioned in question 2, because the lighter body has a weaker opposite force to overcome. Thus in his search for regularities Galileo took acceleration to be the perturbation and the rectilinear uniform motion—reached asymptotically and due, for us, to the resistance of the medium—to be the key regularity in nature. Acceleration, in this view, does not seem to be amenable to mathematical treatment.[26]

Galileo may have found this approach less implausible than we may think it because his preferred medium in which bodies fell was probably water. However, he was concerned with the empirical status of his claim, for he admitted that "at times experience seems rather to point to the opposite." In *De motu antiquiora*, experience and experiment were often a problem to be explained rather than a support to Galileo's views. He also tried to have recourse to parallax effects in taking into account the point of view of the observer, whereby equal portions of the trajectory traversed in equal times subtend different angles if they are seen by a fixed observer.[27]

In the essay E23, Galileo expanded his reflections and investigations. The essay is better organized than the dialogue and appears to be complete. Important differences between the essay and the dialogue include the role of statics in the study of motion, such as the balance and the inclined plane, the attempt to explain experimental results against Borro, and the study of projectile motion. Compared with the dialogue, in E23 we find a more sustained effort toward employing mechanics or the science of machines in the study of motion. In chapters 6 and 9 of the essay Galileo tried to use the balance to explain gravity and levity, in terms of the weight of a body and the weight of an equal volume of fluid. In a balance, the fall of one weight is due not to its entire gravity but only to that portion overpowering the other weight. Similarly, the gravity of a falling body is diminished by the presence of a fluid. Here Galileo tried to combine the two Archimedean traditions deriving from the balance and floating bodies, but his efforts were plagued by a failure to use appropriate terms such as *gravitas in specie*, or specific gravity, an expression used by Benedetti.[28]

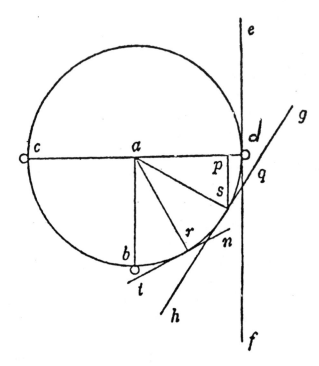

Figure 2.6. Galileo's balance and inclined plane *(De motu antiquiora).* Galileo argued that the balance is in equilibrium in the initial position *cad.* If one arm is moved to *s* or *r*, the body in *s* or *r* is not allowed to move along *ef,* as when it was in *d,* but rather when in *s* it would move initially along *gh,* and when in *r* along *tn.* The body at *r* or *s* exerts less force to descend, and this decrease can be determined by means of the perpendicular to *cad.* A body in *s,* for example, would act as if it were at a distance *ap* from the fulcrum in *a.* The body would descend along *ef* more readily than along *gh* and along *gh* more readily than along *tn,* in the same ratio as it is heavier at *d* than at *s* and at *s* than at *r.* On the basis of simple geometry joined with his views about motion and force, Galileo could conclude that the same heavy body would descend vertically with greater force than on an inclined plane in proportion as the length of the incline is to the perpendicular.

In chapter 14 Galileo introduced another mechanical device into his investigations of motion, the inclined plane. Galileo's initial purpose was to find out why a heavy body falls more swiftly along a steeper plane than along one less steep. In order to answer this question, he addressed the problem of equilibrium and considered a balance in which one arm can be fixed at different angles (fig. 2.6). Though differing from Pappus's solution as adopted by dal Monte, Galileo's solution was fully within dal Monte's project of explaining the inclined plane in terms of the lever, and they may have discussed it together. As in figure 1.2, here the arms of the lever are not along a straight line, although once one of the arms is locked into any position it is no longer allowed to rotate. This example and others we shall examine in the next chapter help identify an important tradition in the history of mechanics where Pappus, dal Monte, and Galileo occupy a nodal position. It is not by accident that Galileo's solution to the inclined plane can be found also among dal Monte's manuscripts.[29]

Galileo was very careful in dealing with the problem of accidents and talked of the "roughness of the moving body or of the inclined plane" and imperfections deriving from the shape of the body. While dealing with the inclined plane, Galileo developed an approach different from Pappus's. While Pappus believed that a force was necessary to move a body along a horizontal plane, Galileo argued that if the plane and ball are very smooth and perfectly hard, "any body on a plane parallel to the horizon will be moved by the very smallest force, indeed by a force less than any given force." Thus Galileo did not shy away from complex cases but rather tackled them by trying to remove or minimize the impediments; and this important passage considers the problem of setting a body in motion on a horizontal plane seen as a limit case of the inclined plane with zero inclination. While presenting arguments in support of his claim, Galileo stated that bodies move down a plane even when the inclination is very small, as can be seen in the case of water. The association of water with the inclined plane will become a characteristic feature of Galileo and his school in the treatment of river flow. The ties with the inclined plane with zero inclination suggest that Galileo conceptualized this motion as parallel to the horizon, since motion in a straight line tangent to the Earth would move upward away from its surface. Thus for Galileo a body set in motion covers a circle arc along the Earth's horizon, rather than a rectilinear segment. In an important marginal note he characterized such a circular motion not as mixed (meaning composed of natural and violent motions) but rather as neutral (that is, excluding both). It is unclear at this stage whether Galileo would consider this neutral motion to continue indefinitely.[30] There is a structural similarity in many mechanical problems pertaining to what we might call the "zero-point." In the equilibrium of the balance, the problem is what happens when the centers of gravity and suspension coincide; with upward motion, the problem is what happens when the motion changes direction; and with the inclined plane, the problem is what happens when the plane has zero inclination.

Galileo established a further proportion between speed and slope of a plane that is fully in agreement with the treatment of fall in a medium based on uniform speed. He argued, "The [uniform!] speeds of the same body moving on different inclinations are to each other inversely as the lengths of the oblique paths, if these entail equal vertical descents."[31] This conclusion follows directly from Galileo's approach and fits well with his attempts to use statics in the investigation of motion. In the light of these remarks Galileo's statement at the outset true?

of chapter 14 that no philosopher had undertaken the study of the ratios of motions along inclined planes appears to be justified. Thus in one single leap Galileo elegantly completed dal Monte's program of reducing the inclined plane to the lever and made his first unsuccessful attempt to introduce motion into the picture by using the inclined plane as a tool of investigation.

At the end of chapter 14 in E23 Galileo mentioned that in theory one should be able to construct an inclined plane with such a slope that two bodies of equal size and different materials would fall along it and down the vertical in equal times. What one needs is for the inclination of the plane to compensate exactly for the difference in specific gravity of the bodies. On making the test, Galileo probably ran into difficulties and tried to find justifications for this fact.[32]

Another important topic treated extensively in E23 concerns the experimental investigation of free fall and of the ratio between, on the one hand, the difference between the specific gravity of the body and the fluid, and, on the other, the body's speed. At the end of chapter 8 Galileo announced "a great difficulty" in that "those ratios will not be observable by someone who makes the experiment [*periculum facit*]", and in that the lighter body seems to move faster at the beginning. The expression *periculum facere* indicates a trial performed intentionally in order to test a theory.[33] The issue of falling bodies is treated more extensively in chapter 22, where Galileo attempted a refutation of Borro's views that the wooden sphere moves faster because it contains three heavy elements, namely air, water, and earth, whereas the lead sphere lacks air and therefore has only two. Galileo stated that taking a sphere of wood and one of lead, "if they are both let fall from a high tower, the lead moves far out in front," adding, "This is something I have often tested." It is only at the beginning of motion that the opposite holds. Thus in all likelihood Galileo repeatedly dropped balls from high places such as the leaning tower of Pisa; lack of additional contemporary records may be attributed to the fact that such public events and disputations were rather common at Pisa and possibly elsewhere. Galileo was only one among several scholars to have performed similar experiments. Giuseppe Moletti, for example, also mentioned wooden and lead spheres, but he argued that they reach the ground at the same time.[34]

We have seen that Galileo attributed the behavior of falling bodies to the dissipation of the force holding them in place. This is also the reason why a lead ball shot from a cannon travels further than a wooden one, namely the motive force is preserved longer in the lead. Galileo also stated, "This same thing is seen if two weights, one of wood, the other of lead, are suspended, each by a cord, and then receive an impetus [by being drawn to one side] the same distance from the perpendicular. If, then, the weights are allowed to swing, the lead will certainly move back and forth for a longer interval of time." This is the first reference in Galileo's writing to the pendulum, indicating that his interest in its oscillations dates back to his time at Pisa.[35]

The closing chapter of E23 deals with projectile motion and answers a question similar to question 5 from the dialogue mentioned above, namely "why objects projected by the same force move further on a straight line" the closer they are shot along the vertical. It is remarkable that Galileo took the initial portion of the trajectory to be rectilinear. He again used the inclined plane to emphasize a paradox. If the cannon is vertical, he argued, the motive force can impress its entire power on the ball, whereas if the cannon is horizontal, the ball escapes before experiencing the full impact of the motive power. Moreover, if the ball is shot vertically, Galileo thought, it has to exhaust all its projecting force before turning, whereas if it is shot at a different angle, the trajectory begins to bend before the body has lost all its projecting force. Galileo seems to be thinking in terms of a conflict between different forces that are operating not in tandem but rather in competition with each other. The diagram too suggests that the initial portion of the trajectory is rectilinear.

Galileo's work was not unusual in this respect. Probably the best-known attempts to deal with the trajectory of cannonballs were Tartaglia's, who in *Nova scientia* (1537) had approached the question using a peculiar combination of Aristotelian physics, geometry, and military experience. Tartaglia divided the trajectory of a cannonball into components in which the projectile's motion was either violent or natural. He argued that no part of the trajectory was traversed with mixed motion (a composition of the two), because in natural motion speed increases whereas in violent motion it decreases; in mixed motion, speed should both increase and decrease, which is absurd. However, he also argued that "every violent trajectory or motion of uniformly heavy bodies outside the perpendicular of the horizon will always be partly straight and partly curved, and the curved part will form part of the circumference of a circle."

The decomposition of the trajectory into three segments allowed Tartaglia to explain the problem using mathematics (fig. 2.7). The initial portion of the trajectory was nearly rectilinear and motion was violent. It was not, strictly speaking, straight, because the weight of the body continually acts on it and bends its path. The final part was also rectilinear, but this time motion was natural. The two segments were joined by a circle arc traversed with violent motion. Strictly speaking, the arc was not circular, but Tartaglia argued that it could be taken as such because the circle did not differ sensibly from the real curve. Tartaglia was trying to use curves he could treat mathematically, namely the straight line and the circumference, though without being able to provide a suitable justification for his choice. In *Quesiti et inventioni diverse*, however, Tartaglia did state that the trajectory was

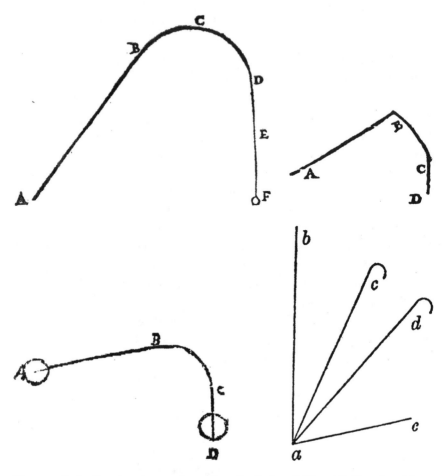

Figure 2.7. Projectile trajectories according to Tartaglia, Cardano, Baldi, and Galileo. *Above left,* Tartaglia *(Nova scientia). Above right,* Cardano *(De subtilitate). Below left,* Baldi *(Exercitationes). Below right,* Galileo *(De motu antiquiora).* Galileo showed different trajectories of bodies shot vertically as in *ab*, obliquely as in *ac* and *ad*, and nearly horizontally as in *ae*.

always curved. He claimed that the longest shot is achieved at an elevation of 45 degrees, but his proof was far from rigorous. The physician Gerolamo Cardano discussed the shape of a trajectory in a way similar to Tartaglia's. For Cardano, however, the central part of the trajectory resembled a parabola. The Urbino mathematician Baldi too saw it in a similar way, except that he did not specify the shape of the central portion, claiming that it was "curva & mixta linea." Both probably relied on Tartaglia.[36]

In the conclusion of the chapter, and of the essay, Galileo explained the paradox of the inclined plane. He argued that a body shot along an inclined plane held obliquely moves further if the plane is closer to the horizontal position than to the vertical, because the plane holds a greater portion of the body's weight. Thus a body shot into the air would behave very differently from one shot along an inclined plane. It is not entirely clear from the text whether at that point Galileo had performed experiments on balls' trajectories on inclined planes, but we know that he did perform them, probably in the early 1590s. We have confirmation of this from a letter by Muzio Oddi to Cavalieri (stating that Galileo had performed experiments on parabolic trajectories with dal Monte), from dal Monte's manuscripts, and from Paolo Sarpi's diary. We have no direct evidence as to whether and when they met, but it is likely that they did—either in January 1590, for example, when dal Monte, who was Visitatore of the Tuscan fortresses, was at Livorno for four weeks, or in 1592 when Galileo was on his way to Venice. An entry in Sarpi's diary dated 1592 or early 1593 indicates that Galileo discussed with Sarpi the experiments he had probably performed with dal Monte.[37]

Despite their protestations about the impossibility of using mathematics to deal with matter, dal Monte and Galileo experimented with motion and the trajectory of projectiles. Dal Monte experimented with crossbows and guns, and also with throwing an inked ball over an inclined plane, satisfying himself that the curve was symmetrical and similar to a parabola or hyperbola. He added that at each point of the trajectory, motion was always resulting from both violent and natural components, thus contradicting Aristotle and Tartaglia.[38] He also compared it to the curve marked by a chain hanging between two nails infixed in a wall, which at first sight resembles a hyperbola or a parabola. In this case too dal Monte argued that the shape of the curve results from a combination of natural and violent components, extending somewhat the Aristotelian dichotomy to a case in which there is no motion. Presumably the chain's tendency to fall toward the center of the Earth was natural, while the nails provided the violent component preventing the chain from reaching its natural position. Dal Monte tried to account for the shape of the projectile's trajectory in terms of the shape of the hanging chain not simply because they resemble each other, but also because they are generated in a similar fashion from a combination of natural and violent components. This analogy must have been appealing because of its potential for linking statics to motion. Notice that the bridge linking them stands on two pillars: the identification of common causes and the similarity in the outcome or the effect. Despite these appealing features, this was a far cry from an Archimedean axiomatic science, and it is not surprising that those early experiments remained unpublished at that time.[39] Although it is not clear at whose behest those experiments were performed, Galileo is the more likely candidate. His less refined speculations in the last chapter of E23 and the range of experimental tests on motion in *De motu antiquiora* point to him. Galileo was already challenging Aristotelian orthodoxy in that work, whereas dal Monte's view of mathematics was not anti-Aristotelian and his approach to motion was more cautious and problematic.[40]

Although Galileo's speculations and experiments with dal Monte were not a science, they reveal an attempt to uncover mathematical regularities. Galileo and dal Monte used a simple machine, the inclined plane, to investigate projectile motion. They used it experimentally, trying to produce physical, measurable curves. It is not clear what Galileo made of those experiments, besides the fact that he later mentioned them in the second and fourth days of the *Discorsi*. The results stated in dal Monte's manuscripts seem more advanced than those announced by Galileo in chapter 23 of E23; therefore it seems reasonable to date his meeting and experiments with dal Monte after *De motu antiquiora*. Subsequently Galileo may have thought that the issue was too complex and decided to start afresh from the simpler case of falling bodies. However, he claimed that he was first stimulated to the investigation of motion by the desire to find the trajectory of projectiles. Both with the catenary and the projected ball on the inclined plane there is an almost seamless transition between experiment and geometry. In both cases the experiment produced a curve that could then be measured and studied.[41]

In *De motu antiquiora* Galileo was caught in a conceptual and terminological maze at the intersection between Aristotelianism and a new science. For example, Galileo still used the Aristotelian term *levitas*, meaning the absolute lightness of a body, despite the fact that in his Archimedean framework all bodies were heavy and "levity" was the effect of their specific weight being less than that of the medium. The problems and contradictions, as well as in many cases the lack of experimental confirmation, led Galileo to leave his work unpublished. Even the attempted elaboration in E10 did not lead to satisfactory results.[42] The similarities to the work by Benedetti, however, testify to the crucial role of Archimedes in the early efforts toward a mathematical science of motion.

Within the framework of Galileo's early reflections, we notice the inclined plane entering the scene and taking on a multiplicity of roles, both conceptual and experimental, as a bridge toward the investigation of motion. Galileo tried to link speed with slope and then to test against each other his assumptions about speed, slope, and buoyancy. In *De motu antiquiora* and in the experiments with dal Monte, the inclined plane was also used to conceptualize horizontal motion and to investigate the trajectory of projectiles. Thus the inclined plane allowed Galileo to slow down the motion of falling bodies and to visualize the trajectory of projectiles. Despite the problems he encountered, his experiments are indicative of his concern with measurement and empirical data.

2.5 Mazzoni, Stevin, and Galileo

During his three years of teaching mathematics at Pisa, Galileo was a colleague of the Platonist philosopher Jacopo Mazzoni, a friend of dal Monte who had frequented the Urbino court. Mazzoni and Galileo were quite close, and in a 1590 letter to his father Galileo stated that he was studying with him. Mazzoni's 1597 *Praeludia* is a philosophical treatise based on a comparison between the philosophies of Aristotle and Plato. Galileo received a copy soon after publication and replied with a celebrated letter in which he stated his Copernicanism. Mazzoni also discussed the role of mathematics in the study of *physica* and defended Plato's use (against Aristotle) of mathematics in the investigation of the shape of the elements in the *Timaeus*. Such views went hand in hand with Galileo's. *Praeludia* dealt with several issues pertinent to motion and mechanics and relied on Benedetti's *Diversarum speculationum liber*, which it often quoted. Mazzoni started from three suppositions: (1) motion stems from a *virtus motrix*, which for bodies moving downward is gravity and which for bodies moving upward is a force (or *vis*) exerted by the heavier medium; (2) each of the four elements, against Aristotle, is weightless in its proper place (for example, water in water weighs nothing); and (3) the weight of a body in a medium is reduced by the weight of an equal volume of medium. On this basis Mazzoni claimed that the speed of falling bodies is not proportional to their weight and inversely proportional to the resistance they encounter, but rather is proportional to the difference of density between the body and the medium. Referring to Benedetti's later views, Mazzoni qualified this statement with the words "insofar as they encounter the same resistance in the medium." Although he did not discuss motion in a void, Mazzoni provided a reasonable account of Benedetti's views. Thus, although it is not certain that Galileo knew Benedetti's views firsthand by the time he composed *De motu antiquiora*, he certainly knew them through Mazzoni's 1597 account.[43] Most likely, however, by then Galileo's speculations were advancing to the point of rendering Benedetti's views of little use.

Benedetti's views came also to Stevin's attention through Jean Taisnier. In his *Appendix to the Art of Weighing* printed at the end of *The Elements of Hydrostatics* (1586), Stevin attacked the views on falling bodies of both Aristotle and Taisnier (the plagiarist of Benedetti's *Demonstratio). As we saw in the previous chapter, Stevin was opposed to using statics or the art of weighing in connection with considerations about motion. He argued, against Aristotle, that for two similar bodies falling in air, speed is not proportional to weight. He also argued, against Taisnier, that these same two similar bodies do not fall in equal times. Stevin does not talk about *gravitas in specie,* but from the subsequent discussion it seems reasonable to interpret the word "similar" as meaning "of the same material" and, therefore, equally dense. Unlike Benedetti, Stevin did not simply provide a theoretical refutation of Aristotle's views but also, as he reported, went to the trouble of testing Aristotle's claims about falling bodies in collaboration with his friend Jan Cornets de Groot, father of the famous jurist Hugo Grotius and Burgomaster of Delft between 1591 and 1595. In order to refute Aristotle, they simultaneously dropped two lead spheres, one ten times heavier than the other, from a height of thirty feet onto a board. From the sound of the impacts, they concluded that the spheres fell with the same speed, not, as claimed by Aristotle, with different speeds proportional to their weights.[44]

Although the experimental refutation of Aristotle seems to be in agreement with Taisnier's account, Stevin performed other experiments to refute the latter as well. From a height of five or six feet he dropped both a single hair or fiber of cotton and a packet of the same material weighing one pound and similar in shape to the fiber. The fiber, whose material is more compact than the packet, takes more than twenty-five times longer to fall than the packet does, whereas according to Taisnier, Stevin claimed, they should have fallen in equal amounts of time. Even more dubious-sounding are Stevin's experiments concerning the ascent of air bubbles in water. Stevin claimed to have observed that large bubbles rise much faster than smaller ones, again against the view that similar bodies move in a medium at the same speed.[45]

Stevin's theoretical arguments rely on two basic principles. First, he argued that every body in motion experiences some impediment, such as that from the contact between the air and the surface of the falling body. For bodies of similar shapes but different sizes, however, volume and surface are not proportional; the former grows as the cube of a linear dimension and the latter as the square. As we have seen, a similar line of reasoning had been independently adopted by Benedetti in **Demonstratio* and *Diversarum speculationum liber.* Stevin's second argument seems to be a peculiar application of the principle of continuity. He started by claiming that there is no proportionality in the speeds of bodies in water if one sinks and the other floats. Restricting the example to bodies that both sink, Stevin imagined an infinite number of lighter and lighter bodies approaching the point where one sinks and the other floats, and he appears to be disturbed by the idea that a proportionality involving speed should end in a limiting case that involves no speed at all. The example is less troubling to us because we can accept that in that limiting case, speed is equal to zero. The discussion is plagued by terminological and conceptual difficulties because there is no clear distinction between absolute and specific gravity.[46] Overall, Stevin was trying to argue that no proportion can be established involving motion—in other words, that the mathematical tool of the theory of proportions is inadequate to describe a complex phenomenon like motion, even in simple cases.

If we can be sure that Mazzoni read Benedetti and that Stevin read Taisnier's rendering of Benedetti, we are on much less certain terrain about Galileo. In *De motu antiquiora* Galileo, like Benedetti, relied on Archimedean hydrostatics as a way to investigate falling bodies. The similarity of approach and subject matter supports the thesis that Galileo knew at least some of Benedetti's works. For example, in arguing that bodies of different sizes but with the same specific weight fall in the same medium with equal speed, Benedetti imagined decomposing the larger body into smaller ones connected by weightless threads. Similarly, Galileo imagined connecting bodies of different sizes in a *reductio ad absurdum*. Following Aristotle, the conjoined body should be at the same time both faster, because it is heavier, and slower, because the smaller body slows down the larger one. Moreover, Benedetti and Galileo wrote in similar terms about the projection of balls by guns or bats, arguing that an impressed force acts more strongly on a body offering substantial resistance than on a body offering only weak resistance. This is why gunpowder exploding inside a sealed container produces a bigger effect than an explosion in an unsealed container.[47]

On the other hand, Benedetti is not mentioned in Galileo's writings, and Alberti and Tartaglia too had stated views on buoyancy similar to those later put forward by Galileo. A close examination of many passages in *De motu antiquiora* leads one to wonder whether Galileo had seriously pondered Benedetti's works, if he had even read them at all. The analogy concerning the bodies connected by weightless threads is not conclusive evidence, for Benedetti *decomposed* a large body whereas Galileo *connected* smaller ones.[48]

Other passages are rather surprising. Galileo refers to Scotus, Saint Thomas, and Philoponus as having held views different from Aristotle's, though not as having refuted him. Galileo continues: "For Aristotle assumes that the speed in one medium is to the speed in the other, as the rareness of the first medium is to the rareness of the second. And no one up to now has ventured to deny this relation."[49] One may also wonder how closely Galileo had studied Philoponus, whose role with regard to motion was already clear to Commandino, and Benedetti, who had indeed challenged Aristotle precisely on that point. A similar reasoning applies to the discussion of resistance to motion. Benedetti had devoted several chapters against Aristotle to this problem, reaching conclusions differing considerably from Galileo's. Benedetti developed his views on this subject from the *Resolutio* onward, correcting himself in the ****Demonstratio* and *Diversarum speculationum liber*, in which he conceived of resistance as proportional to the surface of a body in its direction of motion. By contrast, Galileo in chapter 10 of E23 seems to identify resistance with buoyancy, much as Benedetti had done in the earlier *Resolutio*. In chapter 13 Galileo attempted to refute the opinion of the fourth-century commentator Themistius, who had argued that in a void all bodies fall with equal speed. Benedetti, by contrast, agreed with Themistius.[50] Moreover, several commentators have noted important terminological and conceptual differences between Benedetti and Galileo. When Galileo used the term *impetus*, or *virtus impressa*, he meant a self-dissipating entity comparable to the heat of an iron after it is removed from the fire. By contrast, Benedetti had in mind something that is conserved.[51] Both criticized Aristotle for having stated that rectilinear and circular motions cannot be compared or have no ratio. Although both Benedetti and Galileo invoked Archimedes in their refutations of Aristotle, the former referred to *Quadrature of the Circle*, in which the area of the circle is compared to that of a right triangle with one side equal to the radius and another equal to the circumference. Galileo, however, referred to Archimedes' *On Spirals*, "where a straight line is found equal to the circumference of the circle around the spiral of first revolution." Furthermore, in the same discussion, Benedetti argued that Aristotle would have been right in stating that the circular motion of heavenly bodies is not comparable to the rectilinear motion of terrestrial bodies, because the former is uniform whereas the latter always increases its speed because of the continuous action of its natural propensity toward the center of the Earth. This important observation is related to Benedetti's view that impetus is conserved. Although both Benedetti and Galileo established a link between, on the one hand, the specific densities of the body and the medium and, on the other, the speed of the body's fall, the former conceived the speed to increase indefinitely, whereas the latter considered the speed to be constant after an initial period.[52]

As to the similarity between Benedetti's and Galileo's views on the projection of balls by bats and guns, once again one should not take it blindly as evidence of Galileo's having read Benedetti. Galileo's predecessor at Padua, Moletti, made analogous statements in an unpublished treatise on mechanics dating from about 1576. Discussing projectile motion, Moletti argued that the projectile must be able to yield to the moving force, and it must be able to resist as well. This may sound paradoxical, but, Moletti argued, "resistance is the cause of movement, in so far as resisting and yielding impress the moving power on the mobile with more vehemence." My contention here is not, of course, that Moletti's treatise was a common source for Benedetti and Galileo, but rather that opinions similar to theirs were not uncommon at the time. The same holds true for the claim about the power of gunpowder explosions in small sealed spaces, which is also found in Biringuccio's *De la pirotechnia*, for example.[53]

Another possible avenue leading from Benedetti to Galileo goes through dal Monte. Dal Monte's notebook shows that he was familiar with and critical of *Diversarum speculationum liber*. It seems plausible that dal Monte and Galileo discussed Benedetti's work together when they experimented with projectiles and hanging chains. We have seen, however, that their meeting likely occurred after Galileo had completed *De motu antiquiora*.[54]

In conclusion, direct evidence that Galileo read Benedetti is lacking, and overall it seems implausible that Galileo studied Benedetti's works carefully. The analogies between their views on falling bodies, however, emphasize the significance and impact of the Archimedean tradition as a vehicle for early thinking on the subject. Once again we witness the attempt to employ an established area of the mathematical disciplines to explore and find regularities and perturbations in new domains.

CHAPTER THREE

The Formulation of New Mathematical Sciences

Reflecting over the matters on motion, about which I lacked a
wholly indubitable principle to be set as an axiom to prove the accidents I had observed, I reached a proposition with many natural
and self-evident features.

Galileo

3.1 The Broadening of the Mechanical Tradition

In 1592 Galileo moved from the mathematics chair at Pisa, which he had held
for three years, to the more prestigious and better-paid chair of mathematics at
Padua, a position he retained until his return to Tuscany in 1610. He owed both
jobs to dal Monte's help, but links with the Marquis were not simply a matter of
patronage. Dal Monte's foundational concerns were crucial to Galileo. In the 1612
Discorso on hydrostatics he tried to use the balance to account for the siphon, and
in the second day of the *Discorsi* he relied on the lever in his work on the beam.
In tackling the science of motion Galileo moved away from dal Monte's strict tradition, both because the balance proved an unsuitable starting point and because
dal Monte doubted that a full mathematical science of motion was possible. After
repeated failures, Galileo eventually relied on the inclined plane and pendulum,
but his search for suitable postulates continued after the publication of the *Discorsi*. His search for axiomatic formulations posed the problem of identifying and
justifying new axioms, but his primary ambition was not to find new principles
but rather to find new consequences of established ones.[1]

In this chapter I discuss the sciences of motion, waters, and resistance of materials. It is instructive to compare and contrast them not only with respect to the
problem of foundations but also with regard to other issues. Engineers dealing
with water flow, shipbuilding, and military affairs approached the new mathe-

matical sciences in different ways. As to physical causes, in the *Discorsi* Galileo did not wish to speculate on the cause of gravity, yet he put forward daring speculations on the cause of cohesion. This dichotomy shows that simplistic categorizations of his method are untenable. Galileo's efforts to rely on the lever highlight his reluctance to base the formal presentation of his sciences on experimental foundations. No experiment was invoked in the formal presentation of the science of the resistance of materials, for example, and his use of the balance in his study of buoyancy is revealing of his concerns. Moreover, in *Le mecaniche* and in the second day of the *Discorsi* he presented a new proof that a balance is in equilibrium when the arms are inversely as the weights, showing his need to prove even such an empirically well-tested proposition. The last revision to his science of motion in the posthumous 1656 edition of the *Discorsi* was the addition of a proof of his only axiom, thus removing its partly experimental justification. Of course, axiomatic formulations are only part of the story, and Galileo did rely on experiments in a variety of ways, from heuristic to rhetorical. In order to understand his way of proceeding, it seems helpful to distinguish the various phases of his itinerary, such as the private heuristic stage, the informal presentation in print, and the formal axiomatic formulation. Further, I wish to emphasize the role of his intellectual and material manipulation of tools such as levers, inclined planes, and pendulums. The new mathematical disciplines and the principles underlying them relied on a creative mathematical, experimental, and conceptual handling of those objects.

Manuscript sources, lecture notes, and letters show that during his eighteen years at Padua, Galileo took the decisive steps toward a new science of motion. Section 3.2 relies on *Le mecaniche*, a short treatise linked to Galileo's Padua lectures, and the letters to Contarini in 1593, dal Monte in 1602, Paolo Sarpi in 1604, and Luca Valerio and Antonio de' Medici, both in 1609. Section 3.3 examines the dispute on buoyancy between Galileo and the Aristotelian philosophers, arguing that Galileo's work was a springboard for Castelli's science of waters. Although the *Dialogo* is primarily devoted to the world-systems, it is there that Galileo published some of his findings about motion, and it is through the *Dialogo* and its later Latin editions that most scholars north of the Alps became familiar with his findings. Section 3.4 deals with those portions of the *Dialogo* relevant to motion. In the *Discorsi* Galileo put forward the rudiments of another mathematical science dealing with the resistance of materials, which is the subject of section 3.5. Finally, section 3.6 discusses the formal presentation of the science of motion and the role of experiments in Galileo's work.

3.2 Galileo at Padua and the Science of Motion

At Padua, Galileo belonged to a number of intellectual circles, such as those of Pinelli and Contarini, of technicians such as the instrument maker Marcantonio Mazzoleni (who moved with his family into Galileo's house in 1599 in order to fabricate compasses, receiving in exchange lodging, meals, and a small stipend), and of students of the mathematical disciplines (who often resided in his house too).[2]

In 1593 an exchange on the position of oars in galleys occurred via Pinelli between Galileo and Contarini, who was one of the three Provveditori to the Venice Arsenal. The problem was related to the lever but was complicated by the fact that the whole system is in motion. Galileo's strategy was to shift the oar outward in order to increase the lever's efficiency. His solution, making the oarsmen's task very hard, betrayed a lack of familiarity with practical issues, but the exchange is significant for the parallel with Contarini's previous exchange with dal Monte, and for the link it establishes with the Arsenal.[3]

Within the first few years at Padua, Galileo completed *Le mecaniche,* first published by Mersenne in French (Paris, 1634). We have some preliminary versions dating from 1593 and 1594 before the final version, generally dated between 1598 and 1600. Although simple machines seem to multiply force, Galileo wished to clarify the principles on which they operate with respect to the weight to be moved, the force or power to move it, the distance, and the time required. For example, a given force can move an arbitrary weight, provided the weight can be divided into smaller components to be moved one at a time. Galileo also included three definitions: of gravity, or the tendency of bodies to move downward; *momento;* and center of gravity. By *momento* he meant the tendency to fall or the impetus resulting from the combination of gravity and position, for example the position of a weight on the arm of a balance. The introduction of the term *momento* is of great significance because Galileo used it as a bridge from statics to the study of motion, shifting from *momento* or gravity as a cause to the impetus or the effect.[4] By "center of gravity" Galileo meant, following Commandino, that point around which the parts have equal moments. *Le mecaniche* treats simple machines such as the balance and lever, the wheel and axle and winch, the pulley, the screw and Archimedes' water-raising screw, and the force of percussion. I discuss the screw, which Galileo used in his study of the inclined plane, and the force of percussion. Much in dal Monte's tradition, Galileo used the lever to explain the other machines. He also attempted a reformulation of Archimedes' doctrine based on a series of intuitive operations on a prism attached to the lever. Galileo's reformulation, which was later included in day 2 of the *Discorsi,* does not contain new re-

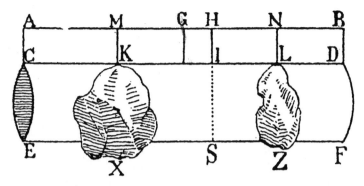

Figure 3.1. Galileo on the equilibrium of the balance *(Le mecaniche).* A solid of uniform density *CDFE* is attached by the extremes *C* and *D* to a balance *AB* suspended from its middle point *G*. If we cut it along *IS* and suspend both parts from the balance by string *IH*, the system will still be in equilibrium. If we now suspend the two unequal parts by their middle points *K* and *L*, by means of *KM* and *LN*, and cut strings *AC, HI,* and *BD*, the two solids will remain in equilibrium. Now the two weights *CS* and *SD*, or *KX* and *LZ*, must be of equal moment. Galileo can show that *NG:GM::CS:SD*. Therefore the balance is in equilibrium if the weights are inversely as the distances from the fulcrum at which they are hung.

sults in mechanics, but it is revealing with regard to his views on the structure of the science of equilibrium and represents a blueprint for his later attempts to formulate a science of motion. In both cases he was not seeking principles based on experiments. The example of the lever is especially revealing because the notion that equilibrium occurs when the weights are inversely as the distances from the fulcrum was universally accepted and would have been common knowledge in every market square since ancient times. Yet this was not enough for Galileo, as it had not been enough for Archimedes. That proposition had to be proved from more secure principles relying on symmetry, such that equal weights hanging from equal distances are in equilibrium, on the example of Archimedes, *On the Equilibrium of Planes* (fig. 3.1).[5]

In his work on the screw, Galileo criticized Pappus's solution to the inclined plane on two grounds. He argued that once the accidental impediments—which are not to be considered by the theoretician—are removed, no sensible force is required to move a weight along a plane parallel to the horizon. This claim posed the basis for a rapprochement between equilibrium and motion. Galileo also criticized Pappus's peculiar way of tackling the problem, arguing that he had not solved it. Galileo's solution derived from *De motu antiquiora,* but this time he removed the problematic proportions involving motion and treated the problem using *momento.* He noticed an interesting symmetry, namely that while the body attached to the bent arm of the lever moves along a semicircle, it is possible to conceive it free from the lever and moving along a bent surface, the two cases being equivalent (see fig. 2.6). This remark is suggestive of pendular motion, a topic Galileo had barely mentioned at Pisa but was to explore more fully at Padua. It is unclear whether at this stage Galileo still saw acceleration as an accident in free fall.[6]

As to the force of percussion, Galileo mentioned the force of a hammer hitting a nail, concluding that a power *(potenza)* that can move a small resistance over a long distance will be able to move a resistance one hundred times larger over a distance one hundred times smaller. For example, a hammer with a resistance equal to four is moved with a force such that, if it were free, it would move ten *passi*. If the hammer hits a beam with a resistance equal to four thousand, it would move it by one-thousandth of the ten *passi*. Galileo continued to speculate on the force of percussion until the end of his life and drafted an additional day of the *Discorsi* on this topic. He performed experiments too at Padua, dropping heavy balls either onto one arm of a balance, trying to determine the counterbalancing weight on the other side, or onto the strings of different bows and measuring their deflections in comparison with the deflection produced by balls simply hanging from the strings. The stronger the bow, the greater the static weight equivalent to the falling one. Galileo's conclusion was that since the force or tension of the bow can be increased indefinitely, the equivalent static weight increases indefinitely as well; therefore the force of percussion is infinite. These investigations represent a link between mechanics and Galileo's research on motion.[7]

Galileo's 1602 letter to dal Monte was part of a larger exchange that sadly has not survived. Writing to his mentor, Galileo announced his discovery of the isochronism of pendular oscillations, claiming to have established it experimentally. He then proceeded to provide to his skeptical correspondent specific indications as to how experiments ought to be performed. The pendulums consisted of two leaden balls attached to thin threads about two to three *braccia* long, one *braccio* being just under two of today's feet. Galileo's insistence that the balls could be of different sizes but must be both made of lead may suggest that he still believed specific gravity to be a factor in the speed of falling bodies. By removing them from the equilibrium position, one by a large amount, the other very little, and counting their oscillations with help from a collaborator, Galileo argued that they were exactly isochronous. He insisted on the truth of this claim for the rest of his life, even though it is observably incorrect, as Mersenne was to claim in the 1630s. Probably Galileo brushed aside a discrepancy of one oscillation in every two or three dozen as a perturbation due to a variety of effects, as he did in many other instances related to physical effects. Galileo also announced his discovery that it

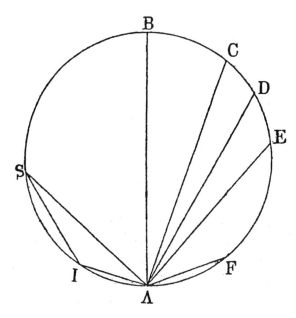

Figure 3.2. Galileo's chord theorem (1602 letter to dal Monte). A body falls in equal times along the diameter *BA* of a circle and along its chords *CA*, *DA*, *EA*, and *FA*. One of Galileo's attempted proofs relies on the idea that the speeds are as the *momenti*. If the *momenti* are as the distances, the speeds are as the distances as well, and the motions occur in equal times. Paradoxically, the proof may also be applied to uniform motions.

takes the same amount of time for a body to fall vertically along the diameter of a circle as for it to fall along its chords, a result known as the chord theorem (fig. 3.2). Finally, Galileo claimed that he had proved that a body falls through the two segments *SIA* faster that along the chord *IA*. He had attained these results "without transgressing the bounds of mechanics," by which he meant that he was relying on the properties of the inclined plane and the equality between *momento* of gravity and speed. Obviously those bounds were a matter of common concern with dal Monte. Galileo was seeking a link to pendular oscillations by proving that arcs *SIA* and *IA* are traversed in equal times, namely the experimental result he had announced earlier in the letter.[8] Here experiments were used heuristically, pointing to a result to be demonstrated from first principles at a later stage.

From this letter it is not clear whether Galileo still adhered to the view that bodies fall with a nearly constant speed, believing the initial increase in speed to be a temporary perturbation, as in *De motu antiquiora*. Paradoxically, the chord theorem could be proved on the basis of the deceptive results based on uniform motion in free fall from *De motu antiquiora*. Although many elements may have led Galileo to abandon his approach, the letter to dal Monte and its accompanying manuscripts provide no clear evidence that he in fact did abandon it, and they do not mention accelerations.[9]

During the early Padua years Galileo freed himself from a key notion from *De motu antiquiora*, namely that the speed of fall depends on the excess of the specific gravity of the falling body over that of the medium. Galileo had probably come to realize that all bodies regardless of density fall with the same speed, either by means of experiments with inclined planes or pendulums or by reasoning about joining bodies with different specific gravities. Similarly, Galileo may have grasped that distances traversed by falling bodies are as the square of the times by means of experiments with inclined planes. The specific path he followed, however, cannot be easily reconstructed.[10]

In his letter to Sarpi of 16 October 1604, Galileo introduced the notion of acceleration, arguing that distances traversed by falling bodies are as the squares of the times and that they increase following the odd-number rule. Acceleration takes center stage here and is no longer considered as a perturbation occurring at the beginning of motion. Galileo was trying not simply to find new propositions on motion but also to establish an axiomatic science of which they would be part. In 1604 he submitted to Sarpi his "indubitable" principle according to which the speeds increase in proportion to the distances traversed—a principle that is deceptively intuitive but incorrect, since the speeds increase as the times elapsed. Galileo claimed that at each point the body has enough impetus to make it raise to *a*, "whence it is manifest that the impetus at the points *d*, *c*, and *b* go decreasing in the proportions of the lines *da*, *ca*, and *ba*" (fig. 3.3). Since Galileo had argued that the projectile "will return downward by the same degrees by which it went upward," the same conclusion will apply to falling bodies as well.[11]

Manuscripts datable to the same period show that Galileo's statement was part of an investigative program that occupied him for years. Koyré argued that Galileo was led to the proportionality between speeds and distances by the intuitively privileged role both of spatial geometry over time and of direct proportionality over more complex relations.[12] Koyré's claims can be clarified with the help of one manuscript that is especially close to the letter to Sarpi and provides an important clue as to what led Galileo to his "indubitable" principle. The manuscript states, "This principle appears to me very natural, and one that corresponds to all the experiences that we see in the instruments and machines that work by percussion, in which the percutient produces so much the greater effect, as the height whence it falls is greater; and once this principle is assumed, I shall prove the rest."[13] Galileo refers here to percussion, thus establishing the link between those investigations and the study of motion I hinted at above. He established a proportion between

Figure 3.3. Diagram from Galileo's letter to Sarpi. Galileo claimed that a body falling through *ad* has at each point a degree of speed proportional to the distance traversed, *ab* in *b*, *ac* in *c*, and *ad* in *d*.

height of fall and the effect of the percussion. It seems plausible that by joining this proportion with another between speed and the effect of percussion, Galileo would have concluded that the height of fall is proportional to speed. Indeed, in several later passages Galileo stated that the speed of a falling body could be ascertained by the percussion it produces. This is an intuitive idea and it is possible to test it experimentally by dropping balls onto a soft target, such as clay or wax, and measuring the indentations.[14]

Galileo's association of speed with distance did not rely only on percussion or the vertical motion of projectiles. In the letter to Sarpi he referred also to the vertical motion of projectiles, associating speed with distance.[15] It is plausible that Galileo was led to the erroneous proportion between speeds and distances by comparing falls along an inclined plane and along the vertical. That equal speeds are reached at equal vertical distances from rest deceptively suggests the incorrect proportionality between speed and distance.[16] All these investigations are mutually reinforcing and show the mechanical roots of Galileo's practice and his reliance on objects such as inclined planes, falling and projected bodies, and their geometrical representations.

In the manuscripts accompanying the letter to Sarpi, we see Galileo beginning

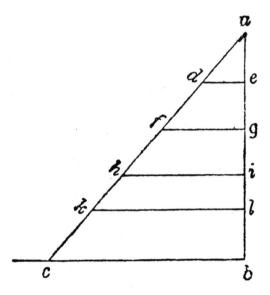

Figure 3.4. Inclined plane and triangle of speeds (elaboration of Galileo's manuscript 179v). For an object falling from *a* along the vertical *ab* or an inclined plane *ac*, the speed of fall is the same at the same vertical distances *ed, gf,* etc., from the vertex *a*. If the vertical represents a distance, we have a spatial representation of fall along an inclined plane, but if it represents time, we have a graph known as the triangle of speeds where *ed, gf,* etc., represent increasing speeds at different times, each known as *gradus velocitatis.* At times Galileo mixed the two, claiming that speeds are as the distances. If *ab* represents time, *cb* represents the maximum degree of speed. The mean-speed theorem states the equality of the distances traversed in equal times with a uniformly accelerated motion from rest, or with a uniform speed equal to the mean degree of speed of the accelerated motion. Galileo replaced the mean degree of speed with half the maximum speed *cb* (see also fig. 3.17). The double-distance rule states that the distance traversed with a uniform speed equal to the maximum degree of speed *cb* acquired with a uniformly accelerated motion from rest is double the distance traversed with that accelerated motion in the same time.

to adopt a new representation of motion and a new concept of speed. Uniform speed in free fall was no longer tenable, and in order to deal with variable speeds he had recourse to different concepts, terminology, and visual geometrical representations. He associated at each point a degree of speed, *gradus velocitatis* or *momentum velocitatis,* represented by the length of a segment and broadly corresponding to the notion of instantaneous speed (fig. 3.4). He took the aggregate of all the infinitely many *momenta velocitatis* with which the body has fallen over a certain distance, corresponding to the area of a triangle, to be still a speed (we consider it to be proportional to the distance traversed). Besides *gradus velocitatis,* other terms such as *latitudo, intensio,* and *uniformiter difformis* show Galileo's reliance on the medieval terminology associated with the *calculatores,* a group of philosophers active at fourteenth-century Oxford and Paris. Their works and methods were broadly known in the sixteenth century, but we still do not know Galileo's exact sources. The visual representation and theorems Galileo used also resemble those of the *calculatores,* as with the mean-speed theorem and the double-distance rule.[17]

Galileo probably became familiar with this tradition when he was professor at Pisa around 1590, although he started relying on it in his own investigations only around 1604. Since medieval scholars had developed theorems about speed and time, the reliance on distance in the letter to Sarpi and the confusion between the geometrical representation of an inclined plane and the graph of the triangle of speed are problematic for those who see the *calculatores* tradition as a direct inspiration for Galileo's science of motion. To my knowledge, the issue of medieval predecessors to Galileo's work did not arise in his time, despite the fact that many of his contemporaries were openly hostile to him and would have welcomed any opportunity to question his originality; nearly three centuries elapsed before historians raised the question. Moreover, the extensive body of preparatory manuscripts shows his struggle primarily with mathematics and experiment rather than medieval sources, whereas overall the *calculatores* tradition was concerned with abstract relations between changing magnitudes rather than with an empirical investigation of nature. It seems that when Galileo abandoned his views on uniform motions in free fall, he lacked the mathematical tools necessary to deal with accelerated motion and began actively using the *calculatores* tradition precisely in this fashion.[18]

Before leaving Padua, Galileo formulated a different rule for falling bodies and rejected the notion of proportionality between distances and instantaneous speeds.[19] Some scholars have argued that despite this change, some features associated with that notion of proportionality remained even after Galileo corrected it; since the product of speeds and distances has no physical meaning, Galileo did not consider the aggregate of all the speeds to represent the distance traversed even when he took speeds to be proportional to time, or $v \propto t$. Cavalieri followed Galileo in this regard, but Cavalieri's thinking did not go through a phase involving the notion of speed being proportional to the distance, or $v \propto s$. Moreover, at some point Descartes accepted the relation $v \propto s$, yet he later took the areas of the triangle of speeds to be proportional to the distances traversed. Therefore there is no necessary correlation between $v \propto s$ and the failure to see the area of the triangle of speeds as proportional to the distance traversed. Others have looked at Galileo's strict adherence to the theory of proportions, whereby nonhomogeneous magnitudes cannot be multiplied: for example, it is not possible to multiply speed and time to arrive at a distance. Galileo, however, may have taken the distances

traversed to be proportional to the areas of the triangle of speeds without necessarily taking those areas as the products of speed and time; this was the route
Borelli followed in *De vi percussionis* (Bologna, 1667), in which he relied explicitly
on Cavalieri's method of indivisibles, a powerful tool for finding the areas or volumes of geometrical figures by considering aggregates of lines or surfaces. Borelli
argued that as the *fluxus* of a line (or the aggregate of infinitely many lines) is no
longer a line, so the *fluxus* of a speed (or the aggregate of infinitely many degrees
of speed) is no longer a speed. Galileo, however, seemed to conceive the aggregate
of all the speeds more as a "sum" and hence as a type of speed itself, rather than
a distance.[20] In conclusion, although we understand the proportionality between
distances of fall and the square of the times to be equivalent to the proportionality between instantaneous speeds and times, that equivalence is hardly self-evident. Galileo attained an understanding of the latter only with great effort and
several years after the former. Few examples could emphasize more effectively the
need to pay close attention to the original mathematical concepts and notation.
Galileo may have attained this result either through experiments or through
mathematical reasoning by recombining some of his previous findings, such as
the proportionality between distance and the square of the time and the double-
distance rule (fig. 3.5). A similar line of reasoning was later incorporated into the
Discorsi.[21]

The theory of proportions proved problematic in the transition from uniform
speed to accelerated motion, and Galileo was puzzled at the paradoxes that
emerged. A body falling along the vertical and along an inclined plane, in a given
time, covers a greater distance along the vertical than along the incline; therefore
it has a greater speed along the vertical. However, drawing a parallel to the horizon, the body reaches it with the same speed along any incline; therefore one could
also argue that the speed is the same along the incline and the vertical. It might
clarify matters for us to write $v = g_1 t_1 = g_2 t_2$, where v is the common final speed,
g_1 and g_2 are the accelerations along different inclines, and t_1 and t_2 are the respective times; but this was not an option for Galileo, who did not use such parameters as g. In fact, it has been argued that Galileo never fully disentangled this
paradox.[22]

Other relevant documents date from 1609. The first is a reply by Valerio in
Rome to a lost letter by Galileo with two principles on falling bodies, showing that
Galileo was still seeking suitable principles for his new science. Valerio was rather
confused about the first one, but according to a reasonable reconstruction Galileo
would have stated that given two inclined planes of equal height, one twice as long
as the other, at each instant the *impetus* acquired is double on the shorter plane.
The second principle states that a body falling along an inclined plane acquires as

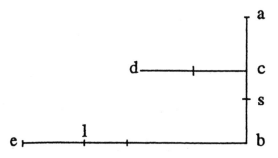

Figure 3.5. The speeds are as the times (elaboration from Galileo's manuscript 91r). Here Galileo superposed two diagrams, one representing distances and the other time. A body released at *a* falls vertically along *acb*. At *c* it will have acquired a speed enabling it to cover in the time of fall the distance *dc* = 2*ac*; at *b* it will have acquired a speed enabling it to cover in the time of fall a distance *be* = 2*ab*. This is the double-distance rule. Thus we have (1) *dc*:*eb*::*ac*:*ab*. The same vertical axis also represents time; *ac* is the time of fall from *a* to *c*, and *as* is the time of fall from *a* to *b*. Let us call them t_{ac} and t_{ab}. From the law of fall, t_{ac}^{2}:t_{ab}^{2}::*ac*:*ab* or ::ac^2:(*ab* $\times$ *ac*). Therefore t_{ac}:t_{ab}::*ac*:*as*, where *as* is the mean proportional between *ab* and *ac*, and also t_{ac}:t_{ab}::*as*:*ab*. Since *eb* is traversed with uniform motion, *eb*:*bl*::*as*:*ac*, or (2) *eb*:*bl*::*ab*:*as*, where *bl* is the distance traveled in the time *ac* with the speed acquired falling from *a* to *b*. Since the times are the same, the moments of speed at *c* and *b* are as *dc* and *bl*. But from (1) and (2), *dc*:*bl*::*ac*:*as* or ::t_{ac}:t_{ab}, namely the moments of speed are as the times.

much *impetus* as if it had fallen along the height of the inclined plane. Both principles would reemerge in different forms at later stages.[23]

The second document is a letter to Antonio de' Medici in which Galileo announced that he had recently found that a ball shot horizontally by a gun in a high position reaches the ground in the same amount of time regardless of whether it falls perpendicularly or is propelled forward a great distance (fig. 3.6). This proposition implies the independence of the vertical and horizontal components of motion, a crucial proposition with implications for the motion of projectiles and cosmology. Galileo also claimed that all the shots reaching the same height take the same amount of time to reach the ground. His illustration shows parabolic trajectories, and related contemporary manuscripts testify to his renewed efforts on this topic; whereas in 1590 the shape of the curve would have been puzzling to Galileo and dal Monte, in 1609–10 Galileo was able to analyze it in its components, showing that the violent motion of a projectile consists of a natural motion downward composed with a uniform motion in any direction (see also fig. 3.18).[24]

Galileo may have also conducted experiments in order to find or confirm that instantaneous speeds are as the times. If a ball falls along an inclined plane set at a given height on a horizontal floor, on a table for example, and is projected horizontally, the time of fall in a parabolic arc from the point where it leaves the table

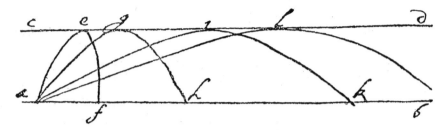

Figure 3.6. The isochronism of projectile motion (Galileo to Antonio de' Medici). If the trajectories of projectiles reach the same height *cd*, all the shots *aef*, *agh*, etc., reach the ground in the same
time.

to the floor is the same regardless of the horizontal speed. Since time is the same,
the distances traversed along the floor are proportional to the speed; therefore the
speed acquired at the end of the fall along the inclined plane can be found by measuring the distances traveled on the floor. Releasing the ball on the inclined plane
at distances proportional to 1, 4, 9, 16, for example, the ball falls along the inclined
plane in times proportional to 1, 2, 3, 4. But the distances covered on the floor will
also be as 1, 2, 3, 4; thus the speeds acquired in the uniformly accelerated fall along
the inclined plane are as the times.[25]

It may be worthwhile here to recapitulate some issues concerning Galileo's use
of experiments in the early phases of his formulation of the science of motion. As
we have seen in previous chapters, sixteenth-century mathematicians from Tartaglia and Benedetti to dal Monte and Stevin were not, or at least were not simply,
closet calculators but men engaged in practical pursuits from military engineering to the testing of simple machines. This was the tradition to which Galileo belonged. At the same time, we must beware of reading modern experimentalism
into the culture and tradition of sixteenth-century mathematicians, who lacked
shared and established criteria for dealing with experimental data and who may
have used such data to test an established theory or to discover new mathematical
relations. If it is difficult to doubt that Galileo, like many of his contemporaries,
performed tests and experiments, it is difficult to ascertain their role and significance.

Distances can be easily measured, but the measurement of times requires
greater ingenuity, and speeds can be measured only indirectly. Galileo may have
inferred experimentally parabolic trajectories (with dal Monte), the relation $s \propto t^2$, as in the inclined plane experiment narrated in the *Discorsi* (see section 3.6),
and that all bodies fall with the same acceleration, but he would have inferred only
indirectly the proportionality between instantaneous speeds and times. The ques-

tion of whether Galileo reached his science of motion on the basis of experimental trials or on that of purely abstract reasoning, both logical and mathematical, has been debated extensively and is harder to settle. Some have claimed that settling this issue would mean grasping a crucial feature of what has been called the Scientific Revolution.[26] The answer to this question seems elusive to me because from Galileo's rough calculations and extemporaneous jottings about an experimental apparatus it is hard to tell whether he was trying to establish or simply to verify a claim. Conversely, from the outline of a proof it is hard to judge whether Galileo was finding there and then the results we read, or whether he was simply giving formal dress to results previously inferred in a different fashion. We do know, however, that Galileo did perform "experiments" at times when he had no firm view as to the outcome, experiments we can thus call heuristic. Two such experiments are those performed with dal Monte on parabolic trajectories and the geometrical shape of a hanging chain.

One case potentially indicative of Galileo's research habits comes from pure mathematics rather than the physical world. Galileo was interested in the cycloid, a curve he thought very elegant. Although he was unable to prove any of its subtle mathematical properties, he sought to find the ratio between its area and that of the generating circle by cutting cardboard shapes and weighing them. He found that ratio to be approximately 3 to 1. As it happens, this is the exact ratio, but Galileo did not believe it to be expressible in such simple terms. Galileo had recourse here to an empirical test as a preliminary aid toward a formal proof, which in this case he never reached. We may take this example as indicative of his empirical bend and of the likelihood that in the study of motion too he may have used, *mutatis mutandis,* empirical tests for similar purposes. For example, Galileo did empirically measure discrepancies between a parabola and a catenary, in addition to trying to prove their identity.[27] Such tests were part of his private investigations and could have not appeared as part of a formal proof. There are also significant differences between the two cases, however. In mathematics the postulates were given and the problem was that of producing a proof. In the new science of motion, however, the problem was that of finding both suitable postulates or principles and theorems deduced from them. Given the vagaries and uncertainties of the physical world, the principles had to appear natural and acceptable, much like Archimedes' in *On the Equilibrium of Planes.* Galileo's frantic search for "indubitable" principles in the letters to Sarpi and Valerio is revealing in this respect. Experience and experiments could corroborate the choice of the axiom, however, and could then show the relevance of the science to the study of nature by showing the agreement between theoretical predictions and experimental results.

3.3 From Buoyancy to the Science of Waters

In the winter of 1609–10 Galileo started making spectacular telescopic discoveries on the four satellites of Jupiter, which he called Medicean stars, the composition of the Milky Way, sunspots, and the phases of Venus. In recognition of his findings, and of his dedication of Jupiter's satellites to Grand Duke Cosimo II, in 1610 Galileo was called to the Medici court at Florence to serve as mathematician and philosopher to the Grand Duke of Tuscany. A good portion of the mathematicians' duties at sixteenth-century courts involved astrology and a range of issues such as water management, gnomonics, and civil and military architecture. Besides the courts, mathematicians could be found at the universities, teaching Euclid and elementary astronomy to physicians, who needed it for medical astrology, and in society dealing with the mechanical arts.[28] During the seventeenth century several trends emerge. The location of mathematics on the map of knowledge and its relation to the medical disciplines were transformed. Whereas in the sixteenth century astrology was the main point of contact, in the seventeenth century astrology declined in importance, while mechanics gained a stronger footing in the university curriculum. Philosophical and anatomical concerns with the mechanical understanding of the body gained a prominent position with Galileo and the Padua anatomy professor Gerolamo Fabrizi, among many others. Galileo's intellectual status and salary were greatly enhanced by his discovery of the Medicean stars, the publication of *Sidereus nuncius* (Venice, 1610), and his move to the Tuscan court. Galileo's new position close to the ruling prince—not to mention his salary, which was huge compared with what he had earned as a mathematics professor at Pisa—gave him greater authority in the debates of the time. While enabling him to argue from a position of strength, however, his higher status did not automatically settle any controversies. Moreover, it is unclear how much Galileo's higher status would have benefited the status of mathematicians elsewhere.

Soon after his call to Florence, Galileo became involved in a dispute over floating bodies, an area at the intersection between mechanics and natural philosophy treated by both Aristotle in the fourth book of *De coelo* and Archimedes. Galileo argued that the topic was not merely philosophical but also concerned the construction of bridges and other machines above water. The dispute originated in a discussion of why ice floats. Relying on Archimedean hydrostatics, Galileo argued that ice weighs less than an equal volume of water, whereas the Aristotelians argued that ice floats because of its flat shape. On his side Galileo counted the Benedictine monk Benedetto Castelli, who had been his student at Padua and had obtained the chair of mathematics at Pisa in 1613, while the dispute was still developing. Galileo drafted a report for the Grand Duke, published under his own name two editions (both in 1612) of a *Discorso* on bodies in water, and collaborated with Castelli in drafting a *Risposta,* which appeared under Castelli's name in 1615. On the opposite side were the member of the Florentine Academy Ludovico delle Colombe, an Anonymous Academician (probably the Provveditore of Pisa University, Arturo Pannocchieschi d'Elci), the professor of Greek Giorgio Coresio, and the professor of philosophy Vincenzo di Grazia, each of whom authored a piece on the topic.[29]

The dispute involved fine terminological distinctions and unfolded through public experiments, court debates, and publications. Galileo's opponents were able to show that a thin layer of a material with a greater specific gravity than water, such as ebony, could float. To counter them, Galileo argued that a wet heavier body will always sink, while one placed on top of the water behaves like an empty vase in that although its matter is heavier *in specie* than water, the combination of its matter and the air it contains or is contiguous to is lighter. The example of the floating vase had been proposed by the Pisa philosopher Buonamici in his attempt to refute Archimedes. In the case of the body on water, Galileo claimed that the body sinks only a small amount, since the air above it diminished its gravity by functioning like the air in the vase (fig. 3.7). While discussing the behavior of fluids, Galileo also linked buoyancy to the constitution of matter and atomism.[30]

One can see an analogy with the case Galileo had faced in *De motu antiquiora* with uniform speed and acceleration, namely the problem of distinguishing the central issue from accidental perturbations. In the 1610s Galileo believed that the behavior of bodies in water depended fundamentally on their respective densities,

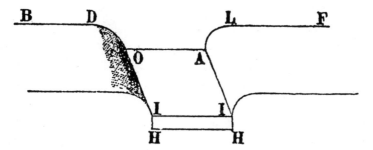

Figure 3.7. Galileo's little banks (1612 *Discorso*). According to Galileo, a thin body floats even if its specific gravity is greater than that of water because its specific gravity combined with that of the air above it make it lighter. Galileo relied on the little banks *OI* and *AI* to save an Archimedean account of buoyancy.

whereas the floating of thin bodies placed flat on the surface of the water involved perturbations or accidents. By contrast, the Aristotelians believed the shape of the bodies to be the crucial factor in their behavior in water. Since surface tension was not fully understood at the time, the issue was not as straightforward as Galileo would have liked.

Galileo introduced a distinction between gravity *in specie* and *absoluta,* a distinction sorely missed in *De motu antiquiora.* He also introduced for the first time in print the notion of *momento* and was led to define it in the second edition of the *Discorso* on bodies in water in connection with motion. He argued that *momento* depends on a body's gravity, its speed, and the inclination of the spaces, or better planes, over which it moves. He took two principles from the science of mechanics. The first states that equal weights moved with equal speeds will act with equal forces and *momenti.* The second principle states the proportionality between the *momento* or force of gravity and the speed of a body.[31] Here motion is considered an integral part of mechanics in the tradition of the pseudo-Aristotelian *Quaestiones.*

Dealing with buoyancy, Galileo once again had recourse to the balance in order to provide new foundations. He had already attempted the same analogy in *De motu antiquiora,* where his efforts had been caught in a terminological maze with gravity *in specie* and *absoluta.* In the *Discorso* on bodies in water Galileo modified the analogy and applied the notion of *momento* to the equilibrium of water in two connected containers, arguing that "the *momento* of the speed of motion of a body compensates for the *momento* of gravity in another" (fig. 3.8). The analogy is with a balance with unequal arms, where a weight of two pounds is in equilibrium with one of two hundred pounds if the former moves over a space one hundred times greater than the latter.[32]

In response to Galileo's arguments, both d'Elci and di Grazia argued that mathematics is ill suited to the investigation of nature, because it does not provide causes and it cannot deal with motion.[33] It is remarkable that the role of mathematics was challenged even in a classical area like floating bodies that had already been treated by Archimedes. The final piece from Galileo's side came with the *Risposta alle opposizioni* (Florence, 1615) drafted in collaboration with Castelli, who signed the dedication.

A few years later Castelli made a major contribution to the establishment of a new mathematical science, the science of motion of waters. His contribution was linked to his previous work on buoyancy. Castelli occupied Galileo's former chair of mathematics at Pisa until 1626, when he moved to Rome, where he later obtained a chair at La Sapienza. The origin of his work is also to be found in a dispute, though not at court with philosophers but rather in the field among *periti*

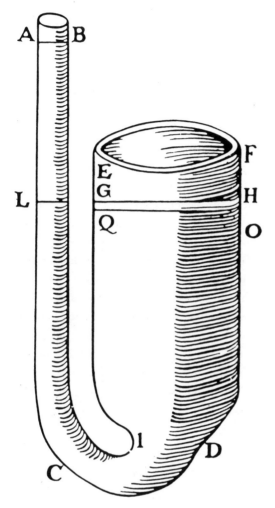

Figure 3.8. Galileo's siphon (1612 *Discorso*). Water in the two containers *EIDF* and *ICAB* is in equilibrium if it is at the same level, regardless of their cross-sections. According to Galileo, the system works like a balance wherein a small motion of the water in the large container, to *QO* for example, makes the water in the thin container rise in the same time from *L* to *AB*. The speeds are inversely as the cross-sections, but since the cross-sections are as the weights in *GD* and *LC*, the speeds are inversely as the weights, just as in the balance. Therefore the different speeds compensate for the different weights, and the moments are equal.

and engineers. In 1625 Castelli was sent with Monsignor Ottavio Corsini to the lands north of Bologna in order to study the flow of the rivers Po and Reno, and at the same time he engaged in a correspondence with Galileo on problems of river hydraulics. A mathematician originally from Brescia, Castelli appeared to be both competent and sufficiently neutral to mediate between the rival cities and views of Ferrara and Bologna. The issue had theoretical aspects similar to those addressed by Galileo's science of motion (such as the relation between slope and speed, linked to the problem of the inclined plane), practical ones (such as the effects of tides and winds), and complex intermediate ones (such as the relation between depth of the river and speed). Castelli also performed experiments with the Florentine Giovanni Ciampoli, who paid for them, to study water flow. They tried to show that two equal pipes fixed to different heights of a barrel full of water discharge different amounts of water. Although Castelli argued that the lower one discharges more, he did not attempt to establish a proportion between the height of the water above the pipe and the speed, as Torricelli was to do a few years later. Castelli, however, began to use the pierced cistern in the context of river hydraulics.[34]

In *Della misura dell'acque correnti* (Rome, 1628) Castelli stated the fundamental proposition of the equality of the rates of flow at different points of the river (fig. 3.9). Unlike cotton wool, water is not compressible; therefore the speeds of the water of a river flowing through two cross-sections are inversely as the areas of their respective cross-sections. This proposition affected the way matters were dealt with at times, i.e., by taking exclusively into account the cross-section of the riverbed, and added as an additional factor the speed of the water through each cross-section. This was a crucial move in the debates on the effects of diverting the Reno into the Po, or any river or canal into another riverbed, because an increased speed decreased the height of the river. Those who have investigated Castelli's sources have looked at the practical engineering tradition of the sixteenth century, including the work of Leonardo. In that tradition it was recognized that, besides cross-section, speed had to be taken into account while measuring the discharge of a river, but we do not find a rigorous formulation for the rate of flow. I suggest that Castelli's principle has additional roots in the previous debates on buoyancy, where

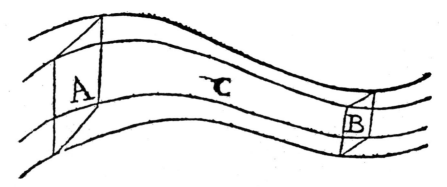

Figure 3.9. Proposition 1 from Castelli's *Della misura.* Castelli showed that the quantity of water flowing through sections *A* and *B* of a river in a steady state must be the same, and that water cannot be compressed; therefore the speeds at *A* and *B* are inversely as the respective cross-sections. Although Galileo studied pipes and Castelli rivers, Castelli's figure looks like a straightened version of Galileo's siphon shown in figure 3.8.

Galileo had argued that the speed with which water moves in a siphon is inversely as the cross-sections of the two branches.[35]

Castelli's engagement with water flow was not exceptional; Galileo's teacher Ostilio Ricci was considered a *peritissimus aquarum mensor*, and Galileo too had been dealing with similar issues for some time. While at Padua around 1604–5 he had been engaged in a dispute with the engineer Gabriele Bertazzolo about the great speed of rivers in floods. The engineer argued that the greater weight of the water was responsible for the greater speed, whereas Galileo believed that the relevant slope for determining speed was not that of the riverbed but that of the river's surface. Thus the greater height of a river during floods would be responsible for the increased speed.[36]

Around 1630 Galileo became involved in another dispute about river flow between various member of his school, such as Mario Guiducci and the cousins Niccolò and Andrea Arrighetti, versus the engineer Alessandro Bartolotti. Bartolotti argued that in order to avoid periodic floods the bed of the Bisenzio River, a tributary of the Arno, had to be straightened. Galileo's contribution, in the form of a letter to a court high officer, opposed the engineer's plan on the basis of notions derived from his as yet unpublished science of motion. For example, one finds statements that the speed of bodies at the bottom of two inclined planes with different lengths and the same height is the same, or also the chord theorem. Galileo's letter is a short treatise showing that the science of motion was his beacon in matters of hydraulics and river flow even when in the end he had to admit that fluids behave differently from solid bodies. For example, contrary to the conclusion he had reached at Padua, Galileo had to admit that the speed of water during a flood is far too great to be accounted for by the increased slope of the river surface. Galileo's views must have seemed rather impractical then, as they do know, given that even his faithful follower Viviani suggested that the course of the Bisenzio be straightened in order to reduce friction, apparently with good results.[37]

Another episode involving hydraulic engineering occurred for the construction of a closed water conduit to the Boboli gardens. Andrea Arrighetti, a former student of Castelli's and superintendent to the fortifications in Tuscany, was a consultant to the project and became concerned because the source of the water was barely above its intended outflow. During the debates Arrighetti informed Castelli that his book was often invoked. In order to test his views Arrighetti had a model built, involving four hundred *braccia* of pipes and imitating the key features of the real case, including the proportion among the length of the parts of the conduit, its many bends, and the tiny difference in the levels of the source and the outlet. The poor results predicted by the model were confirmed in practice, and the project had to be scrapped. The construction of such models in reduced scale was an important feature of hydraulics at the time. Experimental apparatus aimed at testing laws or regularities in nature sought to simplify the problem to the bare essentials. By contrast, engineering models had to retain some degree of complexity in order to provide a realistic estimation of the outcome, especially in the case of such a capricious and unpredictable substance as water.[38]

3.4 Motion between Heaven and Earth

In 1632 Galileo published the *Dialogo,* his ill-fated masterpiece wherein three interlocutors debate the merits of the Copernican and Ptolemaic systems over four days. Salviati, who usually speaks for Galileo himself, and Sagredo, an unprejudiced interlocutor standing for an ideal reader, are named for two (deceased) close friends of Galileo's, whereas Simplicio is a slow-witted Aristotelian whose bookishness is brilliantly, and mercilessly, ridiculed. The book was written in Italian and was addressed to a learned but nonacademic audience. In many cases Salviati does not explain his views to Simplicio but rather poses questions that elicit correct answers, in a Socratic manner. Galileo's own views thus appear to be a natural extension of common knowledge, whereas Simplicio's emerge as the result of a badly digested concoction of hopeless Peripatetic doctrines.[39] Although the *Dialogo* deals primarily with questions of astronomy, the informal style of the work and its many digressions make it wholly unlike any astronomical treatise. The book contains not only many discussions about motion, anticipating results later presented in the *Discorsi,* but also material that Galileo did not discuss elsewhere. Here I present a selection of some relevant themes.

In the first day, the interlocutors discuss motion along inclined planes (fig. 3.10). These discussions introduce a number of themes, such as the idea that motion along an inclined plane with zero inclination is perpetual. Galileo had already unequivocally introduced this important notion in his 1612 letters on the sunspots, in which he had equated a body's state of rest with its state of uniform circular motion on a surface concentric with the Earth. This equation of rest with eternal motion—wherein both are characterized by the term "state"—was a significant departure from traditional doctrines and from his own earlier views. Galileo established a link between the angle of inclination of a plane and the speed of the ball. Since the angle varies continuously, Galileo could plausibly conclude that a falling body goes through all the degrees of speed starting from zero. In the *Discorsi* Galileo defended the same thesis by means of the force of percussion. A body falling from a greater height acquires a greater speed and produces a greater in-

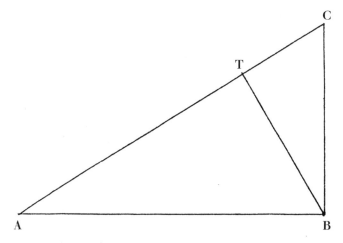

Figure 3.10. Galileo's inclined plane *(Dialogo)*. Given the inclined plane *ABC*, Galileo could prove that in the time it takes a body to fall along the vertical *CB*, another body released at the same time and falling along the incline will reach point *T*. This result derives from the assertion that the time of fall along the inclined plane *CA* is to the time of fall along the vertical *CB* as *CA* is to *CB*. Since the times are as the lengths, applying a well-known theorem by Euclid we have *CA:CB::CB:CT*. Galileo's claims are sound for sliding bodies, but if rolling is involved they become problematic, as Mersenne was to show. See chapter 4.

dentation on a yielding material. Thus the continuity of space guarantees the continuity of speed as measured by the indentations produced.[40]

The second day is largely devoted to the effects of the rotation of the Earth on its own axis and is especially relevant to the science of motion. This section includes discussions of the near-isochronism of pendular oscillations, the double-distance rule proved by means of indivisibles, the law of falling bodies, and the odd-number rule, for example. Here I focus especially on the composition and relativity of motion, the extrusion of bodies set in rotation, and the trajectory of falling bodies on a rotating Earth. The standard barrage of objections to the Copernican system concerned the fate of moving objects on the Earth, from cannonballs to birds and clouds. It is here that Galileo discussed the experiment of the stone dropped from the mast of a moving ship, the behavior of bodies such as butterflies or a fish in a bowl of water in a room below the deck of a moving ship, the trajectory of bodies shot from a moving carriage, and the idea that a ball falling from the sphere of the moon would "anticipate the Earth's vertigo," because it would retain its greater circular motion. Galileo did not consider an eastward deviation for bodies falling from a tower, however.[41]

The problem of extrusion is discussed in some detail. The question Galileo addressed is why heavy bodies are not projected away from the rotating Earth, an ob-

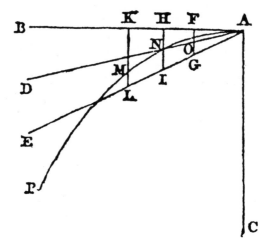

Figure 3.11. Extrusion in Galileo's *Dialogo.* C is the center of the Earth, *AP* is its circumference, and *AB* is tangent in *A. AF, FH,* and *HK* are traversed in equal times. *AE* is a line with an arbitrary inclination to *AB,* and *FG, HI,* and *KL* are the speeds acquired by the body in the corresponding times. The line *AD* with a smaller inclination to *AB* represents the effect of diminishing gravity, reducing the speed of the falling body to *FO, HN,* and *KM.* Galileo's reasoning is that approaching *A,* the distance between the tangent and the curve is smaller than the distance between the tangent and *AD,* representing the speeds acquired as a result of gravity. Curiously, Galileo was here comparing distances with speeds, though this is not the only problem with the diagram. Galileo intended to take the speeds acquired in "all the time" *AF,* or *AH,* or *AK* (*GOF,* 7:225−26: "Il grado di velocità acquistato in tutto il tempo AK sia come la linea KL rispetto al grado HI acquistato nel tempo AH"). Here is the fault in the proof: Galileo should have taken not the instantaneous speeds generated at single instants at *F, H,* and *K,* but rather the aggregate of the speeds acquired in the intervals, or the overall speeds, which are as the distances traversed. Since the overall speeds are as the squares of the times, *AD* should be not straight but parabolic. Now approaching *A,* the distances from the tangent to the circle and to the parabola would be comparable, therefore one can have extrusion. Here over a short distance gravity is taken to act along parallel lines.

vious and pressing problem in a Copernican universe. His answer involved comparing two magnitudes diminishing *ad infinitum.* A body on a rotating support, such as a rotating wheel or a rotating Earth, would be projected outward along the tangent. At the beginning of motion the body would have to pass through all the degrees of speed starting from rest. The key idea is that at the beginning of motion the distance between the tangent and the Earth would be much smaller, in fact infinitely so, than the length of the tangent segment; therefore a body with any finite gravity would not be extruded. Galileo's subtle reasoning is incorrect. He is certainly right in arguing that as we approach the origin of the tangent, the deviation from the tangent becomes infinitely smaller than the relevant tangent segment. The fault in his reasoning is that he should compare the body's gravity with the tendency to escape along the radius generated by the tangential motion, which are of the same order of magnitude.[42]

Subsequently Galileo tried to compare the deviation from the tangent with the fall due to the body's gravity, arguing that the latter is larger than the former. He also tried to take into account the effects of a decreasing gravity. Since the speed of falling bodies is not proportional to their weight, if their weight decreases, their speed would not diminish in the same ratio. Galileo's diagram and reasoning show a similarity with his approach to the problem of horizontal projection, where no amount of power can make the projectile go in a straight line. In the remarkable figure 3.11 Galileo superimposed a triangle of speeds upon a diagram of the Earth, thus combining a graph with a geometrical representation.[43]

In a passage in the second day Salviati puts forward a fancy of his, or a *bizzarria*, concerning the path of falling bodies on a moving Earth (fig. 3.12). The chief idea consists in supposing that the real trajectory of a body falling from a tower is the arc of a circumference from the top of the tower to the center of the Earth, traversed with a uniform speed equal to that of the body at the top of the tower. Thus rectilinear motion and acceleration would be reduced to mere ap-

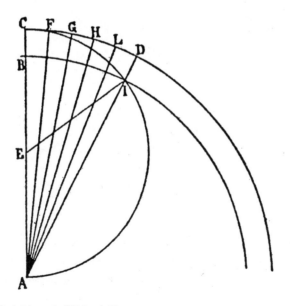

Figure 3.12. Galileo's *bizzarria (Dialogo)*. BI represents the Earth's surface and *A* its center. *BC* is the height of a tower whence a body falls. Its trajectory to the center of the Earth would be the circle arc *CIA* with center *E*. According to Galileo, the body would travel along arc *CIA* with the same speed it had when it was moving along *CFGHLD*, since the arcs *CF, FG, GH*, etc., are equal to the corresponding arcs on *CIA*. Indeed, the angles in *E* are double those in *A*, whereas *CE* is half *CA*, therefore the arcs on the corresponding circumferences have equal length.

pearances, real motions being only uniform and circular. Galileo did not defend or explain this opinion further, but when several correspondents questioned him he claimed that it was presented merely as a speculation. The *bizzarria* is indicative of his concern with accelerations being the same for all bodies, the cause of acceleration, and the mathematical order of nature. Despite his withdrawal of the idea, the *bizzarria* had quite a history, as we shall see below.[44]

The third day of the *Dialogo* presents an analysis of astronomical observations that displays considerable skill in data handling and error theory. The occasion is provided by twelve observations of the 1572 star used by Scipione Chiaramonti, who tried to argue that the star was inside the sphere of the moon. The problem was how to analyze and combine the observations in a nonbiased manner. In the extensive discussion, Galileo offered several comments indicating that he had devoted considerable thought to such matters. For example, he argued that observations of the elevation of a star are necessarily inaccurate, equally likely to be either too high or too low. Small errors concerning very distant objects have a very large effect compared with the effect of errors concerning closer objects. Galileo has recourse here to the notion of significant figure, arguing that there is no point in making calculations accurate to small distances when the errors in the measurements are known to be much larger. Although Galileo's comments concerned astronomical observations, they could easily have been applied to other areas. Yet Galileo generally refrained from detailed discussion of the issues pertaining to error theory.[45]

The *Dialogo* contains the first published account of the pendulum. Galileo stated that oscillations are nearly isochronous regardless of amplitude and that the period decreases when the pendulum is shortened, but he did not specify that the oscillation period is as the square root of the length; Mersenne was probably the first to do so in the 1634 *Les mecaniqves de Galilée.* Even without that relation, Galileo seized on the notion of the pendulum to establish analogies with the oscillation of water in a siphon, the orbital motion of planets and satellites, and the periods of tides.[46]

In the fourth day Galileo tried to explain tides as a result of the combination of the annual and diurnal motions of the Earth. Since a point on the surface of the Earth would move on an epicycle resulting from the simultaneous uniform circular motions of the Earth on its axis and around the sun, that point would periodically accelerate and slow down. Tides would be the result of those changes of speed, much as water in a basin oscillates if the basin is pushed forward and then stopped. Galileo also performed several experiments with moving vessels filled with fluids, showing once again that simple objects were at the basis of his investigations. This subject was not taken up again in the *Discorsi*, since this was a main cause of Galileo's troubles with the Inquisition.[47]

Galileo's *Dialogo* was not the only work to anticipate some of the results later presented in the *Discorsi*. In that same year Cavalieri published *Lo specchio usto-rio* (Bologna, 1632), in which he devoted two chapters to falling bodies and parabolic trajectories relying on his new mathematics. Cavalieri was readying for publication *Geometria indivisibilibus*, his mathematical masterpiece on the method that dominated midcentury infinitesimal mathematics. Cavalieri attributed to Galileo his main results on motion, such as the odd-number rule and parabolic trajectories. As to the latter, he stated that, neglecting gravity, the trajectory of a body shot at any angle would be rectilinear, and he tried to prove the relation $s \propto t^2$ with his indivisibles.[48]

3.5 The Science of the Resistance of Materials

Following his trial for heresy and subsequent house arrest at his villa in Arcetri near Florence, Galileo was forbidden to publish.[49] Despite this prohibition, he brought to completion his masterpiece and had it smuggled out of Italy, possibly by the French ambassador to Rome, François de Noailles, who had been one of his students at Padua. The work was published at Leyden by the prestigious Elzeviers in 1638 when Galileo was seventy-four. The *Discorsi* was an epoch-making work and Galileo's crowning achievement. Despite the status it has enjoyed ever since, however, we still lack a critical evaluation of how it was perceived by his contemporaries.

The *Discorsi* describes a series of discussions taking place over four days among the same three personages as in the *Dialogo*, although Simplicio appears less obtuse here and at times even puts forward views Galileo had held in the past. Galileo added an appendix on the centers of gravity of solid bodies, relying on his work of the 1580s. The wording of the title—*Discorsi e dimostrazioni matematiche intorno a due nuove scienze*—was chosen in part by the Elzeviers, but the emphasis on "two new sciences" was Galileo's. Whereas Tartaglia had provided a rather loose set of principles and propositions, Galileo believed he had provided a rigorous axiomatic formulation on the model of Archimedes.

The discussions of the first two days are entirely in Italian and are devoted to the science of the resistance of materials; this section was largely completed in Padua and was originally intended for a separate treatise.[50] In this section Galileo supposed matter to be inalterable, so as to overcome the objection that no mathematical science of matter can be formulated because of its contingency. (We encountered this issue earlier, in a passage from the correspondence with dal Monte quoted in section 1.3.)[51] The opening praise for the Venice Arsenal as a philosophical venue and for the knowledge of the *proti* or senior technicians working there was not mere rhetoric. Recent research suggests that work undertaken at the Arsenal on the problem of constructing larger and larger galleys was one of Galileo's likely sources.[52]

models

The first day's discussion begins with Galileo challenging the view that small models are reliable means for testing real machines built in scale, namely machines with all their linear dimensions increased in proportion to the dimensions of the model. Rather, he argued, if models of a ship or bridge are extended in scale, eventually they will be unable to sustain their own weight, let alone any weight placed on them. Galileo attributed his knowledge of this proposition to a *proto* and took this as the starting point of his new science, which is dealt with more rigorously in the second day. We can grasp Galileo's point if we consider that if all linear dimensions are doubled, weight alone, which is only one effect of an extension in scale, becomes eight times greater. The rest of the first day is devoted to many digressions on the composition of the continuum, atomism, the pitch of musical strings, and resistance to motion. Galileo argued that there are three ways to raise the pitch of a string. To make it sound a higher octave, its length must be halved, or its tension quadrupled, or its cross-section reduced to one-quarter.[53] Galileo associated resistance to the speed of a body, its surface, and the difference between its specific gravity and that of the medium. He argued that in a vacuum, all bodies would fall with the same acceleration, whereas in a medium, lighter bodies would be more affected than heavier ones.[54]

cf James / re Pascal

The many digressions of the first day are subtly linked to topics addressed in the subsequent days in more ways than one. For example, in day 1 Galileo discussed the physical causes of the resistance of materials, the same topic investigated in a mathematical fashion in day 2, and the related problem of the composition of the continuum, a key theme for days three and four. Galileo tentatively put forward the view that the cohesion of bodies depends on infinitely many nonextended interstitial *vacua* opposing separation and therefore holding together infinitely many nonextended particles of matter. Galileo tried to provide an explanation for why water cannot be raised higher than eighteen *braccia* in the same terms. In this case too, Galileo identified a technician as his source of information. The opening of the second day summarizes the results achieved previously: "There was then a search for the cause of that cohesion, which is extremely strong in some solids; and the chief cause proposed was that of the void."[55]

The second day's topic can be characterized as statics of the rigid body, where balances and beams are not ideal geometrical entities but bodies with their shapes,

Figure 3.13. Galileo's beam *(Discorsi).* Galileo tried to determine the resistance of a beam *ABCD* infixed at right angles in a wall, or cantilever, by means of the lever. *B* is its fulcrum, *BC* is one of its arms, and *BA* is the other at right angles to it. The moment of the weight *E* in *C* is counterbalanced by the moment of the beam's resistance to breaking, proportional to the number of fibers in its base or to the area of the cross-section. Since the fibers are equally spaced, they work as if they were all in the middle point of *AB*, by analogy with a system of weights. Thus the resistance to breaking is as the cross-section times half the height.

dimensions, and at times weights. The text presents a series of propositions based on the doctrine of the lever, which provides the foundations for the new science. This time the lever works in a novel way, namely the two arms are perpendicular to each other, and one of them counterbalances the other by means of the number of fibers with which it is attached to the wall. Galileo was concerned not with giving values for the resistance of individual materials but rather with comparing the resistance of the beam to being pulled along its axis with its resistance to rupture when it is pulled down transversally, as in figure 3.13. The former is used purely as a unit measure of the resistance of the fibers. Galileo's conclusion was that the ratio is as the length of the prism to half its height.[56]

Galileo asked questions such as the following: How much does the resistance of a horizontal beam protruding from a wall decrease if one increases its length? Taking its weight into account, the answer is that it decreases as the square of the length. How much does the resistance of a cylinder increase if one increases its

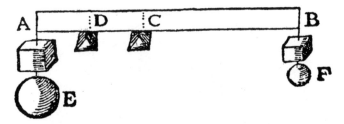

Figure 3.14. A beam supported at both ends *(Discorsi).* In order to study a beam supported at both ends and loaded with a weight at some point on its axis, Galileo considered the case of a beam supported at some point on its axis and loaded at both ends. He argued that the smallest forces *A, B* required to break the beam in *C* are to the smallest forces *E, F* required to break the beam in *D* as *AD* × *DB* is to *AC* × *CB.*

thickness? The answer, if this time we ignore its weight, is that it increases in the third power of the diameter. Galileo mentioned additional problems, such as "given a prism or cylinder with its own weight, and the maximum load it can bear," to find "the maximum length beyond which it would break under its own weight."[57] Galileo was thus able to show why human constructions such as ships and buildings, as well as nature's productions such as trees and animals, cannot exceed certain dimensions (fish are a different case, because their weight is sustained by water). The text combines geometrical theorems with beautiful insights into the mechanical constraints to natural and artificial constructions. It also reflects a concern with deeper understanding and mathematical elegance, but Galileo was after practical results too. Accordingly, the investigation now shifts to the resistance of a beam supported at both ends (fig. 3.14).[58]

Galileo's subsequent propositions are devoted to the task of reducing the weight of a beam by removing portions where it is stronger than necessary, an important consideration in the construction of buildings and, especially, large ships, where weight is a key factor. The opening lines regarding the Venice Arsenal return to mind here, but Galileo's treatment of this problem is rather confusing, because without warning he shifts to the cantilever, which is not central to shipbuilding. Galileo proved that a parabolic beam loaded at one end offers the same resistance at any distance and weighs only two-thirds of the corresponding prism. His solution is singularly inelegant, for it is based on the assumption that the beam is weightless, but in this case there would be no advantage in giving it a parabolic profile. On the other hand, a heavy beam would have a different profile. Moreover, Galileo's diagram, reproduced in figure 3.15, is inaccurate in that the parabola should be tangent to *BX* and his is not.[59]

Galileo also provided practical and easy ways of drawing parabolic lines, such as throwing balls over an inclined plane and marking out their trajectories, or hanging a thin chain between two nails fixed in a wall. Thus the speculations and experiments carried out with dal Monte emerged as part of the intersection between two new sciences, one dealing with the resistance of materials and the other with motion. In the same manuscript dal Monte had discussed the resistance of a rope to being broken by a hanging weight, arguing that length was not a factor. The same example and conclusion can be found in the second day of the *Discorsi*.[60]

In concluding the science of the resistance of materials, Galileo addressed the resistance of hollow solids, such as cylinders. He argued that two cylinders of equal weight and length, one full and the other hollow, have a resistance proportional to their diameters. The conclusion follows from Galileo's assumptions, since the number of fibers is the same for the two cylinders; the only difference is the length of the arm of the lever, namely their height.[61] Once again Galileo referred to human constructions and nature's operations, as in the bones of birds and the stalks of plants that can support a weight much greater than their own. Thus his new science presents a rich interplay both with architectural or engineering concerns and with animal and plant anatomy.

Galileo's science of the resistance of materials contains some inaccuracies and lacunae that were already evident to Viviani, but it broke new ground and established links to other areas of knowledge and the engineering tradition. Moreover, Galileo's new science provides us with a valuable tool for exploring crucial themes such as the interplay between physical causes and mathematical explanations.

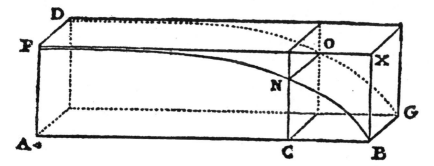

Figure 3.15. Parabolic beam *(Discorsi)*. The beam *ABGDF* with a parabolic profile, infixed in the wall at *AFD* and loaded at *BG,* offers the same resistance at any point *C* along its axis. In the parabola, *BA* is to *BC* as *AF*^{2} is to *NC*^{2}, but the resistance of each cross-section is as *NC*^{2}/*BC*, which is constant, therefore the resistance is constant at any point *C* along the axis.

3.6 The Science of Motion

Galileo's science of motion, addressed in days 3 and 4 of the *Discorsi*, has been read in different ways since its publication and therefore presents great interpretative challenges. Through a combination of reasoning and experience, Galileo had long established the laws of falling bodies and parabolic trajectories. Even with the correct proportion between speeds and times and other related propositions, however, Galileo's task was far from complete. One of the difficulties was to formulate the new science by means of the theory of proportions in classical axiomatic fashion, selecting suitable definitions and axioms. Galileo's axiomatic formulation differs widely from the heuristic path followed. Without understanding these constraints, we would fail to grasp the complexity of his achievement. Galileo's procedure looks unnecessarily cumbersome to the modern reader familiar with the "Galilean laws" of falling bodies whereby $s = \frac{1}{2}(gt^2)$ and $v = gt$. Writing Galileo's results in this anachronistic form, however, profoundly obscures the methods he adopted and the difficulties he encountered. Galileo did not use algebraic equations, but rather proportions between homogeneous magnitudes and indivisibles. Each requires different procedures and involves specific problems of formulation. Galileo faced problems with the mathematical formulation, and in some cases he relied on unorthodox rules. Although Cavalieri with his method of indivisibles attempted a rigorous formulation of an area of mathematics relevant to the new science of motion, some problems and contradictions remained.

Galileo's science of motion consists of three books, on uniform motion, on naturally accelerated motion, and on the motion of projectiles. Book 1 deals with an apparently elementary issue, namely the relation between speed, time, and distance in uniform motion, with all its six proportions including compound ones. While providing an example of the theory of proportions, Galileo was also supplying theorems he would use in books 2 and 3. Since he had no mathematical tool for dealing in general with accelerated motion, in a number of cases he tried to "transform" accelerated motion into an equivalent uniform motion, a procedure fraught with dangers that was intended to allow him to employ the theorems of book 1. Thus references to book 1 in the later books are useful indicators of his methods and the difficulties he encountered.[62]

Galileo has Sagredo discuss the cause and nature of gravity, only to be sidelined by Salviati arguing that he wishes to investigate only some "passions," or features of accelerated motion. This attitude stands in sharp contrast to the science of the resistance of materials, where Galileo had embarked on an elaborate investigation of physical causes. This dichotomy suggests that the shift away from physical causes was not part of any general program on Galileo's part. Indeed, his reflections on resistance and buoyancy suggest that he saw weight as a physical cause and that he refrained from expanding on it because of internal difficulties rather than because of a methodological shift. Notice also that the *Discorso* on bodies in water ends with an emphatic claim that the difference in the specific gravity of the body and the medium is the "only, true, proper and absolute reason" why some bodies float and others sink.[63]

One of the major problems Galileo encountered was that of establishing the foundations of the new science, as shown by his letters to Sarpi in 1604 and Valerio in 1609. In previous cases, from simple machines to hydrostatics and the resistance of materials, he had relied on the lever, but this approach seemed to be unfeasible for the science of motion. At the outset of *De motu naturaliter accelerato* Galileo claimed that his entire construction was built on a definition and a principle. The former states that uniformly accelerated motion is that motion, starting from rest, acquiring equal moments of swiftness in equal times. Galileo briefly revisited his own mistake of the 1604 letter to Sarpi in which he had claimed that speed is proportional to distance, making Sagredo put forward that view and having Salviati refute it. This problematic attempt to refute it is indicative of troubles with Galileo's views about speed and indivisibles. Galileo argued that if the speeds of a falling object are proportional to the distances, the time of fall is the same for any distance. He justified his claim by arguing that if a body falls two *braccia* in time x and another falls four *braccia* in time y, for each speed of the first body there is a corresponding double speed of the second. In comparing two infinite sets, Galileo seemed to imply that if one were to find the uniform speed with which the first body travels in time x two *braccia*, the uniform speed with which the second body travels in time y four *braccia* would be double; therefore, since the speeds are uniform and are as the distances, the times x and y are the same. But the two bodies would also fall together over the first two *braccia*, therefore the second body would have to travel the remaining two *braccia* instantaneously, which is absurd. This reasoning was problematic in that Galileo was applying a theorem valid for uniform motions to motions that are not uniform but accelerated.[64]

Galileo's postulate assumes "that the degrees of speed acquired by the same moveable over different inclinations of planes are equal whenever the heights of those planes are equal." His attempt to justify this assumption is twofold. On the one hand, Galileo has Sagredo state that his *lume naturale* (something like an innate intuition) indicated that the assumption was true. On the other hand, Salviati proposed an *esperienza* with a pendulum falling "little short" of a necessary demonstration (fig. 3.16). He argued that the bob of a pendulum acquires enough *momento* to return to its original height. The height does not change if one inserts a nail in the wall forcing the bob to move along different arcs. Therefore the speed acquired, or lost, depends on the height of fall, or ascent. Salviati proposed to accept the postulate provisionally, before having it subsequently confirmed: "Hence let us take this for the present as a postulate, of which the absolute truth will be later established for us by our seeing that other conclusions, built on this hypothesis, do indeed correspond with and exactly conform to experience." Galileo struggled for decades to reach a suitable formulation and continued to struggle with his followers even after 1638, as in the second 1656 edition of the *Discorsi*, in which he attempted to prove his postulate.[65]

Soon after the first two propositions, whose contents I discuss below, Galileo presented another experiment whose purpose was different from the one we have just considered. This experiment was not meant to justify an axiom or the coherence of his theory; it had no foundational role. The mathematical theory would remain valid as such even if nature behaved differently. Galileo wanted to show that the mathematical result he had proved—that the distances traversed by a body with uniformly accelerated motion are as the squares of the times—holds true in nature. In order to show this, he described an experiment with an inclined plane:

> In a wooden beam or rafter about twelve braccia long, half a braccio wide, and three inches thick, a channel was rabbeted in along the narrowest dimension, a little over an inch wide and made very straight; so that this would be clean and smooth, there was glued within it a piece of vellum, as much smoothed and cleaned as possible. In this there was made to descend a very hard bronze ball, well rounded and polished, the beam having been tilted by elevating one of its ends above the horizontal plane from one to two braccia, at will.[66]

Time was measured by weighing the water flowing through a thin tube at the bottom of a large bucket. The result was that the distance had to be reduced to a quarter in order to halve the time. Galileo accompanied his account with claims of remarkable precision, less than a tenth of a pulse-beat.

In theorem 1 Galileo encountered problems with the composition of the continuum. He wanted to prove that the time required by a falling body to cover a given distance from rest with uniformly accelerated motion is the same as that required to cover the same distance with a uniform motion whose degree of speed is half the maximum degree of speed acquired in falling.[67] In this way an accelerated motion can be treated as uniform. In his proof, Galileo had recourse to the areas of the triangle and parallelogram as aggregates of infinitely many lines representing speeds at all the different instants of time (fig. 3.17), but he did not try to establish a rigorous new mathematical theory, as Cavalieri had done with his method of indivisibles. Rather, Galileo tried to justify his procedures pertaining to

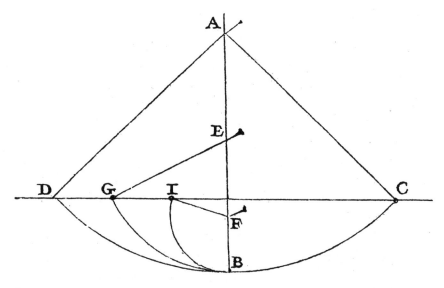

Figure 3.16. Galileo's pendulum *(Discorsi)*. Galileo's *esperienza* consists in inserting a nail at different heights such as *E* and *F* on the vertical of an oscillating pendulum. The bob will always reach
the same height *D, G,* or *I* along smaller arcs, and from there it will reach back up to *C*. Hence it
has the same speed in *B* as long as it falls from any point on the line *CD*. Here Galileo considered a
circle arc as an infinite series of inclined planes with different inclinations.

acceleration, the nature of the continuum, and infinite aggregates of lines on a
one-by-one basis, attempting to make them look plausible. At the end of his Padua
years Galileo conceived a treatise on the continuum, *De compositione continui,* and
in the *Postils to Rocco* of the mid-1630s he discussed issues of infinitesimal mathematics. In both cases, however, his chief preoccupation was a defense of his physical theories. Thus overall Galileo saw mathematics not as an end but as a tool enabling the formulation of a science in Archimedean fashion.[68]

    In the following two theorems too Galileo had recourse to book 1 on uniform
motion. Theorem 2 states that if a body falls from rest with uniformly accelerated
motion, the spaces are as the square of the times. Galileo relied on the theorem
whereby the spaces traversed with uniform motion are in the compound proportion of the speeds and the times. Falling bodies do not move with uniform motion,
but through theorem 1 we can find a uniform motion covering the same space in
the same time, namely with half the final speed. Since in uniformly accelerated
motion the speeds are as the times, half the speeds will be also as the times. Hence
the spaces traversed are as the times squared.[69]

    Theorem 3 states that if the same body falls along an inclined plane and then
along its vertical, the times are as the lengths of the plane and the vertical, re-

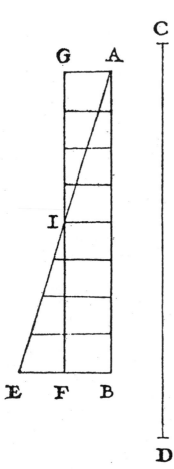

Figure 3.17. Theorem 1 from the *Discorsi*. The time *AB* with which a falling body covers a given distance from rest with uniformly accelerated motion is the same as that required to cover the same distance with a uniform motion whose degree of speed *FB* is half the maximum degree of speed *EB* acquired in falling. Here all the lines drawn from every point on *AB* parallel to *EB* represent the increasing degrees of speed from *A*. *FG* is parallel to *BA*. Extending all the lines from *AEB* to *IG*, Galileo compared two infinite aggregates offering a reformulation of the mean-speed theorem: "The aggregate of all the parallels contained in the quadrilateral *[AGFB]* equals the aggregate of those included in the triangle *AEB*"; therefore the moments of speed in the two cases are the same, and the distances traversed are also the same. Galileo did not take the areas of the triangle and parallelogram to be proportional to the spaces traversed, but introduced for this purpose the segment *CD*.

spectively. Galileo drew any parallel lines to the horizontal basis of the plane and, on the basis of his postulate, could claim that the degrees of speed acquired at the extremes of those lines, on the incline and along the vertical, are equal (see fig. 3.4). Thus, he argued, the incline and the vertical are traversed with the same degrees of speed. Once again, at this point he felt authorized to rely on book 1 to conclude that if the speeds are the same, the distances are as the times. Here Galileo was again comparing infinite aggregates of speeds where, in addition, the aggregate along the incline appears to contain more points than the aggregate along the vertical. Sagredo provided a more solid proof by arguing that both the motions along the vertical and the incline are uniformly accelerated and their final speeds are the same. Therefore, for theorem 1, they can be replaced with uniform motions with a common speed. Now one can apply the theorem from book 1 leading to the desired result.[70]

Although an analysis of the rest of book 2 would lead us too far afield, brief mention must be made of Galileo's failed attempt to find the curve of fastest descent between two points on a vertical plane. Galileo believed that the circle was the curve of fastest descent and that it was isochronous; that is, a body released at any point on a vertical quadrant will take the same amount of time to reach the lowest position. Theorem 22 and its scholium argued—correctly—that a body's descent is faster along two conjugate segments inscribed in the circular arc than along the chord. From here Galileo concluded—incorrectly—that the circle is the curve of shortest descent. All that he could conclude was that descent along the circle arc was faster than along any inscribed polygon, not that the circle was the line of fastest descent among other curves.[71]

A barrage of difficulties of a different nature emerges in day 4, in the discussion of parabolic trajectories. The entire construction is based *ex suppositione* on the notion that a horizontal uniform rectilinear motion and a uniformly accelerated downward motion do not interfere with each other; thus one can unmask a projectile as a falling body and erase the distinction between violent and natural motions (fig. 3.18). Here the horizon is taken to be straight, although it is really

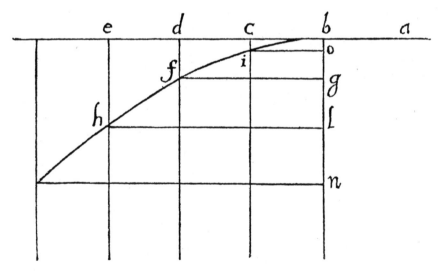

Figure 3.18. Parabolic trajectory *(Discorsi).* Galileo shows uniformly accelerated motion *bogln* lurking inside the trajectory *bifh* of a projectile, where *bo, og, gl, ln,* are as 1, 3, 5, 7. The horizontal line *ab* represents a plane or line supporting the body, but at *b* the support ends and the body begins to fall. The prolongation of *ab, bcde* represents the flow of time. Galileo can switch from a spatial representation to a graph involving time because in uniform motion the distances are as the times. He can then show that the distances fallen are as the squares of the times, and the trajectory is a semiparabola.

curved, and the lines of gravity are taken to be parallel among themselves, whereas they converge toward a center. Galileo dealt with these difficulties in part by appealing to Archimedes, stating that the Greek mathematician had considered the directions of a balance's weights to be parallel, and in part by claiming that his approximation does not represent a serious problem because the errors thus introduced are very small. Here we encounter again the problem raised by Tartaglia, dal Monte, and Benedetti. Finally, the resistance of the medium is neglected. Galileo argued that the impediments of the medium are so variable and unpredictable that no science can be formulated. He devoted several theorems to the problem of composition of motion but did not provide a proof of parabolic trajectory for oblique projection, where the impetus seems to increase or decrease depending on the inclination to the horizon and composition of motion becomes trickier. Galileo was possibly thinking in terms of inclined planes with zero inclination, which led him to see the horizon as a privileged direction.[72] At the end of the *Discorsi*'s fourth day he presented numerical tables for shooting based not on empirical data but on mathematical computations.[73]

At the end of the fourth day Galileo revisits the analogy between projectile trajectory and a hanging rope. Salviati explains to Sagredo that neither a hanging rope nor a projectile trajectory could ever be horizontal because both result from the combination of two forces, one horizontal from the gun's projection or those who pull the rope, and the other downward due to the weight of the ball and the rope. In both cases the curve would be parabolic, although Galileo claimed that a hanging chain is *almost,* or "quasi *ad unguem,*" parabolic. In fact, the curve of a hanging chain is not parabolic, as Huygens was to show a few years later.[74] Galileo shifted from a rope to a small chain, namely the same device used by dal Monte. Unlike his predecessor, however, Galileo could prove that the trajectory of projectiles is parabolic, rather than resembling a parabola or hyperbola. Moreover, dal Monte would have seen the curve as resulting from "natural motion progressively overtaking violent motion," whereas for Galileo the two components were independent and horizontal motion was always conserved, neglecting the impediments. Galileo and dal Monte used the term "violent motion" in different ways.[75]

These issues lead us to reflect on the way Galileo discussed and presented experiments in the *Discorsi.* The case of the science of resistance of materials was peculiar, in that it was believed to rely on established Archimedean postulates and, despite its practical and engineering contents, was not directly dependent on experimental confirmation. On the one hand, Galileo formulated the science using proportions, wherein numerical parameters do not occur; on the other, he stated that he was considering ideal materials without cracks or blemishes. The science of resistance of materials had empirical implications in claiming, for example, that a law of scaling based on simple proportions does not work. In the science of waters Castelli's proportion was derived from purely mathematical considerations, and in the *Discorso* on bodies in water Galileo established a link with the doctrine of the lever. Once again, the science had empirical import but was not dependent on empirical data for its validity.

The case of the science of motion was different for a number of reasons. First, its mathematical formulation was not implied by a priori considerations, such as in Castelli's proportion. Galileo's claim that the proportionality of speed and time was the simplest looks to be a rhetorical justification and was challenged by contemporaries, as we shall see in the following chapter. Galileo also claimed that one could initially take his construction as a purely mathematical exercise and only subsequently verify its empirical adequacy. On the basis of the knowledge available at the time, bodies could have fallen following different rules, if they followed one at all. Thus to make the science of motion empirically significant, one had to show that it accurately represented the actual fall of bodies, whether projected or not. As a friend of dal Monte, Galileo would have known that unlike statics, the science of motion could not be expected to provide absolutely exact results. In print Galileo was usually reluctant to provide any data about experimental results. In the third day of the *Discorsi*, in the discussion of the experiment with balls rolling down a groove in an inclined plane, for example, Galileo stated that in the measurements of time he had never found a difference even of a tenth of a pulse-beat. This sounds more like a rhetorical device than a presentation of experimental error, even though Galileo was reporting an experiment that in all likelihood he had performed.[76] In another passage in day 1, dealing with the speeds of falling bodies, Galileo gave some numbers and introduced an important notion. Simplicio is about to accept Salviati's conclusion that bodies of the same material fall in the same time, regardless of their weight, but admits his reluctance to accept it. Salviati replies:

> But I do not want you, Simplicio, to do what many others do, and divert the argument from its principal purpose, attacking something I said that departs by a hair from the truth, and then trying to hide under this hair another's fault as big as a ship's hawser. Aristotle says: "A hundred-pound iron ball falling from the height of a hundred *braccia* hits the ground before one of just one pound has descended a single *braccio*." I say that they arrive at the same time. You find, on making the experiment, that the larger anticipates the smaller by two inches; that is, when the larger one strikes the ground, the other is two inches behind it. And now you want to hide, behind those two inches, the ninety-nine *braccia* of Aristotle, and speaking only of my tiny error, remain silent about his enormous one.[77]

Galileo here claimed not absolute precision but rather much greater precision than Aristotle by means of the notion of order of magnitude. Moreover, Galileo also attempted to explain away small differences in terms of secondary factors, such as a difference in specific gravity or a difference in size affecting air resistance. He tried to argue that buoyancy affects the speed of fall in that a body's gravity is diminished by a factor depending on the density of the medium. For a body whose density is one thousand times greater than air, for example, the diminution is one part in a thousand, whereas for a body whose density is only ten times greater than that of air, the diminution is one-tenth. Speeds of fall, other things being equal, would be affected in the same proportion: "For example, an ivory ball weighs twenty ounces, and an equal amount of water weighs seventeen; therefore the speed of ivory in air is to its speed in water approximately as twenty is to three."[78] The numbers provided by Galileo were not the results of experiments, but rather seem to be simply plausible values suitable for didactic purposes. Falling bodies were less problematic than the motion of projectiles, where parabolas seemed to diverge more and more from the point where, if unhindered, they were believed to reach, namely the Earth's center.

Galileo's reflections show how much more sophisticated he had become in dealing with perturbations and accidents compared with dal Monte, for example, who expected perfect agreement between nature and his simple machines. The balances built for him in the Urbino workshops to prove indifferent equilibrium when the centers of suspension and gravity coincide went hand in hand with his reluctance fully to engage in the study of motion. Moreover, Galileo extended dal Monte's approach to the equilibrium of fluids and the resistance of materials, but he also created a new science of motion departing, albeit reluctantly, from the principle of the lever. Finally, Galileo tentatively and somewhat problematically ventured into new mathematical terrain with the theory of indivisibles. Although he broke new ground in all these areas, his science of motion remained plagued by difficulties in its formulation and results.

Our perception of Galileo's achievements comes from a very different perspective from that of his contemporaries and immediate successors. Thus it is impossible for us to envisage which of his assumptions, methods, and results were accepted, doubted, or challenged without going back to the sources of the years immediately following the appearance of his major works.

Novel Reflections and Quantitative Experiments

We have found that while [a ball] falls perpendicularly from a
height of five feet, it falls only one foot on a plane inclined by
fifteen degrees, although it should fall sixteen inches.

Mersenne

4.1 Different Readings of Galileo

This chapter explores the investigations on mechanics and motion and the
many different ways in which Galileo's masterpieces were read to about 1650. This
was a period of intense research and debate, including such major works as Mersenne's *Harmonie universelle* (1636), *Cogitata physico-mathematica* (1644), and *Novarum observationum tomus III* (1647), Baliani's *De motu naturali* (1638 and
1646²), Gassendi's *De motu impresso* (1642) and *De proportione motus* (1646), and
Torricelli's *Opera geometrica* (1644).[1] Moreover, scholars engaged in extensive
epistolary exchanges stretching from France and Italy to the Low Countries.

The impact of Galileo's *Dialogo* and *Discorsi* would have been enormous even
without the condemnation of 1633, which made Galileo the most famous intellectual in Europe. The banning of the *Dialogo* was fairly effective in Italy but
largely ineffective in Catholic France, not to mention Protestant Europe. In addition to prohibiting the book and confining its author to house arrest, the censors
banned him from putting anything further into print. The following year, however, Mersenne published a paraphrase of Galileo's *Mecaniche* as *Les mechaniques
de Galilée* (Paris, 1634), and Latin translations of the *Dialogo* appeared in 1635 at
Strasbourg, in 1641 at Lyons, and in 1663 at London. When the *Discorsi* appeared
in Leyden in 1638, Galileo had to claim that the printing had been undertaken
without his knowledge. Soon thereafter Mersenne published a French paraphrase

of the work, *Les nouvelles pensées de Galilée* (Paris, 1639). A second Italian edition appeared at Bologna in 1656 with his other works excluding the *Dialogo*.[2]

There are important differences between the *Dialogo*, in which portions of the science of motion are presented piecemeal, and the *Discorsi*, which puts forward an axiomatic science. Even so, it is striking to notice how differently Galileo's science of motion was read by other scholars. On the title pages of *Les mechaniques* and *Les nouvelles pensées* Mersenne presented Galileo as an engineer, despite the fact that in the *Discorsi* Galileo is described as "Filosofo e matematico." While some tested Galileo's claims experimentally, others focused on reformulating his axiomatic structure, on its methodological implications, on the composition of the continuum, or on physical causes. The inclined plane, the pendulum, and falling and projected bodies became prominent objects of research. The debates following the *Dialogo* and *Discorsi* show that the link between experience and the new science was a complex matter in the eyes of contemporary observers. Although Galileo inspired reflections and experiments on a wide range of themes, many of his contemporaries rejected his views, or at least important components of them. Several claims were subjected to quantitative investigation, such as the existence and nature of differences in the speeds of different falling bodies, the exact distance fallen by a body in a given time, the length of a seconds pendulum, and the trajectory and other features of projectiles. In the process, experimental results were reported in print and compared with theoretical predictions. Galileo seems to have been only marginally interested in this aspect. Galileo, with his disciples Viviani and especially Torricelli, was primarily concerned with the axiomatic foundations of the science of motion. Scholars such as Fabri, Baliani, Gassendi, and others tackled the issue of the continuity of speed and time, proposing alternatives to Galileo's formulation. Descartes above all attempted to formulate a new worldview and questioned Galileo's more limited approach and his reluctance to deal with physical causes. Of course, these scholars' interests were not limited to this or that theme but often extended to more than one and shifted over time.

In the eyes of seventeenth-century scholars, views on space, time, and motion were connected to other issues. Copernicanism is the most obvious example, and, significantly, Mersenne, Gassendi, Cabeo, Roberval, Baliani, and Riccioli all dealt with the problem of tides.[3] The composition of matter and the continuum, rarefaction and condensation, and many other themes were also linked through a web of intellectual threads to many surprising areas. For Mersenne, for example, mechanical principles enunciated in the pseudo-Aristotelian *Quaestiones mechanicae* were the basis for one of his proofs of the existence of God, while the Society of Jesus banned several propositions on the composition of the continuum relevant to motion.[4] The material presented in this chapter covers only a selection of the themes and debates centered on mechanics and motion.

Section 4.2 examines the reception of the *Dialogo,* paying special attention to Mersenne's *Harmonie universelle* and the way he mined the *Dialogo* for problems to be examined mathematically and experimentally. Section 4.3 examines the reflections on foundations by Galileo and Torricelli. I then move in section 4.4 to the reception of Galileo's works among a number of scholars, largely in France and Italy, such as Gassendi, Fabri, and Baliani. I am especially interested in the interplay among conceptual analyses of motion, the mathematical representation of falling bodies, and the philosophical debates on causes. Section 4.5 deals with debates and experiments on parabolic trajectories, the pierced cistern, and the resistance of materials. Finally, section 4.6 examines the experiments performed by Riccioli and some of his fellow Jesuits.

Galileo was not very forthcoming in providing detailed accounts of how his experiments were performed and of the numerical values he had obtained: he was more interested in proportions among variables than in absolute values, such as the distance traversed by a falling body in one second, for example. At times he even argued that it was not necessary to perform them, as when in the *Dialogo* he discussed whether a body dropped from the mast of a moving ship falls at the foot of the mast. In a previous letter Galileo had claimed that he had performed the experiment, however, and some of his readers also tried it. Baliani and his collaborators performed it aboard a galley in the Genoa harbor in September 1639, and Gassendi arranged for balls to be dropped from the mast of a ship moving at full speed in the Marseilles harbor. Both Baliani, in a letter to Galileo, and Gassendi, in print, provided some quantitative values. Baliani stated that the ball had fallen in three seconds from a mast of forty *braccia* while the galley moved sixteen *braccia* in the same time. Gassendi provided a long list of experiments on relativity of motion, but he provided only one quantitative value for the case of the stone dropped from the top of the mast of the moving trireme, namely that the ship traveled four miles in a quarter-hour. The reader can calculate that the trireme was making well over twenty feet per second. These values show that the experimenters were looking for macroscopic effects that should have been visible even in such a crude test. If the choice was between two alternatives, the falling body was clearly moving with the ship, since the opposite view implied that it should have been left behind by dozens of feet. The result indicated that the vertical fall and horizontal projection were independent, thus removing a standard objection against Copernicanism.[5]

Many of Galileo's contemporaries were much more forthcoming than he in their accounts, not only giving details of how, where, and in front of whom experiments were performed but also providing numerical values for the results they had obtained. The emergence and growing importance of this trend warrants our attention. The works by Mersenne and the Jesuit astronomer Gianbattista Riccioli stand out as marking an important change in the way experimental data were reported in print.

4.2 Mersenne's *Harmonie* and the *Dialogo*

Mersenne's views on falling bodies shifted dramatically over time, but at each stage he claimed that they were based on experience. Initially he believed that bodies in free fall slow down. Then he argued that they move with a uniform speed. He was quite adamant about his views, despite Beeckman's claims to the contrary. Ultimately, after having seen Galileo's *Dialogo*, he published a pamphlet, *Traité des mouvemens* (Paris, 1634), in which he claimed that bodies accelerate following the odd-number rule whereby the spaces traversed are as the squares of the times. His subsequent investigations on motion were largely inspired by Galileo, and he examined several features of Galileo's views both experimentally and conceptually. Mersenne took Galileo very seriously and spent a considerable amount of time investigating even those claims Galileo had put forward as mere examples or curiosities. Despite many criticisms and increasing doubts, he always referred to Galileo with great respect. In his 1634 *Traité* Mersenne provided numerical data extrapolated from his observations on free fall, arguing that in five seconds a body falls three hundred feet, falling twelve feet in the first second. His concern with reporting experimental data and providing extensive numerical extrapolations would remain a consistent feature of his work.[6]

The little *Traité* was incorporated into *Harmonie universelle,* a sprawling work and a bibliographic nightmare of which, it has been alleged, there are probably no two identical copies. There are three separate publishers, the pagination is in disarray, and the numbering of propositions is often erroneous. For these reasons it has become customary to refer to the modern facsimile edition based on Mersenne's own copy, inscribed with his corrections and marginal annotations.[7] The intellectual organization of the work is no better than the quality of its printing. Information on similar matters is scattered across hundreds of pages and is often difficult to retrieve despite an imposing subject index.

Before venturing into this labyrinth, we must explore the relevance of Mersenne's enterprise to the history of the science of motion and mechanics. His work deals with topics ranging from the nature of sound to the motion of bodies and includes a lengthy disquisition on musical instruments that discusses such matters as pendulums and vibrating strings. The vibration of strings had been investigated by mathematicians going back to the Pythagorean school of ancient times. In the sixteenth and seventeenth centuries several scholars studied the properties of vibrating strings, including Galileo's father Vincenzio, Benedetti in *Diversarum speculationum liber*, Galileo in the first day of the *Discorsi*, Beeckman in his discussions with his visitors, especially Descartes, and Mersenne in several publications, notably *Harmonie universelle*. They examined the relationships among length, thickness, tension, and pitch of chords, for example. As Mersenne's title indicates, these investigations belonged to the science of music or harmonics, one of the four classical mathematical disciplines of the quadrivium, together with arithmetic, geometry, and astronomy. During the seventeenth century, however, harmonics progressively vanished as a separate discipline and became a part of mechanics closely connected with elasticity.[8]

Harmonie universelle investigates the motions and properties of strings in books 1 (dealing with the nature and properties of sounds) and 3; the latter is followed by a short treatise by Roberval on the equilibrium of weights on inclined planes and other matters. As the title indicates, Mersenne's primary focus was universal harmony, the Augustinian belief that harmony pertains to all of creation. Motion and mechanics are valuable tools for uncovering musical harmony because, as one reads at the end of book 1, sound is local motion, and this explains why Mersenne devoted so much space to these topics.[9]

One of the most characteristic features of Mersenne's work is the large number of tables, whose purpose is not always immediately clear. One of them is meant to teach a deaf man how to tune a string instrument. Here musical harmony is revealed to the deaf by columns of neatly arranged numbers (fig. 4.1). Ironically, the table is incorrect, but even this Swiftean attempt is revealing of Mersenne's love for the order and harmony of numbers arranged in neat columns. His tables often served the same purpose as Galileo's geometrical diagrams. Of course, Mersenne too used geometrical diagrams, but arithmetical presentation was especially well suited to his fascination with musical harmony, while Galileo's preferences were for geometry and visualization.[10]

Book 2 is devoted to motion, but additional material on motion is presented in the unnumbered pages following the table of contents. Mersenne states that those pages had been left blank and that he therefore filled them with additional observations. (He often inserted such prefaces, printed after the rest of the book, to include corrections and additional material.) Some of the problems addressed concerned the distance fallen by a body in a given time, fall along inclined planes, and the length of the seconds pendulum. These problems were obviously linked, for a pendulum that marked out one second exactly would have been useful in experi-

Tablature harmonique pour les sourds.

Les 8 sons, ou notes de l'Octaue.	Les 7 degrez de l'Octaue.	Table I. La tension des chordes proportionnées selon la raison doublée des interualles.				Table II. La grosseur des chordes proportionnée selon la raison simple des interualles.		Table III. La longueur des chordes proportionnées selon la raison simple des interualles.			Table IV. La Tension des chordes proportionnées selon la raison simple des interualles.			
		liures.	onces.	gros.	grains.	parties de ligne	dixièmes.	pieds.	poulces.	lignes.	liures.	onces.	gros	grains
1 VT		1	0	0	0	10		4	0	0	2	0	0	0
	ton mi.													
2 RE		1	4	15	54	9		3	7	2⅖	1	12	12	58
	ton mai.													
3 MI		1	10	9	0	8		3	2	4⅘	1	9	9	43
	fem.mai.													
4 FA		1	14	3	32	7½		3	0	0	1	8	0	0
	ton mai.													
5 SOL		2	6	4	0	6²		2	8	0	1	5	5	24
	ton mi.													
6 RE		2	14	3	32	6		2	4	9;	1	3	3	14
	ton mai.													
7 MI		3	11	12	18	5;		2	1	7;	1	1	1	5
	femi.maj.													
8 FA		4	4	0	0	5		2	0	0	1	0	0	0

M

Figure 4.1. Mersenne's table of tuning for the deaf *(Harmonie universelle).* Here Mersenne relates musical intervals to properties of the strings. Table I gives the tension of strings of equal length and thickness. Table II gives the thickness of strings of equal length and tension. Table III gives the length of strings of equal thickness and tension. Table IV gives their tension when both thickness and length are variable. Table IV is inconsistent with II, however, because in II by thickness Mersenne means diameter, whereas IV makes sense only if thickness means cross-section, as it does elsewhere in the book.

ments with falling bodies and in studying other phenomena. I start by discussing the problem of determining the distance fallen by a body in a given time.

The interlocutors in Galileo's *Dialogo* discussed the time required by a body falling from the moon to the center of the Earth. In the context of this discussion Salviati stated that, "per replicate esperienze," a ball of one hundred pounds would fall one hundred *braccia* in five seconds. In 1632, following the publication of the *Dialogo*, the Genoese patrician Baliani tried to find out from Galileo how he had obtained that result, since he lacked a sufficient height from which to conduct the experiment. At the time Galileo had other concerns, but in 1639, following a second request by Baliani, he replied that he had used the inclined plane; thus he had not performed the experiment directly. As to time, with the help of some friends he had counted the number of oscillations of a pendulum of arbitrary length during a twenty-four-hour period, defined as the time necessary for a star to return to its original position in the night sky, and then found the seconds by means of the relation between periods of oscillations and distances. Despite giving all these details, Galileo was reluctant to commit himself to the value he had provided, arguing that it might need revision. Indeed, in a marginal notation in his own copy of the *Dialogo* Galileo gave different values, having Simplicio say that a lead ball of one hundred pounds would fall more than one hundred *braccia* in four pulse-beats, a value Mersenne would have much welcomed. It is possible that after Baliani's 1639 letter Galileo had the experiment performed directly from a high tower, as opposed to by extrapolation from an inclined plane. In the meantime, in 1638 Baliani published *De motu*, a treatise in which he reached conclusions about falling bodies similar to Galileo's, despite his doubts about the degree of acceleration. In a letter of 1627 Galileo had told him that bodies fall according to the odd-number rule, though he did not provide a proof; thus Baliani's work was not entirely independent of Galileo.[11]

In *Harmonie* Mersenne had taken the Florentine *braccio* to be equal to 20 royal inches, but in a corollary he added that the learned scholar Fabri de Peiresc had given him a value of 21½ royal inches. In 1644 during his Italian journey Mersenne was finally able to find the correct value of 23 Parisian inches when near St. Peter's at Rome he found a shop selling samples of units adopted in Italy.[12] The story of Mersenne's struggle with metrology effectively highlights some of his concerns.

The experimental determination of data such as the distance fallen in a given time was not the only topic on his mind, however, and Mersenne discussed mathematical problems too, such as Galileo's triangle of speeds and the idea that speed increases uniformly going through infinite degrees of slowness, an idea Mersenne seemed to accept quite readily at this stage. Discussing the triangle of speeds, Mersenne indicated that all the lines representing the degrees of speed make up the triangle, which he took to be proportional to the space traversed by the falling body, something Galileo had not done. Mersenne seems to have taken the horizontal lines to represent both instantaneous speeds and the spaces traversed with those speeds, conflating somewhat the notion of indivisible (a line representing an instantaneous speed) and an infinitesimal (a tiny surface representing a space traversed). Roberval, a mathematician in Mersenne's circle, when talking of indivisibles understood them not as lines but as infinitesimal rectangles.[13]

Table des cheutes.

	I	II	III	IV	V
I	3	3	5	5	3
2	·9	12	8	13	5
3	15	27	13	26	8
4	21	48	21	47	13
5	27 33	75	34	81	21
6	39	108	55	136	34
7	45	147	89	225	55
8	~~51~~	192	94	319	89

Figure 4.2. Mersenne's table for free fall *(Harmonie universelle)*. On the left side Mersenne indicated half-seconds. Column I shows the spaces traversed according to the odd-number rule during each interval (multiplied by 3), whereas II gives the total distances. Column III gives the spaces traversed during each interval according to Wendelin's rule, and IV gives the total distances. Column V gives values according to Wendelin's rule with a different initial value. For short amounts of time, up to four or five half-seconds, the values in columns I and III are very close, as are those in columns II and IV. The written corrections are Mersenne's.

Although Mersenne accepted the odd-number rule, he cautioned against reaching hasty conclusions based on only a few observations. For example, he discussed another proportion suggested by the Copernican astronomer Gottfried Wendelin, Canon of Condé and Tournai, and produced a numerical table of the rival rules showing that although the values tend to diverge substantially, over short amounts of time the differences are not huge (fig. 4.2).[14]

Mersenne further discussed motion in a resisting medium and whether falling bodies reach a constant speed. Without mentioning Descartes by name, he reported the latter's views on this matter, according to which at successive time intervals the increase in the speed is reduced by the same fraction; therefore the body never reaches a final speed.[15]

Several passages dealing with motion and mechanics in Galileo's *Dialogo* attracted the attention of scholars, such as his speculations about the trajectory of a body falling from a tower on a moving Earth. Galileo's spokesman Salviati presented this reasoning as a mere fancy, a *bizzarria*, but readers reflected extensively on this issue. Mersenne was the center of a correspondence network involving scholars such as Fermat, Pierre Carcavy, and others. He devoted two propositions from *Harmonie universelle* to refuting Galileo's fancy. During this period it was generally believed that if it were possible for falling bodies to continue moving beneath the surface of the Earth, they would reach its center. Mersenne showed that, according to Galileo's fancy, the time of fall to the center of the Earth is independent of the height of fall, thus a body falling from the moon and a body falling from a tower on the Earth would reach the center of the Earth in the same amount of time, namely about six hours.[16] In this respect Galileo's speculation looks implausible, for it is counterintuitive. Mersenne tried to examine some problems with empirically testable consequences, but in his calculations he erroneously extended to the entire fall an approximation valid only for very short amounts of time. Despite this inaccuracy, his efforts toward finding numerical values are worth noting. At the end of his analysis of Galileo's speculation, Mersenne provided his own analysis of the motion of a falling body on a moving Earth (fig. 4.3). Following Fermat, Mersenne accepted neither that motion is rectilinear nor that it is uniform, and argued that the trajectory is a spiral composed of circular motion whose speed decreases with the radius combined with a uniformly accelerated motion toward the center along the radius.[17]

Mersenne examined Galileo's claims about extrusion, reporting the original geometrical diagram and reasoning. Galileo's attempt had been marred by his failure to take into account parabolic trajectories. Subsequently, however, Mersenne shifted from geometry to arithmetic and by using numbers he adopted the correct proportion, concluding that by increasing the Earth's rotational speed, extrusion would occur. His strategy was to show that, by selecting increasingly shorter times, the distances between a point on the Earth and the tangent from a nearby point become comparable to the distances fallen by a body in free fall. Although Mersenne reached a sound result, it is doubtful whether he understood the problem with Galileo's proof.[18] Even so, his attempt was sounder than that of the Jesuit Nicolò Cabeo, who in 1647 claimed that extrusion does not occur and that, neglecting gravity, projection would occur not along the tangent but along a curved line. Cabeo also denied that that impetus lasts perpetually, even neglecting impediments, since all things naturally come to an end.[19]

In *Harmonie universelle* Mersenne explored motion along inclined planes with two main objectives in mind: to check whether falling bodies go through all degrees of speed and to test Galileo's claim that, given the height, the time of fall is inversely as the length. From his account it seems that Mersenne did not measure time but simply compared the distances traversed in equal times by bodies falling

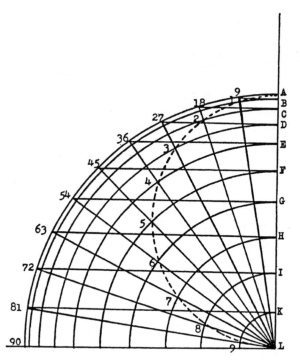

Figure 4.3. Mersenne on Galileo's *bizzarria (Harmonie universelle)*. In Mersenne's rendering of Galileo's fancy, a body falls from *A* to the center of the Earth in *L* along the dotted line. At successive instants marked 9, 18, 27, etc., the body is in 1, 2, 3, etc., respectively; 9−1 is equal to *AB*, 18−2 is equal to *AC*, etc., which are as the versed sines of the angles of rotation. In modern terms the versed sine is 1 − cosωt, where ω is the Earth's angular speed and t is the time. At the beginning of fall, ωt is very small, therefore the versed sine can be approximated by $(\omega t)^2$, and fall is proportional to the square of the time. Mersenne found empirical problems with Galileo's *bizzarria* in that from the value of the Earth's radius (17,181,818 feet) he calculated that in three seconds a body would fall only just over four inches, rather than the distance of 108 feet shown by experience. Mersenne applied the proportion whereby the distances are proportional to the square of the times. Since in six hours there are 21,600 seconds, we have the following proportion: 17,181,818:21,600::distance fallen in three seconds:3². Hence the distance fallen in three seconds is about four inches. But the distance fallen is proportional to the square of the time only as an approximation when the time of fall is very small. For longer times the versed sine can no longer be approximated in the same way.

vertically for five feet and descending along a corresponding inclined plane with different inclinations. He calculated the distances a ball should descend along the inclined plane with different inclinations by extrapolation from free fall and then compared those distances with the values he had obtained experimentally. Mersenne found that in the same amount of time it takes a ball to fall perpendicularly from a height of five feet, the ball "falls only one foot on a plane inclined by fifteen degrees, although it ought to fall sixteen inches." He provided a long list of data on the discrepancies between expected and empirical data. The inclinations ranged from 15 degrees to 75 degrees. He consistently found balls to fall shorter distances on the inclined plane than expected, but he could detect no clear pattern. For example, despite allegedly having repeated the trial several time with great accuracy, he found that a body falls three feet on a plane at 45 degrees and two feet nine inches on a plane at 50 degrees, clearly an anomalous result. Despite his fondness for numerical tables, in this case Mersenne did not present in tabular form his experimental results and the discrepancies with Galileo's rule. His tables privileged regularities in nature over discrepancies between experimental and theoretical values. Overall, the discrepancy with Galileo's values was such that he doubted whether Galileo had indeed performed the experiments, but in a marginal note to his own text he referred to Galileo's inclined-plane experiment in the *Discorsi.*[20] Despite such problematic results, Mersenne still relied on the inclined plane to find out whether falling bodies go through all the degrees of speed or slowness. Following Galileo, he reasoned that one could arrange an inclined plane with such small slope that a body would take years to fall, and any given time could be increased by reducing the slope.[21]

Mersenne studied the motion of strings, including their length, tension, frequency, and forces. He claimed that the vibrations of a string are isochronous regardless of their amplitude, because pitch does not change even while the string's motion decreases, a result already known to Beeckman, who had communicated it in a letter to Mersenne. Mersenne tried to experiment with very long strings, where frequency is very low and the vibrations can be counted visually, to show that frequency is inversely proportional to length. As to tension, Mersenne claimed that for two strings of equal thickness and material but different lengths to vibrate in unison, the tension must be as the square of the length. In fact, Mersenne added that in this and related cases one needs to adjust the ratio so that if the length is doubled, the weight generating the tension in the string is not 4 times the original weight but 4¼. This correction is presented as a purely empirical rule, one that we would attribute to the stiffness of the string, but Mersenne was not so explicit in his explanations. With regard to thickness or cross-section, Mersenne claimed that two strings of equal length are in unison if the force or tension is as the thickness. He also addressed the vibration of air columns in organ pipes, where variables like length, width, shape, and material provide opportunities for theoretical and experimental research and analogies with strings.[22]

Mersenne also provided an analysis of the motion of the string, using an approach that, as we have seen, characterized many seventeenth-century investigations of mechanics and motion, namely approaching new problems by identifying similarities to familiar ones. Mersenne tried to conceptualize the motion of strings by establishing analogies with the behavior of other objects, such as the oscillations of a pendulum or the motion of projectiles.[23] His analogies are quite different from what the modern reader might expect. For example, he considered the pendulum to be analogous to a vibrating string attached to only one of its extremes, in that both the string and the bob tend toward a central position at rest. Mersenne believed that in both cases the oscillations were either rigorously or pretty close to isochronous. Mersenne's analysis was still largely Aristotelian, as evidenced by his analogy between vibrating strings and projectiles. In his view, force was proportional to the displacement and to speed. Thus a string removed from the equilibrium position would move fastest upon being released and would slow down approaching the equilibrium position. The same is true with projectiles, which travel more slowly as they become further removed from their origin. Both examples are instances of violent motion tending toward rest. Mersenne seems to defend here the notion of a self-dissipating impetus. In his view, the string passes the equilibrium position because of air disturbance. Elsewhere he argued that the string may keep vibrating because whatever moves would move eternally. He seems to be noncommittal between these two seemingly incompatible explanations.[24]

Scattered across the pages of *Harmonie universelle* are several propositions on the pendulum that count among the earliest in print, although the theme had already been broached in correspondence with Beeckman and Descartes.[25] Mersenne conceptualized the pendulum's bob as moving on a series of inclined planes. Since bodies descend along inclined planes of all inclinations in such a way that the space is as the square of the time, pendulums follow the same proportion. As in the *Addition* to *Les mechaniques* of Galileo, Mersenne stated that the lengths are as the squares of the times. He interpreted this proposition both in terms of the periods of different pendulums with different lengths, and as a relation between the time and the vertical component of the distance fallen for the same pendulum, arguing that if the bob is displaced to the horizontal position and is then released, the vertical components of the bob's trajectory would be as the square of the times.[26] Mersenne returned to the problem later in the *Harmonie* in his treatise on string instruments, where he claimed that the oscillations of a pendulum are not strictly isochronous, the first taking slightly longer than the others. He maintained this same view later in life, as when in 1647 he objected to Baliani that a pendulum might be isochronous in a vacuum but certainly not in air.[27]

In 1646 Mersenne circulated to his correspondents, including Cavalieri, Torricelli, Baliani, Descartes, Huygens, and Roberval, a question on the determination of the center of oscillation of a compound pendulum, that is, a rigid body suspended from an arbitrary point. Whereas in a simple pendulum all the mass is concentrated in a point in the bob, in the compound pendulum the mass occupies a finite volume. The center of oscillation is that point on the line from the suspension point to the center of gravity, whose distance from the suspension point is equal to the length of a simple pendulum with the same period. This problem cannot be solved visually but requires sound principles and computations analogous to those required to determine the center of gravity. The Jesuit Honoré Fabri wrote a short treatise on the subject, showing the links between the center of oscillation and the center of percussion, a problem formulated by Baldi in his commentary on *Quaestiones mechanicae* in terms of the blow of a sword; the problem consisted in determining the point in the sword where the blow has the greatest effect. From the available evidence there seems to have been a division along national lines in the response to Mersenne's problems, with Italian mathematicians apparently showing no interest in it. By contrast, several scholars and correspondents across the Alps tackled it and put forward their solutions. Fabri and Mersenne argued that center of oscillation and center of percussion coincide. While Huygens found the problem too difficult at that stage, Descartes and Roberval circulated their attempted solutions in their letters to Mersenne, Charles Cavendish, and others. Although they could solve several specific cases, a general solution eluded them. Unlike Descartes, however, Roberval understood the need to take into account the direction of the speed of all the portions of the oscillating body.[28]

4.3 Rethinking Galileo's Axiomatic Structure

The results and methods of Galileo's *Discorsi* were not generally accepted by his contemporaries. Even before publication, Galileo and his allies discussed several issues and submitted preliminary drafts to knowledgeable readers in order to receive feedback. Cavalieri discussed mainly those aspects having to do with the doctrine of indivisibles. At Padua the Consultor of the Venetian Republic, Fulgenzio Micanzio, passed on drafts of the first day to the Padua professor of mathematics Andrea Argoli, to Paolo Aproino, a former student of Galileo's, and to the French engineer and mathematician Antonio De Ville, working for the Venetian Republic.[29] Several features of Galileo's formulation proved problematic, but attention initially focused on the science of motion rather than the resistance of materials. A notable exception was Baliani, who expressed to Galileo his doubts about the ratio between the moment of the force and the moment of the resistance of a beam infixed in a wall, and in particular the factor ½. Galileo argued that the entire matter amounted to an application of the principle of the lever.[30]

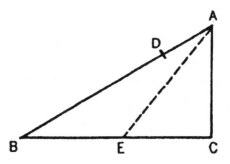

Figure 4.4. Galileo's 1656 addition to the *Discorsi.* Here we have two inclined planes with the same height *AC. BEC* is parallel to the horizon. The proof is based on the incline *ABC* and can then be extended to all others, such as *AEC.* The key idea is to rely on the degree of speed at *D,* on the perpendicular from *C* to *AB.* The degree of speed at *B* is to the degree at *D* as time through *AB* is to time through *AD,* from the definition of uniformly accelerated motion. From corollary 2 to theorem 2, the ratio of the times is as *AC* to *AD.* The impetus acquired in the same time along *AC* is to the impetus along *AB* (to *AD*) as *AB* to *AC,* or *AC* to *AD.* Since the impetuses or the moments are as the spaces, the body reaches *C* and *D* in the same time with degrees of speed that are as *AC* to *AD.* Now the degrees of speed at *B* and at *C* have the same ratio as *AC* to *AD* to the speed at *D;* therefore they are equal.

The foundations of the science of motion were also problematic. Galileo's assistant Viviani expressed doubts on the science's only axiom and led his seventy-four-year-old blind master to a proof of the axiom that was included in the posthumous 1656 Bologna edition. Both Viviani's doubt and Galileo's heroic effort testify to a concern that the foundations of the science of motion had been based on probable arguments and experiences rather than firm and evident principles. Ultimately, Galileo and his followers were reluctant to rely on a new axiom and thought that the science of mechanics would have provided a more solid basis. The statement to be proved was that the degrees of speed acquired by the same body over inclined planes with different inclinations are the same when the planes have the same height. The kernel of Galileo's proof consists in extending to motion a mechanical theorem about equilibrium (fig. 4.4). The moments of gravity of the same body over inclined planes with the same heights but different inclinations are inversely as the lengths of the planes. Extending the principle from equilibrium to motion, Galileo argued that the moments or the initial impetuses are in the same proportion as the degrees of speed acquired in the same time. By "moment" or "impetus" Galileo meant the tendency of a body to descend, equal to the minimum resistance required to hold it.[31]

According to Stillman Drake, in Galileo's addition to the third day there is an interpolation by Viviani, missing in Galileo's dictated text, stating that it is impossible that heavy bodies would descend spontaneously unless their center of gravity approaches the center of all heavy bodies. Drake notes that this passage echoes Torricelli's principle. This would suggest that the attempt to systematize Galileo's science of motion was a primary concern to Galileo's successor as mathematician to the Grand Duke of Tuscany.

In 1644 Torricelli published his *Opera geometrica*, which included *De motu*, a treatise in two books, one on falling bodies and the other on the motion of projectiles. Torricelli relied on Galileo's previous work and did not provide definitions or discuss uniform motion because he was writing for a learned audience. His reasons for revisiting the science of motion were to provide a new and more solid foundation, to expand on Galileo's results, and to provide a more extensive treatment of parabolas. Galileo's notion of *momento* was central to his foundational efforts and to his attempt to find a mechanical basis for the science of motion. Torricelli admitted some discomfort with Galileo's axiom, which lacked evidence and had to be underpinned with inaccurate pendulum experiments.

Torricelli's systematization started from a powerful axiom, namely that two connected heavy bodies cannot move by themselves unless their common center of gravity descends. Unlike Galileo's axiom, Torricelli's was immediately evident and was merely illustrated by examples of machines such as balances or pulleys. Torricelli could then prove Galileo's axiom as a theorem in proposition 5, which concluded the main foundational portion of his treatise. In his proof Torricelli relied on a claim made in the *Discorsi*, in the second part of proposition 6 of *De motu naturaliter accelerato* (book 2, third day), that the moments of equal weights on inclined planes of equal length but with different inclinations are proportional to the heights of those planes. Torricelli questioned Galileo's claim that this relation was known from the elements of mechanics and proved it in proposition 3 of *De motu*. This example, brief as it is, gives us a sense of Torricelli's concerns about interpreting Galileo and formulating a science of motion in a rigorous axiomatic fashion.[32]

Much like Galileo, Torricelli was trying to establish the science of motion on the principles of mechanics. Following Galileo, Torricelli argued that when "the *moles* or quantities of matter are the same, the speeds in different inclined planes are as the moments." This passage became the subject of a debate with Mersenne, who was then traveling across Italy and while in Florence had received a copy of Torricelli's treatise from the author himself. Mersenne was not convinced and asked for a justification of the postulate independent of experience. Torricelli's formulation highlights a lack of symmetry in the way the components of *momento* operate. If an inclined plane becomes steeper, for example, speed and *momento* increase, but an increase of *momento* due exclusively to a larger weight leaves speed unaffected, hence the need to specify the equality of *moles*. Torricelli identified gravity or *momento* as the cause of motion and speed as the effect, arguing that all bodies accelerate in the same way because gravity is greater for heavier bodies in proportion to the resistance they offer. Mersenne objected that in a bent bow the force is not as the speed, to which Torricelli replied that the force moving the heavy body is intrinsic, natural, and uniform, whereas the force of the bow is extrinsic, violent, and acting for a very short time.[33]

These debates highlight the uncertainty and questions surrounding the foundations of the science of motion, and especially the question as to whether the principles of mechanics sufficed, or whether new ones had to be found. But other aspects were highly problematic too.

4.4 Continuity and the Law of Fall

In 1642 Gassendi published *De motu impresso a motore translato*, a work in the form of two letters to the king's librarian, Pierre Dupuy. Several features of his analysis of falling bodies deserve attention. Instead of Galileo's triangle of speeds, Gassendi adopted a visually and conceptually different representation based on a lattice of triangles (fig. 4.5). He claimed that time and speed increase uniformly, but ambiguities between continuous and discrete representations remained. Instead of all the lines, as in the method of indivisibles, Gassendi referred to all the triangles, representing both different degrees of speed and distances traversed by the falling body. Moreover, initially he took the increase of speed in free fall to occur through successive instants, by which he meant tiny time intervals during which the body traverses a very short distance with uniform speed.[34]

Gassendi's physical explanation of the odd-number rule was itself rather odd. He assumed that the falling body would conserve one degree of speed from the first to the second instant of time. Since in the second instant the body had three degrees of speed, he was left with two degrees to account for. One was accounted for by the new action of gravity, the other by an impulsion by the air from behind. At the beginning of motion the body moves only one unit of space, whereas once it has acquired its initial motion, air comes into play to propel it one degree of speed further. Thus without the action of the air, the speed of the falling body would increase only in an arithmetic progression 1, 2, 3, 4, etc. Whereas for Galileo air retarded motion, at this stage for Gassendi it contributed to it. In the course of his correspondence with the Jesuit Pierre Le Cazre in the mid-1640s, Gassendi recognized his mistake and claimed that the action of air was unnecessary because the continuous action of gravity alone was sufficient to account for the odd-number rule. Since speed increases continuously, the degree of speed the body has

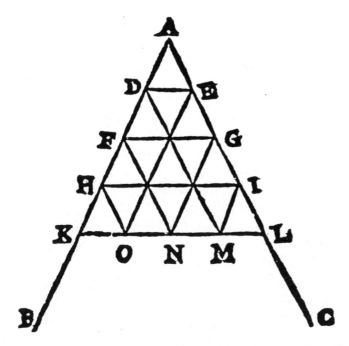

Figure 4.5. Gassendi's triangle of speeds *(De motu impresso)*. From the vertex *A* lines *AB* and *AC* represent time. The parallels *DE, FG, HI, KL,* etc., mark equal time intervals or instants of time, namely *AD, DF, FH,* etc., and the triangles equal to *ADE* represent the degrees of speed or distances traversed, 1 in the first instant, 3 in the second, 5 in the third, etc. Thus speed increases following the odd-number rule. Despite the graphical continuous representation, initially Gassendi took speed to grow discretely and to be constant during each instant of time. Therefore in each instant speed is proportional to the distance traversed, and the odd-number rule applies both to distances and to speeds. Gassendi showed simultaneously speed and the distance traversed by the falling body as the area of the triangles of speeds *ADE, AFG, AHI,* etc.

at the end of the first instant would enable it to cover in the second instant two elements of space, in accordance with the double-distance rule. Thus Gassendi was left with only one element of speed to account for, and this was due to the renewed action of gravity. While writing his 1646 *De proportione* he realized that he had not just to represent but also to take speed to grow continuously in his analysis, and the degrees of speed were represented no longer by triangles but by segments, such as their bases. Gassendi also corrected his previous views about the action of the air. Despite these corrections, a curious tension remained between discrete and continuous actions. In the course of the dispute Le Cazre defended the geometric progression 1, 2, 4, 8, etc., for the distances traversed by a falling body in successive time intervals.[35]

Galileo considered motion and time to be continuous, but others preferred a discrete approach, for various reasons. As we shall see in chapter 5, Descartes thought that the cause of gravity was a series of impacts from particles of the vortex; therefore a continuous increase in speed could not represent physical reality. The Jesuit Honoré Fabri and Baliani believed that Galileo's law was an approximation, because in reality the body's motion consists of discrete intervals with constant speed changing suddenly at the end of each interval. Baliani, for example, tried to show that if the discrete intervals with constant speed become sufficiently small, the difference between Galileo's law and his own decreases (fig. 4.6).[36] Fabri, who was probably Baliani's source, considered the discrete approach to be superior to Galileo's because it was simpler, and because Galileo's approach could not give the cause of acceleration, whereas his own could, in that it explained free fall in terms of a series of impulses. Fabri's work was published under the name of his student Petrus Mousnerius as a collection of the teacher's lectures, a tactic Jesuits often used to avoid the internal censorship of their order. In fact, several of his propositions about the composition of the continuum and indivisibles were forbidden by the Jesuit *Ratio studiorum* because of perceived inconsistencies with the doctrine of the Eucharist.[37] If the time unit is Fabri's "instant," the spaces traversed in each successive instant are as the numbers 1, 2, 3, etc. Fabri was aware that his rule about falling bodies was not invariant if one changed the unit of time, as we shall see below. He was also aware that by taking very short instants, his rule would approach the odd-number rule.[38] It is easy to see this transformation if we use the familiar diagram used by Galileo to represent the odd-number rule together with a diagram representing Fabri's rule. In his representation, Fabri considered the aggregates of speeds to be proportional to the space traversed by the falling body.[39]

The debate about mathematical representation was linked to methodological concerns about the relationship between mathematical theory and experience. Galileo's science of motion had a problematic relation with experience and experiments. His starting points or postulates were posited as natural, but they were also justified by means of experiments. In addition, his whole structure of propositions and theorems was presented as being in agreement with experience. Fabri argued that sensory data lack sufficient precision to provide rigorous mathematical foundations; bodies may fall following a slightly different rule whose difference from the odd-number rule cannot be detected by means of experiments. His fellow Jesuit Cabeo too objected that Galileo's odd-number rule had not been demonstrated; rather, Galileo had "taken from experience" *(assumpsit ex experientia)* that which he ought to have demonstrated. Cabeo, however, misrepresented Galileo's odd-number rule, arguing that for Galileo speeds, rather than distances, would increase as the odd numbers.[40]

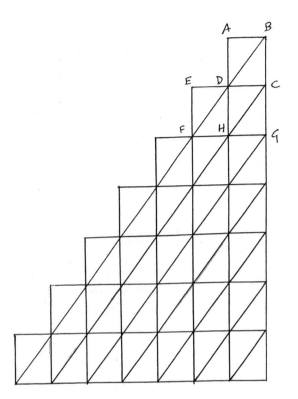

Figure 4.6. Galileo's, Fabri's, and Baliani's rules. According to Galileo, at each successive interval of time the spaces traversed increase as the odd numbers. Following Fabri, at each successive time interval *BC, CG,* etc., the spaces increase as 1, 2, 3, etc., or *ABCD, ECGF,* etc. If we now divide Fabri's spaces in half and choose half a space *BCD* as the new unit, the spaces will increase as the even numbers, 2, 4, 6, etc. Neglecting one unit of space for each time interval, such as *ABD, EDF,* etc., we obtain Galileo's rule. This approximation becomes more accurate the smaller the instant of time and the corresponding space unit. Following Baliani's example, in equal intervals of time the body travels distances proportional to 1, 2, 3, etc. If we now choose a unit of time ten times smaller and a correspondingly smaller unit of distance and apply the same law, in the first of the original intervals the body would travel 55 units (1 + 2 + ... + 9 + 10), in the second 155 (11 + 12 + ... + 19 + 20), and in the third 255 (21 + 22 + ... + 29 + 30). Now 55, 155, and 255 are to each other as 11, 31, and 51; that is, they do not deviate much from the odd-number rule. If we divide the original intervals into one hundred parts instead of ten, we obtain the numbers 5,050, 15,050, and 25,050, which are to each other as 101, 301, and 501. Thus the smaller the time interval, the better the approximation to the odd-number rule.

Baliani's early views differed radically from those of his Jesuit contemporaries. In an important 1639 letter to Galileo, for example, Baliani argued that in astronomy, music, mechanics, perspectives, and the other sciences, the principles are experiences.[41] This empiricism seems hard to reconcile with his later preference for a different law of fall from Galileo's. If the two laws are empirically equivalent because their differences are too small to be detected by experiment, what reason does one have to select one instead of the other? In reality, Baliani's empiricism is mitigated by his search for a causal explanation, a search that became more prominent in the second edition of his *De motu naturali*. As he put it in 1646, "But truly my mind is not entirely satisfied if it does not grasp, or at least investigate, the prior causes from which these effects finally result."[42] In his view his own account of the law of fall is to be preferred because it provides a causal explanation of gravity.

Against this backdrop of proliferating rules of fall, whereby a range of different mathematical proportions were presented as being compatible with experience, the young prodigy Christiaan Huygens, son of the Dutch diplomat Constantijn, represents an exception, defending the correctness both of the odd-number rule and of parabolic trajectories. It appears that Huygens was not responding directly to Galileo, whom he had not yet read, but to a work by the theologian and mathematician Juan Caramuel in which it is stated that the spaces traversed are proportional to time. It also appears that Huygens had reached Galileo's results by the age of seventeen, before having read the *Discorsi*, which he did in the fall of 1646.[43]

Huygens was seeking not empirical confirmation but rather an abstract condition that the law of fall had to satisfy, namely that it must be independent of the unit of time chosen; if a law is valid with any arbitrary time unit, it must also be valid with double that unit. In a letter to Baliani, Castelli reported that Galileo had proposed as a probable argument a similar line of reasoning based on the invariance of the rule with respect to the units of distance. Huygens argued that if in successive identical intervals of time a body falls through the distances AB, BC, CD, DE, etc., then AB is to BC as AC is to CE.[44] Huygens then examined the question of whether there is any geometric or arithmetic progression satisfying this condition. He concluded that there is no geometric progression, and that the claim that bodies fall following the progression 1, 2, 4, 8, etc., is, in his words, "fort ridicule."[45] Next Huygens turned to arithmetic progressions and found that the only one satisfying his condition is the odd-number rule. None of the other progressions work, in particular the progression of the integers, advocated by Fabri and Baliani, which he described as "absurde."[46] As we have seen, however, Fabri did not consider his rule to be valid for any arbitrary unit of time, and he also stated that his units of time were not constant.[47] Moreover, Huygens's condition leads to the odd-number rule only in the restricted context of arithmetic or geometric progressions, while in more general contexts other solutions are possible. For example, the condition would be satisfied if the distances traversed were proportional to any power of the time of fall. Finally, Huygens's apparently innocuous condition was based on the assumption of a constant gravity, since if gravity depended on position, the relation between distance and time would become too complex.

Huygens brought to his analysis of geometric and especially arithmetic progressions not only his powers of very subtle reasoning but also the new algebraic notation whereby the unknown is represented by the letter x and known quantities are represented by letters from the beginning of the alphabet. Huygens, a student of the leading mathematician Frans van Schooten the younger, had studied Descartes' *Geometria*, translated and edited by his teacher, and it is not surprising that he adopted Cartesian methods and notation.

Huygens's argument was not new to Mersenne, who had already heard something analogous in 1643 from the physician Theodore Deschamps and two years later from Torricelli. In June 1645 Torricelli had argued that if the spaces covered by a falling body were as 1, 2, 4, 8, etc., then doubling the time would change the proportion to 3, 12, etc., thus violating the rule. Mersenne's doubts about Galileo's law as expressed in his first letter to Huygens in October 1646 suggest that he had not found Torricelli's reasoning convincing. In a letter of 1647 to Giovanni Battista Renieri, Torricelli also refuted the rule whereby the spaces traversed by falling bodies are as 1, 2, 3, 4, etc. In 1647 Mersenne examined the views on falling bodies put forward by various scholars, including (though they are not mentioned by name) Le Cazre and Fabri. While claiming that the progressions 1, 2, 4, 8, etc., and 1, 2, 3, 4, etc., had been refuted by Gassendi and Le Tenneur, the latter in a book as yet unpublished, Mersenne expressed great uncertainty as to the reason why heavy bodies descend, whether their motion is natural or violent, and whether they are attracted, have a tendency to descend, or are pushed down by some external force, such as the subtle matter advocated by Descartes (see the following chapter).[48]

Mersenne seems to have been influenced by anyone he considered to be a leading scholar of the age. His attitude toward Galileo's views on falling bodies and parabolic trajectories was never one of unreserved acceptance, and it became less sympathetic with time as his earlier beliefs were eroded by his contacts with Roberval and Descartes. The section on *Ballistica* of his *Cogitata* deals both with projected motion and with the forces of bows and strings. He doubted that falling bodies go through all degrees of speed and argued as probable that they have a finite initial degree of speed.[49] While in Rome in 1644–45 Mersenne experimented with dropping bodies from St. Peter's dome, establishing that whereas a heavy ball takes five and a half seconds to fall, pebbles or nuts take about seven seconds. These experiments led him to ask at what point falling bodies stop accelerating, a phenomenon depending on the body's weight.[50] Roberval too doubted the odd-number rule, arguing that falling bodies reach a terminal speed. In a letter to Torricelli he denied on experimental grounds that projectiles shot vertically reach the ground with their projection speed and claimed that, subtracting all impediments, the rectilinear motion of a body would come to a halt, thus challenging Torricelli's contrary statement in proposition 2 for projections in all directions.[51]

4.5 Trials with Projectiles, Pierced Cisterns, and Beams

In theorem 3 of *De motu* Torricelli stated that without the action of gravity, a projectile's motion in any direction would be rectilinear and uniform, and he proved that the trajectories of projectiles are parabolic even for oblique projection, a result stated but not proved by Galileo.[52] At the end of the book (written in Latin) Torricelli included several tables, based on tables of sines and tangents, and a description of a military square (written in Italian). Despite his claims that his square was intended "for philosophers," he probably composed that part of the book with an audience of practical men and gunners in mind. The claims put forward at the end of *De motu* attracted the attention of a diverse group of men at Genoa, including Giovanni Battista Renieri (brother of the professor of mathematics at Pisa), Baliani, and some *bombardieri*, who reported that the results of their firing tests were in striking disagreement with Torricelli's tables. The exchange with Renieri is particularly valuable because Torricelli was often reluctant to engage in disputes over experimental matters, and he stated that if anyone objected to his propositions on motion, the whole doctrine could be taken as a purely mathematical exercise—an implausible claim, given his inclusion of the tables for shooting with Italian commentary.

Writing to Renieri, however, Torricelli indulged in a rare and careful analysis of the origins of experimental errors. The issue concerned the range of a gun shooting at an angle of 45 degrees, which he took to be 2,300 *passi*.[53] Horizontal shots, however, were found to have a range of 400 and even 500 *passi*, as opposed to Torricelli's theoretical estimate of 245. Renieri was surprised and troubled by this considerable discrepancy. In an attempt to solve the puzzle, he proposed that motion is indeed parabolic, as Galileo and Torricelli had claimed, but that the axis of the parabola may not be perpendicular to the horizon. Following this line of

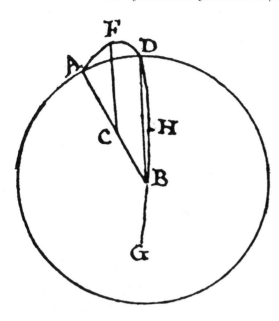

Figure 4.7. Renieri's trajectory. The figure from Renieri's letter to Torricelli shows a parabolic trajectory from point *A* on the Earth's surface *AD. CF* represents the parabola's axis, which is not perpendicular to the surface. According to Renieri, a body projected from *A* would move to *F* and then *D*, and if allowed to move under the Earth's surface, it would proceed to *H*, to the center of the Earth *B*, and to *G*, and after oscillating about *GH* would stop in *B*.

reasoning, if it were possible for a projected body to continue falling through the Earth, it would end at the Earth's center rather than become further and further removed from it (fig. 4.7). Renieri was thus trying to solve both the experimental problem and the theoretical one already mentioned by Galileo in the *Discorsi*, whereby a parabolic trajectory would not lead further and further away from the center of the Earth.

Renieri's letter suggests that Baliani was involved at some level in those experiments. In the second edition of *De motu* (Genoa, 1646^2), Baliani had claimed on theoretical grounds that the motion of projectiles is not parabolic. He accepted that at the beginning of motion the trajectory of a body shot horizontally has a horizontal uniform rectilinear component. At each successive instant, however, one of the components is no longer horizontal but tangential to the curve, resulting in a trajectory deviating more and more on the inside of Galilean parabolas. The problem with Baliani's reasoning is that by considering the tangent to the curve after the first instant, he was already including the action of gravity up to that point; therefore the action of gravity in successive time intervals had to be still proportional to 1, not 4, 9, 16, etc., or the square of the time.[54]

Renieri's letter did not mention Baliani's faulty reasoning, and Torricelli's prompt and lengthy reply ignored Renieri's suggestion, focusing instead on the experimental results. Torricelli restated his two basic assumptions, namely the odd-number rule and the uniformity of horizontal motion, defending the former with the principle of scalar invariance. He continued that Galileo's *Discorsi* mentions some reasons why experience may differ from theory, the main factor being air resistance. Torricelli also identified three possible reasons why the Genoa experiments may have gone wrong: the gun was not level, the plane of fire was not horizontal, or the gun moved slightly upward at the time of firing. He suggested appropriate countermeasures, such as (for the first problem) using many squares to level the gun or (for the second) firing along the seashore. As to the third problem, he suggested setting up a series of screens at different distances from the gun, which would reveal the ball's trajectory at given distances from the firing point. A similar device had been suggested by Niccolò Cabeo in his commentary on Aristotle's *Meteorologica*, a book Torricelli knew well.[55]

In his reply, Renieri communicated the results of further experiments conducted following Torricelli's suggestions, including the use of screens. The results had been contrary to the expectation of the *bombardieri* and also not in accordance with parabolic trajectories. Torricelli again went through the possible causes of error, pointing out that vertical fall can only occur following the odd-number rule, recalling the reasoning outlined above, and that the projectile falls along lines parallel among themselves. Strictly speaking this is not true, but the error is negligible. This remark may have been a veiled comment on Renieri's suggestion that the axis of the parabola is not perpendicular to the horizon. Losing his patience somewhat, Torricelli argued that he was writing for philosophers rather than *bombardieri*, in order to avoid precisely this sort of controversy, and proceeded to point out an incongruity in the trajectory found by the *bombardieri*.[56]

The exchange between Torricelli and Renieri shares common features with the correspondence on pulleys between dal Monte and Contarini discussed in section 1.3. In each case the scholar had problems convincing a learned and sympathetic audience of the soundness of a theory in the face of apparent experimental refutation. These cases show how problematic the links between mathematics and nature were.

Torricelli also studied the motion of waters and formulated a rule for the speed of water flowing from a hole at the bottom of a tank filled to a constant level (fig. 4.8). He probably tackled this problem because Castelli had performed experiments with Ciampoli on water gushing out of a hole in a barrel, and Galileo had discussed the use of a pierced cistern as a time-measuring device in the third day of the *Discorsi* (sections 3.3 and 3.6 above). Torricelli claimed that the jet of water exits the cistern with the same speed it would have acquired falling from the height of the water's surface in the tank. He argued that if a taller container were attached to the orifice of the cistern, the water would reach the same level in both, thus suggesting that the water jet would plausibly reach the same level. Speed

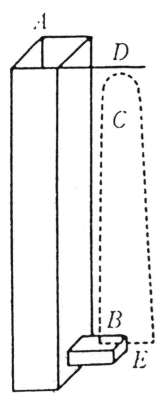

Figure 4.8. Torricelli's pierced cistern *(De motu)*. In tackling the problem of the pierced cistern, Torricelli identified features analogous to the equilibrium of water in conjoined containers and to the motion of projectiles. He arranged for the jet of water spurting out at *B* to be projected upward and showed that it almost reached the level *AD* of the water in the tank. Since in falling to *E* after having reached its maximum height *C* the jet goes through the same degrees of speed as in rising, at *E* the water must have the same speed it had at *B* when coming out of the orifice; hence the water can be imagined as falling inside the container almost as if it were a body falling in air, and its speed at *B* is as the square root of the height of the water in the container.

could not be measured directly but could be inferred either by directing the jet upward and observing the height it reaches, or by measuring the amount of water collected in a given time. Torricelli chose the former method, arguing that the water jet would acquire a speed sufficient for it to reach the level of the water surface in the tank. By analogy with falling bodies, the speed was as the square of the height, provided that the orifice was much smaller than the width of the container. Torricelli showed experimentally that the water rises back almost to the surface level and offered an explanation for the small discrepancy observed based on air resistance and the interference of the first drops falling back. When Mersenne informed Torricelli of his doubts about the motion of waters, Torricelli replied that he considered everything he had said to be true and reiterated the conditions under which his propositions apply, concluding abruptly that if someone rejected his whole doctrine, "quamquam vera sit," he would not care. His dislike of Mersenne is palpable.[57]

In the same year as Torricelli, Mersenne too, probably stimulated by Galileo, discussed the pierced cistern in *Cogitata physico-mathematica*. Mersenne had already discussed the problem in correspondence with Descartes from 1639. Unlike Torricelli, Mersenne provided numerical data as to the amount of water pouring from an orifice and emphasized empirical tests over conservation principles. Mersenne reached conclusions similar to Torricelli's, but he was doubtful as to why the first drops of water reached the level of the water surface, since they patently had not fallen from there.[58]

Mersenne investigated other aspects of Galileo's work in *Cogitata* as well as in *Novarum observationum physico-mathematicarum tomus III.* Since the latter can be seen as a continuation and refinement of the former, they are best considered together. *Cogitata* consists of six parts, including sections on hydraulic-pneumatic, mechanical, and ballistic phenomena. In his investigations, objects and problems associated with them occupy center stage and are key tools of investigation. Besides the pierced cistern, Mersenne studied the rarefaction and condensation of air and discussed the operation of a wind gun or *scopletum pneumaticum,* a metal weapon charged by means of a syringe. Thus the elasticity of air was coming under scholarly purview.[59]

In his section on mechanical phenomena, Mersenne examined simple machines as well as the resistance of materials and the force of percussion. At the time, the resistance of materials attracted less scholarly attention than the science of motion. The former relied on common mathematical procedures and applied the principles of Archimedean statics, which were widely known and well established. Mersenne tested the resistance of materials in several experiments in which he established that the material used to make wires dramatically affected their resistance to breaking. Whereas Galileo was primarily interested in establishing geometrical ratios and in speculating about causes for the resistance of materials, Mersenne, as usual, gave numerical values, claiming that wires with a diameter of one-sixth of a line (where a line is the twelfth part of an inch) break with different loads depending on their material: 7 pounds for gut, 23 for gold, 18½ for silver and bronze, and 19 for iron. Mersenne also discussed the causes of resistance, ranging from the *horror vacui* to atomism, finding all of them problematic. He added that Galileo did not consider the fact that wood splinters while iron, bronze, and other metals bend in an arc before breaking, suggesting that the physico-mathematician solving this problem would be as the "Inventor quadraturae."[60]

Here I wish to mention briefly the problem of the force of percussion, which was the topic of an additional day in Galileo's *Discorsi* that appeared only posthumously in the eighteenth century. The problem, however, was very much alive among his followers, and both Torricelli and Borelli discussed it, the former in the *Lezione accademiche* (also published posthumously in the eighteenth century), the latter in *De vi percussionis* (Bologna, 1667), which concluded that the force of percussion cannot be compared to static weight.[61]

4.6 The Experiments and Tables of Riccioli

The last scholar considered here is the Jesuit Riccioli. In his massive *Almagestum novum* (Bologna, 1651) one finds extensive experimental reports on pendular motion and falling bodies. These topics are discussed in two sections of the work: in part 1, close to the beginning, in *De sphaera elementari*, and in part 2 in the discussion of the mobility of the Earth. Riccioli's approach is largely experimental and relies on tables of numerical results. His tables are quite different from Mersenne's. Mersenne seems to have been mainly concerned with displaying the harmony of the universe and gaining a sense for the numbers involved; his numbers were not the direct result of experiments but rather extrapolations from experimental data showing the numerical values from a given proportion. By contrast, Riccioli provided several tables of actual experimental data. At times there was a perfect coincidence with theory, but other times Riccioli presented data that did not easily fit into any mathematical rule. He showed no interest in Galileo's axiomatic structure.[62]

At the outset Riccioli tried to establish the reliability of the pendulum as a time-measuring device. He reported that oscillations of 40 or 45 degrees are slower than smaller ones, but smaller ones were isochronous if performed well. He counted the number of oscillations during consecutive runs of a sand clock and found an excellent agreement, the discrepancy being one or two oscillations only over several minutes. He attributed this discrepancy to his own carelessness. He also checked the number of oscillations between the meridian transits of stars. The observations conducted on three different nights, 19 and 28 May and 2 June 1645, were in spectacular agreement, yielding 3,212 oscillations in two cases and 3,214 oscillations in the third, the tiny discrepancy being attributed to some perturbation such as wind. Riccioli did not succeed in constructing an exact seconds pendulum, but he was satisfied that the pendulum is an exceedingly reliable tool for measuring time. He argued too that its accuracy should be assessed not in absolute terms but in comparison with other ways of measuring time, none of them perfectly reliable.[63]

In the next chapter of part 1 of the *Almagestum* Riccioli established two important findings: that falling bodies follow the odd-number rule and that not all bodies fall in the same time, size and weight affecting the result. He also criticized Aristotle's claim that speed is proportional to weight. Riccioli returned to these same problems in part 2, in which he expanded the range of his experiments, determining that a body falls fifteen feet in the first second of free fall. Riccioli also criticized previous investigations by fellow Jesuits, including Roderigo de Arriaga, who in his enormous *Cursus philosophicus* had stated not only that all bodies fall with the same speed but also that they do not accelerate, a claim likely to have discredited his entire account in the eyes of contemporaries. Not surprisingly, he was at pains to account for the different effects of the force of percussion, arguing that a greater force of percussion is not necessarily the result of greater speed. Arriaga's account of his experiments was remarkably vague. At one point he mentioned pieces of bread crust dropped from high places, leaving himself open to Riccioli's jibe that he was dropping bodies from his dinner table.[64]

It is in part 2 of *Almagestum* that one finds extensive tables of experimental data. For example, we find a table comparing twenty-one pairs of different balls dropped from the Asinelli tower from a height of 280 feet (fig. 4.9). At four different times over a period of ten years, before a host of reliable witnesses largely from the Society of Jesus, the speed of balls of different weights and sizes was compared by ascertaining which one reached the ground first, and by how much. Never did two balls reach the ground at the same time. In general, the heavier ball anticipated the lighter one by some feet, but there were exceptions: when the two balls weighed the same, the one with the greater specific gravity fell faster (example 12 in Riccioli's table), a walnut ball was marginally faster than a beech one more than twice as heavy (17), and a lead ball was faster than a much heavier clay one (18). Riccioli analyzed the data in some detail and in the midst of contrasting views about falling bodies he tried to highlight the reliability of his experiments by providing a description of how they were performed.

Other tables include data on bodies falling in water. Riccioli concluded that in this case too, bodies fall with a constant acceleration, albeit a smaller one than in air. Then Riccioli turned to bodies rising in water, such as bodies made of beech or wax, glass vessels filled with air or water, and inflated bladders. These experiments seemed highly significant to Riccioli, and he claimed to have established that raising bodies too accelerate, though by only a small degree. Since, according to Aristotle, speed increases in natural motion but decreases in violent motion, Riccioli argued against Galileo that the rising motion of bodies in water is natural. Hence levity would be dependent not on gravity but rather on a positive property distinct from gravity.[65]

Riccioli also included a table recording data on bodies falling along an inclined plane thirty-five feet long. He set the table at different inclinations, the maximum one with a height of approximately twelve and a half feet. He used clay, wood (which went down, as he put it, *saltitando*), and water (used to draw conclusions on river flow).[66] Further tables include measurements of the amount of piercing

Ordo Experimentorū.	Globi eodem momento dimissi ex altitudine Pedum 280.			Distantia tardioris à pauimento iā percusso à velociore.
	Globi eiusdem Magnitudinis	Pondus Globorū Vn-ciæ / Dra-chm	Globus qui fuit velo-cior	
1	Argillace⁹ vacuus / Argillaceus solid⁹	10 0 / 20 0	Grauior	*Pedes* 15
2	Argillaceus vacu⁹ / Argillaceus solid⁹	20 0 / 45 0	Grauior	16
3	Argillaceus / Ligneus	9 0 / 2 4	Grauior	20
4	Argillaceus / Cereus	20 0 / 15 0	Grauior	12
5	Ligneus / Cereus	4 6 / 6 7	Grauior	15
6	Cereus / Ferreus	1 5 / 11 7	Grauior	30
7	Argillaceus / Argillaceus	5 0 / 4 0	Grauior	5
8	Argillaceus / Argillaceus	21 4 / 11 4	Grauior	12
9	Argillaceus / Argillaceus	27 3 / 14 1	Grauior	15
10	Argillaceus / Cereus	18 7 / 1 0	Grauior	35
11	Argillaceus / Plumbeus	2 0 / 2 4	Grauior	25
12	Plumbeus / Ligneus	2 4 / 2 4	Plumbeus	40
13	Argillaceus / Argillaceus	7 0 / 62 0	Grauior	10
14	Argillaceus / Argillaceus	10 4 / 23 0	Grauior	13
15	Argillaceus / Argillaceus	53 0 / 6 4	Grauior	8
16	Argillaceus / Argillaceus	53 0 / 7 1	Grauior	9
17	Ligneus Iugladinus / Ligneus Faginus	2 1 / 4 7	Iugladinus	2
18	Argillaceus / Plumbeus	11 0 / 1 7	Plumbeus	1
19	Argillaceus / Plumbeus	33 0 / 1 0	Argillace⁹	2
20	Argillaceus / Plumbeus	38 0 / 1 0	Argillace⁹	3
21	Plumbeus / Plumbeus	1 0 / 0 4	Grauior	5

Figure 4.9. Riccioli's fall experiment (*Almagestum novum*). In this table, too long for a folio page, Riccioli compared the speeds of descent of pairs of spheres of different weights and materials, providing ammunition for the experimental refutation of the claims put forward by Galileo, Baliani, Arriaga, and Cabeo.

or sinking (the Latin term is *perfossione*) of water produced by a sphere falling from various heights.

At the end of an imposing list of experiments that can only be summarized here, Riccioli presented a list of "theorems" based on his results. The term "theorem" sounds peculiar in this context, and Riccioli's use of it was probably meant to emphasize the certainty of his experimental endeavors. His findings were not deduced from axioms but distilled from experiments. The first "theorem" states the existence of levity, as described above. Others state that heavy bodies accelerate according to the odd-number rule, but that not all bodies fall with the same acceleration. The last theorem states against Aristotle that although heavy bodies are faster than lighter ones, the difference in speed is much less than the difference in weight. Riccioli seemed perturbed by this result, since Aristotle's proposition looked probable given that the effect (speed) was meant to be proportional to its cause (weight). Some people, Riccioli argued, claim that in the absence of accidental perturbations all bodies would fall with the same speed, but he himself disagreed, because if two bodies have different weights, one of them may contain light elements such as air or fire, which would alter its speed of descent. In a rather tentative and convoluted conclusion he suggested some ways to reconcile the new evidence with Aristotle.[67]

The confirmation of Galileo's claim that heavy bodies accelerate according to the odd-number rule is given an anti-Copernican twist in Riccioli's chapter on Galileo's *bizzarria*. Riccioli seems to have been among a minority rejecting composition of motion and argued instead that one motion prevails over the other. His main argument was that if the Earth rotated, falling bodies would move with a uniform speed along a circular arc and therefore they would hit the Earth with the same speed regardless of their height of fall. But since they do accelerate, the Earth does not move. Riccioli's support for the odd-number rule had a sting in its tail.[68]

Riccioli's accurate experiments were widely known during the second half of the century and helped forge a consensus on the empirical adequacy of some aspects of Galileo's work, especially the odd-number rule and the notion that heavy bodies fall with similar accelerations and speed is not proportional to weight. His limited agreement with Galileo was significant, coming as it did from an unsympathetic reader who had gone so far as to include the text of Galileo's condemnation in his own publications.[69]

CHAPTER FIVE

The Motion and Collision of Particles

I recognize no difference between the machines that artisans
make and the different bodies that only nature composes, except
that ... the pipes and springs which cause the effects of natural
bodies are generally too small to be perceived by our senses.

Descartes

5.1 The Rise of the Mechanical Philosophy

Around the middle of the seventeenth century the mechanical philosophy
emerged as a leading player on the intellectual scene. According to this school of
thought, small particles—whether atomic and moving in a vacuum, as for Gassendi, or indefinitely divisible and moving in a plenum, as for Descartes—were
the constituent blocks of the universe. Action at a distance and the Renaissance
notions of sympathy and antipathy were herewith banished from the world. With
few exceptions, if any, phenomena had to be explained on the basis of contact actions among bodies. Mechanical or machinelike analogies and explanations became more and more common in accounting for aspects of the world, both living
and nonliving, and subtle material fluids were devised in an attempt to explain all
manner of strange beliefs, such as that a cadaver's wound would bleed at the approach of the assassin.[1] Descartes' account of gravity, based on a vortex of tiny
particles pushing heavy bodies toward the Earth, proved particularly influential.
His explanation for magnetism was so strikingly illustrated that it has become an
icon of the mechanical philosophy (fig. 5.1). In his drawing, magnetic particles
shaped like screws twirl around and pass through appropriately shaped holes in
the magnet.[2]

Important, though partial, aspects of this view of the micro-world started to
make an appearance well before the major works by Descartes and Gassendi. Classical roots can be found in the atomists and in passages of Plato's *Timaeus*, especially where Timaeus associates the elements earth, water, air, and fire with the

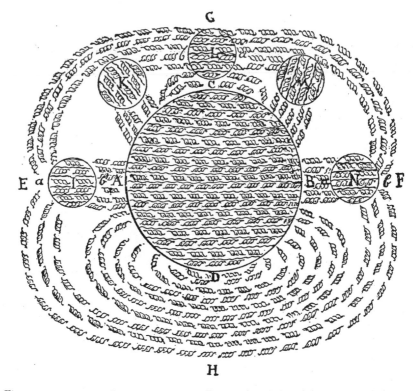

Figure 5.1. Magnet from *Principia philosophiae.* Descartes' rendering of the magnet and of magnetic effluvia highlights one of the key features of his worldview, namely his recourse to invisible
corpuscles and pores to provide causal explanations of phenomena of the visible world.

Platonic solids. Earth is formed by cubes, suggesting stability and solidity, for example, whereas fire is formed by much smaller tetrahedra, suggesting by their
small size and acute angles the ability to penetrate other bodies. This association
powerfully links the shape and size of elementary bodies to their macroscopic features.[3] In the seventeenth century the association between microscopic geometric
shape and macroscopic features became a common way to explain phenomena and
an influential feature of the mechanical philosophy. Fire, for example, was believed to be corpuscular and was associated with tiny fast-moving particles with
sharp sides that Galileo, in *Il saggiatore,* called *ignicoli.* In the same work he also
put forward the distinction between fundamental properties of bodies and mere
sensations, a distinction often referred to in the terms used by John Locke: "primary" versus "secondary" qualities. The former were taken to be real properties
of matter, such as size and shape. The latter, such as color and taste, were consid-

ered to be merely subjective: they would vanish if the perceiving agent disappeared. Galileo compared them quite colorfully to the sensation of tickling generated, for example, by the gentle motion of a feather. In *Principia philosophiae* Descartes too discussed tickling in a similar context. Descartes visited Italy, including Rome, between 1623 and 1625, in the aftermath of the publication of *Il saggiatore,* and indeed it is not difficult to identify several common themes in their works.[4]

Significant as *Il saggiatore* may have been, there is no question that Descartes encountered it well after his conversations and correspondence with the Dutch scholar Isaac Beeckman, which began in November 1618 in Breda. Beeckman was an atomist whose views greatly influenced Descartes. By linking phenomena to the tiny particles of the micro-world, Beeckman and, later, Descartes were able simultaneously to give causal and, at least potentially, mathematical explanations of those phenomena. This was a major transformation in the use and role of the mathematical disciplines. Now a whole new range of phenomena, from gravity to magnetism, was potentially open to both causal and mathematical explanation. A key tool in this program was the formulation of quantitative rules of impact, an area in which Beeckman had begun to work in the late 1610s.[5] The literature on the mechanical philosophy is vast; therefore the following remarks focus on some key themes relevant to the transformations of mechanics.

Despite the heterogeneous character of the mechanical philosophy and the fact that several scholars contributed to its establishment, there can be little doubt that while Descartes emerged as the dominant figure, Pierre Gassendi, Canon at Digne and, since 1645, a professor of mathematics at the Collège Royal in Paris, was the key figure in the revival of ancient atomism. Gassendi wrote very much in the humanist tradition, and his ponderous masterpiece, the 1658 *Syntagma philosophicum,* is a work of extraordinary learning replete with quotations and commentaries on a great number of authors. Nothing could be further removed from the style of Descartes, whose studied reluctance to quote his sources often left his commentators puzzling over his familiarity with a host of texts.

Descartes emerged on the public scene in 1637 with a collection of essays opened by his celebrated *Discours de la méthode* and including *La dioptrique, Les météores,* and *La géométrie.* The *Discours* and *Les météors* in particular begin to outline his views on nature. In 1641 and 1642 Descartes published first and second editions of *Meditationes de prima philosophia,* but it is in the 1644 *Principia philosophiae* that one finds the most comprehensive account of his physics. The *Principia,* building on the earlier aborted treatise *Le monde,* was issued in 1647 in a modified French translation carried out by Claude Picot partly under Descartes' own supervision. Several works by Descartes were published posthumously, such as important portions of his correspondence starting in 1657 and *Le monde* in 1664.

It should come as no surprise that the largest portion of this chapter is devoted to Descartes. Section 5.2 explores the different perspectives of the mechanical philosophy and mechanics. Section 5.3 reviews some of the debates between Descartes and Beeckman as an introduction to Descartes' reading of the *Discorsi*. Thus this section can be seen as complementing and extending themes explored in the previous chapter. Descartes' reading of Galileo can be best appreciated in the context of his own views on nature and motion, which is the subject of section 5.4. Section 5.5 is devoted to the collision of bodies and highlights the importance of the early collaboration between Beeckman and Descartes. I then compare the impact rules formulated by Marci and by Descartes. Descartes' impact rules remained at a very abstract level, and for dealing with the real world he developed different conceptual tools and notions. Section 5.6 looks at the Cartesian universe from a less abstract perspective, focusing on the ways in which Descartes investigated collision in the real world and on his notions of solidity and of the role of the structure and composition of matter, including elasticity.

5.2 Mechanics and the Mechanical Philosophy

A key feature of the mechanical philosophy was the miniaturization of machines. Descartes claimed that the only difference between artificial and natural bodies was one of size. In artifacts made by craftsmen the machines are large and visible, whereas in nature they are small and often invisible, but the motion of a clock is in principle no different from the growth of a tree from a seed to the production of fruits and more seeds. In the French translation of *Principia philosophiae* he provided examples of those machines such as pipes and springs or *ressorts*, thus suggesting a prominent role for elasticity.[6] In this sense the mechanical philosophy resembled not only atomism but also the approach originating in antiquity and so thoroughly pursued by dal Monte, Galileo, and other contemporary mathematicians of trying to uncover macroscopic masked levers operating in other machines such as winches, pulleys, beams, inclined planes, and siphons. In a similar but far more radical fashion, the mechanical philosophy aimed at uncovering different types of machines, often microscopic ones, in the folds of and behind natural phenomena, enabling the investigator to explain how and at the same time why they occurred. This does not mean that Descartes was a follower of dal Monte, however, since Descartes specifically rejected the latter's attempt to reduce the pulley to the lever and gave a prominent place to new principles and laws. Traditionally, causal explanations were reserved for physics, whereas mathematics could provide only description of phenomena. The causal explanations provided by the mechanical philosophy were at least in principle amenable to mathematical description and justified the use of a new term, "physico-mathematics."[7]

There are significant differences having to do with disciplinary configuration between Descartes and later scholars. In some of his writings, such as his *Traité de la mechanique,* Descartes subscribed to the traditional view of mechanics as the science of simple machines relying on a general principle that the effect must be equal to its cause, or that the force to raise a weight to a given height is the same as that to raise, for example, half the weight to double the height. The laws of nature and rules of impact Descartes announced in *Principia philosophiae* bore a problematic relation to mechanics. On the one hand, the notion of weight relied on an explanation of gravity in terms of the collision of particles pushing a body toward the Earth. On the other hand, Descartes' investigation of collision relied on the combination of size or mass and speed of bodies, namely the same notions occurring in the pseudo-Aristotelian analysis of the balance. Descartes' view of the colliding bodies as competing, with a winner and a loser, echoes the exemplar of the balance. This circularity is present also in Descartes' text, since in the preface to Picot he announced his future plan to deal with greater exactitude with medicine, ethics, and mechanics, yet in part IV concluding his treatise he claimed that he had relied on the shape, motion, and size of particles and on the laws of mechanics. There is also a further twist in the story. Some of the more characteristic notions and objects of the Cartesian universe, such as the behavior of bodies in vortices of water, the sling, and pebbles rounded by the motion of the sea much like Cartesian elements (see section 5.6), are discussed in the pseudo-Aristotelian *Quaestiones mechanicae* and its commentaries. Therefore the tradition of the most classical work in mechanics was a key and especially relevant resource for Descartes' worldview. Within a few decades of Descartes' death in 1650, his laws and rules had come to be considered an integral part of mechanics more broadly conceived. Thus if for some it may be problematic to include Descartes' laws and rules in a history of mechanics, it would be even more problematic to exclude them, and they certainly belong to a history of the transformations of mechanics in the seventeenth century. As Dan Garber has recently argued on the basis of detailed textual analyses, Descartes redrew the boundaries of mechanics and transformed its contents.[8]

The relationships between the mechanical philosophy and a mathematical description of nature are quite complex. Since those descriptions involved at the most fundamental level the study of particles in motion, there certainly was a large intersection with mechanics and the mathematical study of motion. On the one hand, both Gassendist and Cartesian philosophical assumptions appear to pave the way for a successful and radical mathematical description of phenomena. Descartes' belief that body is extension opens the door for a geometrization of nature, and he contributed important chapters to the study of optics and motion. On the other hand, his worldview led him to envisage huge numbers of interactions among bodies in such a way as to open in some cases a gulf between the potential and actual geometrical description of phenomena. While some historians have emphasized Descartes' success in providing mathematical formulations, such as in the rules of impact, others have portrayed the mechanical philosophy as a whole as a stumbling block to the mathematization of nature. Despite its program and intentions, the mechanical philosophy often provided qualitative rather than quantitative accounts. Descartes was able to contribute a good deal more to physico-mathematics, and indeed he did so in short specimens he sent to Mersenne in response to his queries. Mersenne was even authorized to publish them, provided he kept their author's name hidden. This episode highlights the tension between Descartes' rigid philosophical program and the mathematical description of phenomena.[9] Gassendi's belief in the void and atoms may have eased the bewildering complexity of the Cartesian world, but his philosophical bent was more skeptical and empirical, and he did not pursue a program to investigate nature in a mathematical fashion, with the exception of some established areas having to do with motion and astronomy. We shall encounter this tension between potential and actual mathematization in several instances below.

5.3 Beeckman, Galileo, and Descartes

After 1610 Galileo was one of the most visible European intellectuals, and there can be no question that Descartes knew of him. We do not know whether he knew of Galileo's works published before the 1630s, although as I suggested above he probably read *Il saggiatore*. In August 1634 Beeckman visited Descartes and lent him for a few hours a copy of Galileo's *Dialogo*. Descartes leafed through it and in a letter to Mersenne was able to comment briefly on the theory of tides, with which he disagreed, and the law of falling bodies, which he believed coincided with his own. Indeed, Beeckman and Descartes had been debating issues related to falling bodies since 1618. The matter was also discussed in subsequent correspondence with Mersenne. Reviewing those exchanges poses a problem to the historian attuned to the nuances of different formulations. Our views and attention have crystallized in a way that leads us to be especially alert to certain themes, though scholars of the time perceived matters differently.[10]

Descartes' work on the problem of falling bodies was instigated by Beeckman.

His attitude toward it changed greatly over the years from 1618 to the 1630s. Initially he tackled the problem mathematically, but later, as his worldview matured, he shifted to a more physical approach wherein the role of mathematics was problematic. It is instructive to explore briefly Descartes' early views as expressed in his diary and correspondence for the light they shed on the problem of falling bodies at the time when Galileo's views were still unpublished. Apart from some cryptic comments on falling bodies in the *Discorso* on bodies in water, Galileo's mature views first saw light in the *Dialogo*. There are striking similarities between several aspects of Galileo's conceptualization and Descartes'. Descartes conceived motion to be continuous and used the triangle of speed much as Galileo had done. In 1618 Beeckman asked Descartes how far a body would fall in one hour, knowing how far it would fall in two hours (fig. 5.2). Several themes running through Descartes' debates with Beeckman and Mersenne resemble those we encountered in chapter 4. They include the study of motion, the analysis of the physical origin of the acceleration of gravity, and the resistance offered by air or other fluids. Beeckman knew that bodies accelerate, but a mathematical study of their behavior was fraught with difficulties. He believed that gravity was caused by the Earth's attraction via a stream of particles, yet the presence of a fluid obstructed the motion of falling bodies. Hence he considered the dubiously coherent case of a body falling in a vacuum but pushed behind by a stream of particles. Given its physical cause, gravity had to act in discrete impulses, yet Beeckman also used a continuous mathematical representation and claimed that the distances traversed were as the squares of the times. As to the role of the medium, Beeckman adopted an approach inspired by Archimedes. He argued that the body's speed increases until the quantity of matter it displaces in an instant of time equals its own quantity of matter. Descartes criticized Beeckman on this point, arguing instead for an asymptotic increase of speed.[11]

It is instructive here to examine Descartes' reaction to Galileo's *Discorsi*, much as we did with a number of other scholars in the previous chapter. Our main source is Descartes' correspondence with Mersenne. In a letter dated 11 October 1638 Descartes put forward his views, praising Galileo's use of mathematics in the study of physical matters. He was less happy with the *Discorsi*'s many digressions and the fact that Galileo seemed not to explain matters fully. Descartes claimed that Galileo had built without solid foundations, namely he had not provided causal explanations. One of the most revealing comments concerns Galileo's reasoning about fall in the void: "Everything he says about the speeds of bodies descending in the void, etc., is built without foundation, for first he should have determined what weight is, and if he had known the truth, he would have known that it is nothing in the void."[12]

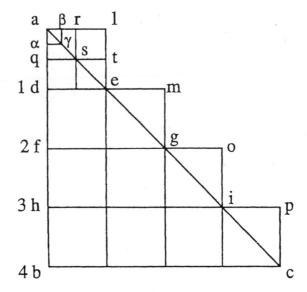

Figure 5.2. Descartes' triangle of speeds (memorandum for Beeckman). Descartes examined the question posed by Beeckman—namely how far a body would fall in one hour, knowing how far it falls in two hours—by means of a geometrical diagram. Initially he considered finite time intervals *ad, df, fh,* etc., but then they shrank to a point. In this case the vertical axis *ab* initially represents time, but Descartes seems to shift between time and space, and later in the text he seems to associate the segment *ab* with distance. The horizontal lines *de, fg, hi,* etc., represent speed. Descartes considered that since the area *afg* is ⅓*fgcb,* if a stone is attracted in a vacuum from *a* to *b,* it will cover the second half of the distance *fb* in one-third of the time it takes to cover the first half *af.* His reasoning seems to be based on the theory of proportions, whereby if the distances are the same, the speeds are inversely as the times. Since the triangle of speeds suggested to Descartes that the speed in the second half was three times that of the first, he reached this conclusion about times. The issue is more complicated than Descartes envisaged, because if *ab* represents a distance, the curve of speeds is not a straight line, but if *ab* represents time, the middle point *f* does not correspond to half the distance. On the basis of the quadratic relation between distances and times, in uniformly accelerated motion the times would be in the ratio of the square root of the distances.

Descartes did not think much of the science of the resistance of materials, claiming that the subject was rather trivial and did not merit a new science. Galileo had attempted to provide a causal explanation of the resistance of materials based on infinitely many interstitial vacua, but Descartes dismissed Galileo's recourse to the infinite.

As to the science of motion, Descartes challenged Galileo's views on uniformly accelerated motion, though he admitted having previously held similar views. Of course, in a universe in which gravity is due to a subtle fluid, a body cannot accelerate uniformly indefinitely. Descartes also dismissed the postulate that the degrees of speeds acquired by a body at the end of its fall along inclined planes with equal elevations are equal.[13] Having challenged the two pillars of Galileo's science, Descartes thought he could make short shrift of the whole. He also doubted that the bodies fall through all the degrees of speed, although he admitted that it would not be impossible. Once again, if acceleration is due to the impacts of the particles of a fluid, one would expect the increase in speed not to be continuous.

A letter written to Constantijn Huygens in February 1643 reveals a significant shift in Descartes' way of dealing with the problem of falling bodies. While describing an experiment on the speed of water flowing out of a container, Descartes discussed the action of gravity and argued that the speed of falling bodies was nearly proportional to the time, and claimed that in the corresponding triangle of speeds the area represents the distance fallen.[14] Descartes' change of heart concerned not fundamental principles but only empirical adequacy. As such, I believe it was representative of the attitude developing in many quarters. Whether a falling body goes through all the degrees of speed or not, Galileo's rules seemed to describe with reasonable accuracy the behavior of falling bodies, and they were slowly gaining acceptance largely on those grounds, as we saw in the previous chapter in the case of Riccioli.

I wish to introduce Descartes' *Principia* by comparing it with Galileo's *Discorsi*. Despite the occurrence of common themes, the two works differ radically as to structure and intended audience. Both were published in the Low Countries by the Elzeviers, the former at Amsterdam and the latter at Leyden. Although the *Discorsi* deals with philosophical issues, such as the nature of the continuum, it is much more in the genre of a work in the mixed mathematics addressed to a diverse audience consisting of men of learning, mathematicians, and even technicians and engineers, such as De Ville. The diversity of the intended audience is evident from the fact that the book is in Italian, whereas the treatise on the science of motion in days 3 and 4 is in Latin and is structured according to classical axiomatic form. Further, in day 2 Galileo gives practical instructions on how to cut parabolic beams, in day 3 some of his reflections on the inclined plane are relevant to problems of river flow, and in day 4 one finds a table of elevations for artillery shots. These three features suggest an audience interested in civil, hydraulic, and military engineering, respectively.

By contrast, Descartes had a different and broader audience in mind and overall more ambitious plans than Galileo. *Principia* was nothing less than an account of the system of the world aimed at replacing scholasticism. Although Cartesianism was often banned in the universities, it spread both indirectly through his works and at several venues arranged by Descartes' followers, shaping the intellectual life in the second half of the century. Treatises based on *Principia philo-*

sophiae, such as Jacques Rohault's *Traité de physique* (Paris, 1671), were actually used as textbooks for teaching natural philosophy both in France and in England.[15]

Although Galileo too played a crucial role in the demise of Aristotelianism and the emergence of a new philosophy of nature, his *Discorsi* could not have fulfilled the same role as Descartes' work. This shift of audiences is an important feature of my story. It helps explain why a mathematical discipline with a limited presence and generally low status in European universities attained such a major position in intellectual affairs. Although one should not conflate mechanics and the mechanical philosophy, the latter often served as a vehicle for the former because in several instances the mechanical philosophy relied on mechanics for its explanations of how the world worked.

In his dedication to Elizabeth, Princess of Bohemia, in *Principia philosophiae*, Descartes characterized her as an ideal reader, indeed the only one who had understood all his previous publications. Descartes also stated in the dedicatory letter to Picot that his book should be read as one would read a novel, without paying too much attention to any difficulties one might encounter. On a subsequent reading one should note those difficulties, marking them with a pen, but, curiously, Descartes claimed that these would be best overcome by yet subsequent readings of the whole rather than by minute analysis of the work section by section.[16] Such a reading strategy would be inappropriate for Galileo's *Discorsi*, especially those portions structured as a mathematical treatise, where the tight deductive structure requires a thorough understanding of each step. Despite the logical and deductive flavor of some portions of his treatise, Descartes did not give it an axiomatic structure. Even those portions that would have potentially been amenable to this treatment, such as the laws of nature and the impact rules, were presented as a list of propositions rather than as theorems. Indeed, even his mathematical work, the 1637 *Géométrie*, is not presented in axiomatic form.

There is another aspect of Descartes' system that I wish to discuss here by way of an introduction to the following section, namely the justification for his laws of nature compared with the justification usually provided by mathematicians for their axioms. Galileo's axiom and also Torricelli's principle—published in the same year as Descartes' *Principia*—were meant to be easily granted by the human mind. By contrast, Descartes invoked theological reasons for his laws. All three laws, according to Descartes, are based on the immutability of God and his operations. This shift from axioms to laws of nature justified with the help of theology represented a major change of worldview in the seventeenth century. God's role remained tied to the concerns of many mathematicians such as Newton and

Leibniz, though others such as Huygens eschewed theological discourse and worked in a tradition closer to that going from Archimedes to Torricelli.[17]

5.4 Motion and Its Laws

In Descartes, Gassendi, and a large portion of the new philosophers in the second half of the seventeenth century, the study of motion took center stage. Gassendi in *De motu impresso a motore translato* (1642) and Descartes in *Principia philosophiae* claimed that rectilinear motion continues indefinitely, a claim later known as the law of inertia. The seventeenth-century use of the term "inertia," meaning inactivity or laziness, dates to Kepler, who understood by it the tendency all bodies have to come to rest once they are set in motion. Kepler's notion shares common traits with the Aristotelian view that motion, once started, comes naturally to rest. However, according to *De coelo*, celestial bodies move in circular motion, which is eternal, and generally they do not follow the same laws as sublunary bodies. Kepler departed from Aristotelian philosophy in his belief that planets and terrestrial bodies behave in the same way, thus breaking the dichotomy between heaven and Earth. The term "inertia" is used by Descartes in a sense profoundly different from Kepler's, namely meaning a body's reluctance to be set in motion. This reluctance depends on both its quantity of matter and its surface. Thus, if we want to adhere to seventeenth-century usage, it would be misleading to call the law according to which bodies in rectilinear uniform motion continue to do so unless they are impeded "the law of inertia."[18] I shall reserve that term for Newton and his followers.

We have seen that for Galileo the pendulum and the inclined plane constituted useful tools in the effort to conceptualize problems about motion. Galileo, however, took uniform motion left to its own devices to be circular. When dealing with the motions of planets, he also took circular motion to be natural. The proponents of the mechanical philosophy, by contrast, reflected more critically on motion and the origin of gravity. They imagined bodies moving through empty space without interference or action by other bodies; they arrived at the notions of the conservation of rectilinear uniform motion and the relativity of motion; and they analyzed the implications of these notions for the science of motion and the system of the world. Descartes' claim was to some extent paradoxical, because he did not believe in the existence of empty space; his statements about motion in a void thus required a considerable power of abstraction. Gassendi's formulation went hand in hand with his atomistic views and belief in the existence of empty space. To the question about what would happen to a stone placed in empty space and set in motion, he replied that it probably "would move with a uniform endless motion" in a straight line, moving the faster the greater the impetus impressed on it. With respect to the real world, however, Gassendi apparently still adhered to the notion that circular motion is natural.[19]

In *Le monde,* composed before Galileo's condemnation of 1633, and then in *Principia philosophiae,* Descartes examined motion and formulated laws of nature dealing with it. According to Descartes, everything in nature had to be explained in terms of the size, shape, and motions of particles. Size, shape, and motion were conceived as modes or properties of bodies. Motion consisted of speed and determination. The latter was a mode of motion, or a mode of a mode, and corresponds only approximately to the idea of direction.[20] Despite its centrality in the Cartesian system, motion was not defined in a clear and unambiguous way. In *Le monde* Descartes defined it in terms of change of place, whereas in the *Principia* he defined it in terms of change with respect to the surrounding bodies. Interpreters, today as in the seventeenth century, have puzzled over this new account of motion, wondering whether Descartes changed his presentation in the aftermath of Galileo's condemnation because he needed to save his system from the charge of Copernicanism.[21]

Besides dealing with matters of definition, Descartes described properties of motion in terms of laws and rules. The first law states that "each thing, as far as it is in its power, always remains in the same state; and that consequently, when it is once moved, it always continues to move." Thus motion is comparable to rest and shape, all being modes of a body: "Thus, if some part of matter is square, we are easily convinced that it will always remain square unless some external intervention changes its shape." Similarly, a body at rest will remain at rest and a body in motion will not alter it unless something external intervenes. Descartes provided an explanation of the first law on the basis of the experience of the motion of projectiles, which stop only because they are retarded by the air or other fluid through which they are moving. This was a major departure from the strict Aristotelian tradition, according to which motion comes to an end of its own accord, although the impetus theory adopted by later commentators did envisage that impetus and motion are not self-dissipating.[22] The second law states "that all movement is, of itself, along straight lines; and consequently, bodies which are moving in a circle always tend to move away from the center of the circle which they are describing." The justification for this law too is two-pronged, based on both God and experience. As to the first, Descartes stated, "This rule, like the preceding one, results from the immutability and simplicity of the operation by which God maintains movement in matter." Next Descartes argued that although no motion occurs in a single instant, at each instant of its motion a body has a determination to move in one direction in a straight line. In a later passage he took "instant" to mean a tiny amount of time, but then he continued stating that bodies that stop for an instant continue their motion without interruption.[23] Since God "maintains [movement] precisely as it is at the very moment at which He is maintaining it," the motion he is preserving is rectilinear. Descartes seems to have in mind here the idea of a tangent to a curve, where the curve represents a body's motion and the tangent its determination at one instant. He explained himself further with the help of a common object and a common experience, namely the rotation of a sling (fig. 5.3). The stone starts moving in a straight line from the point at which it is released.[24] Kepler, in *Epitome astronomiae copernicanae,* had already used the example of a body attached to a rotating wheel and then released, but he had not made of this example a building block for a new understanding of motion. Here Descartes departed from Galileo and Beeckman too, who still adhered to the notion that curvilinear motion is natural. His analysis of circular motion, entailing three tendencies at the same time, one along the circle arc, another along the tangent, and the third along the radius, was to shape the understanding of the problem for several decades to come.[25]

While dealing both with light and with circular motion Descartes introduced the notion of *conatus* (in Latin) or *effort* (in French), generally rendered in English as "endeavor." When he talked of a body's endeavor from the center, he meant not a conscious effort but an inanimate tendency resulting from the laws of nature. Descartes often associated endeavor with constrained motion. A body rotating in a sling is prevented by the sling from flying off along the tangent, but the sling, while constraining motion, does not prevent the body from exerting an endeavor in the form of a tension in the sling. In the Latin version Descartes used the expression "istius vis quantitatem," but the French translation replaced the term *vis* (force) with *effort* (endeavor). This notion would be adopted and transformed by several scholars in the second half of the century.[26]

The third law deals with impact and states that when a body collides with a "stronger" one, it "loses none of its motion," whereas "upon coming in contact with a weaker one, it loses as much as it transfers to that weaker body." Since by "motion" Descartes understood speed times extension, "stronger" and "weaker" could plausibly mean with a greater or smaller amount of motion. In his text, however, Descartes again provided, as for the previous laws, examples drawn from common experience. He claimed, "When any hard bodies which have been set in motion strike an unyielding body, they do not on that account cease moving, but are driven back in the opposite direction; on the other hand, however, when they strike a yielding body to which they can easily transfer all their motion, they immediately come to rest."[27] Thus the strength of a body is associated with matter

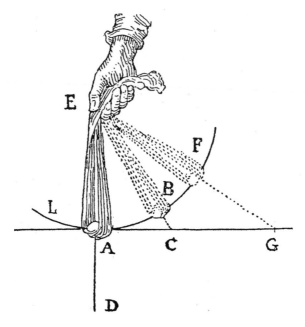

Figure 5.3. Descartes' sling *(Principia philosophiae)*. In part II of his *Principia* Descartes explained
the law that motion is rectilinear using the example of a common object, the sling, held by the
hand at *E*. When released, the stone *A* travels along the straight line *ACG*. Since the stone is constrained, it cannot fly away along the tangent, but it has a tendency to do so. It would seem that according to Descartes the outward tendency could be quantified by measuring the distance of a point
on the tangent from the corresponding point on the curve, such as *BC* and *FG*. One could also measure the force with which the stone tends to escape from the center by the tension of the string. In
part III Descartes returned to the example of the sling to explain how a body can tend to move in
different ways at the same time. He argued that the stone *A* tends to move from *A* to *B*, if one considers all the causes involved. If one considers only the force of motion, *vis motus*, the stone tends
toward *C*. Finally, if we consider only that component hindered by the sling, the stone tends toward
D, along the straight line from the center *E*. In order to explain the outward tendency and its force,
Descartes also used the examples of an ant on a rotating stick and a ball in a rotating tube. He concluded that the force increases with the speed of rotation, though it is not clear whether he had a
direct proportion in mind or a more complex relation.

theory. Notice that the law implicitly provides criteria for classifying bodies as soft
or hard. The third law is valid for all bodies, whether hard or soft, and deals with
their interactions, whereas the first two laws dealt with the behavior of individual
bodies. In *Le monde* Descartes had introduced three laws of nature, broadly corresponding to those of the *Principia*, except for the inversion of the second and
third. He did refer to some rules of collision but did not spell them out. The impact rules are based on a conservation principle derived from the immutability of

what is motion that God conceive it?

the works of God, namely the conservation of motion. This principle differs from its modern version, known as the conservation of quantity of motion, in two important respects. First, by quantity of motion we understand a notion involving direction, whereas Descartes meant a notion independent of direction. Thus a body changing direction by rebounding has the opposite quantity of motion according to our view but the same amount of motion according to Descartes'. Second, Descartes identified body with extension or size of the particles, whereas we do not assume all bodies to have the same density and rely on the notion of mass instead.[28]

5.5 From the Balance to Impact: Beeckman, Marci, and Descartes

The problem of impact is significant both in its own right and also because it lies at the intersection of a network of themes such as the notion of bulk or mass and the classification of the properties of solid bodies. Although practitioners may have employed different names, most of them considered mass, or bulk, or weight to be of central significance. While relying on a rather intuitive notion of bulk at the beginning of their accounts, they often clarified the concept and made it more precise by means of their study of impact. Phenomena of impact are dependent on the material features of the colliding bodies and therefore provide a helpful framework for understanding notions such as hardness, softness, and especially elasticity.[29] At the beginning of the investigations on impact, colliding bodies were classified in elaborate ways and conservation laws were either lacking or defined in ways different from those with which we are familiar. The investigations of impact were based on principles or laws, experiences or experiments, and also the application of tacit assumptions. In some cases experiments were carried out and results were carefully monitored and measured. Quantitative experiments and abstract principles went hand in hand with the mathematization of a domain apparently so unlike a rigorous universal law that hardly anyone would have dreamed it feasible in 1600. The notion that an exact mathematical law described the impact of wood, ivory, and steel bodies as well as wax, clay, and wool ones would have seemed bizarre. Moreover, whereas traditionally the theory of proportions was the main mathematical language used to describe nature, impact phenomena cannot be easily treated in that way but are best dealt with by equations.

Descartes probably developed his views on impact during his early collaboration with Beeckman. In several entries in his diary Beeckman explicitly used the example of simple machines, and especially the balance, as a springboard for the investigation of impact. In a passage from the fall of 1618 Beeckman examined different types of impact. He considered a moving body hitting one at rest and studied the different outcomes due to the bodies' bulk or weight. His bodies did not rebound, and if they were equal, for example, they would move after the impact with half the speed of the impinging body. His justification was the science of machines, whereby "it is observed in all machines that double the weight, raised by the same force, ascends twice as slowly as the previous weight."[30] Beeckman examined other cases too, but my main concern here is to highlight the mechanical roots of his views at the time of his collaboration with Descartes. These views are confirmed in a later passage from the fall of 1629, in which Beeckman considered the case of colliding bodies with speeds inversely as their bulk. He concluded that the bodies would come to rest on the example of the balance: "Sic etiam ratiocinandum de bilance."[31] Beeckman abstracted from the material constraints of devices such as balances or other simple machines to formulate a principle valid for the interaction among bodies free of those constraints.

In 1639 the Bohemian physician and experimentalist Marcus Marci published in Prague *De proportione motus*, a treatise on several aspects of the science of motion with interesting copper engravings, including a title page replete with mechanical devices (fig. 5.4). Marci had seen Galileo's *Dialogo*, and his treatise reflects this to some extent. The most original feature of *De proportione motus* was its investigation of impact, the first ever published. It is possible that Marci's work on this problem stemmed from a combination of interests in percussion and the game of billiards, which is mentioned in the text.[32] Marci provided a classification of different materials on the basis of their behavior in impact. He considered soft bodies, such as clay, wax, wool, or lead, and hard bodies, which are either "absolutely hard" or "fragile," such as glass, terra cotta, or tuff. Absolutely hard bodies can be further subdivided into those that are sound-producing, whose atoms are set in motion by the impact, and those that are not sound-producing. Marci's analysis of impact is especially devoted to absolutely hard bodies and central impact. Marci provided eight "porisms," a notion found in Pappus's *Collectiones* whose meaning is not entirely clear, being either a form of statement intermediate between a theorem and a problem or a type of problem leading to a partly indeterminate solution.[33]

Marci's concepts and notions do not correspond on a one-to-one basis to ours; therefore unqualified translations of his terms are problematic. For example, the key notion he used was that of *impulsus*, which he claimed was proportional to *gravitas* and speed. Marci specified that *gravitas* was different from *moles*, meaning that *gravitas* corresponded to *pondus*, or weight, not to the volume of the body. Marci put forward four suppositions. The first states that the impulse of a body has

Figure 5.4. Title page of Marci's *De proportione motus*. Marci's title page depicts a remarkable range of devices and experiments relevant to the investigation of motion. In addition to the clock and astrolabe, note the man dropping spheres from the tower at the far left, the cannon firing against a row of spheres standing on a horizontal support, the billiards table atop the colonnade, the man on a swing underneath it, and the man playing racketball. The putti seem to be adjusting the weights of a balance.

to be greater than another body's weight to set it in motion. Thus impulse represents a body's activity, whereas weight is resistance; Marci did not worry that impulse and gravity were not homogeneous magnitudes. The second supposition argues that impulse depends on a body's weight and determines the speed after impact. The third supposition is that impact occurs not instantaneously but requires some time. The last supposition defines a perfect impact as that in which a body's impulse is completely destroyed.[34]

Porism 1 states that given two equal bodies, one *A* in motion and the other *B* at rest, after the impact *A* will come to rest and *B* will be set in motion with *A*'s original speed. This is an example of a perfect impact, wherein *A*'s impulse is destroyed. Marci later discussed the case of a row of identical spheres at rest being hit by an impinging one. The outcome is that only the last sphere at rest moves forward with the speed of the impinging one, all the others remaining at rest.[35]

Porism 2 extends 1 to the case where the impinging body *A* has a greater weight than the body *B* at rest. In this case *B* will be set in motion with *A*'s original speed, and *A* will continue with the residual impulse.

Porisms 3 and 4 share many common conditions. In both cases *A* has a greater weight than *B*, *A* is initially at rest, and *B* is in motion. The distinguishing feature is *B*'s impulse in relation to *A*. In porism 3, *B*'s impulse is greater than *A*'s gravity, and that excess is capable of setting *A* in motion with a speed less than *B*'s. After the impact *B* either stops or is reflected, depending on the speed it imparts to *A*.[36] In porism 4, by contrast, *B*'s impulse is smaller than *A*'s gravity, therefore *A* does not move and *B* is reflected with its speed unchanged. In both cases *A* is conceived as offering a resistance to the impinging body. In porism 3 Marci claimed that the "excessus" of *B*'s impulse over *A*'s "gravitas seu pondus" is the cause of the motion of *A*, "principium motus majori." Applying the same principle to porism 1, where the two bodies are equal, it seems that *A*'s impulse is greater than the resistance offered by *B* because *A* is in motion.[37] Porism 4 is followed by two "Problems" related to impact. The first consists in finding the conditions whereby a ball at rest on a plane is hit by another ball with arbitrarily large violence but does not move. The accompanying illustration shows the arrangement and suggests some degree of experimental investigation on Marci's part. The second Problem concerns the motion of balls on a billiards table (see fig. 5.4) and the skipping of stones across water.[38]

Porism 5 can be seen as a generalization of 1. In both cases the colliding bodies are equal. Here both *A* and *B* are in motion before the impact, and after the impact there is a permutation of their speeds.

Porisms 6, 7, and 8 deal with collisions between moving bodies. In all cases *A* has a greater weight than B, but their relative speeds vary, and the different outcomes depend on those differences. In porisms 3 and 4, one body was at rest, and Marci thus compared the impulse of the body in motion with the gravity of the body at rest. In porisms 6, 7, and 8, both bodies are in motion, and one would thus expect Marci to compare their impulses. However, this is not the case; he compares the impulse of the smaller body B with the weight of the larger body A, not with A's impulse. In porism 6, B's impulse is greater than A; after the impact, they both reflect. In porism 7, B's impulse is smaller than A; therefore B is reflected and acquires A's speed before the impact, while A's speed is accordingly diminished. In porism 8, B's impulse is equal to A, A comes to a stop, and B is reflected with an increased speed.[39]

Marci's principal aim was to predict the outcome of an impact by determining whether a body rebounds, stops, or continues its motion, rather than accurately determining its speed. This indeterminacy may explain why, following Pappus, he called his propositions porisms. Overall, they appear to be loosely inspired by the notion of the conservation of impulse, without regard to its sign, but of course it is difficult to have a precise notion of conservation if the speeds after the impact are indeterminate. Marci also considered special cases, such as the impact between pendulum bobs, but he did not use the pendulum systematically in his investigations of impact, as others would do later in the century.[40]

Marci sent a copy of his work to Galileo, but no reply survives. Several years later a copy was sent to Huygens, who commented quite severely on Marci's treatment of impact.[41] The circulation of his work in the years immediately following its publication appears to have been limited.

Five years after Marci's work Descartes published *Principia philosophiae* with his seven rules of impact. Some of Marci's porisms and Descartes' rules correspond quite closely, but overall the latter were based on different principles and were conceived within a different and vastly more elaborate philosophical framework. Given that Marci's work appears to have been little known, it is unlikely that Descartes had access to it, and this makes their similarities all the more interesting. As we have seen, Descartes provided his impact rules as part of his effort to explain the third law. These rules apply to perfectly hard bodies in isolation from anything around them. Thus they cover a scenario so abstract from Descartes' standpoint that no real-world situation could even approach it.[42]

Rule 1 states that if two bodies B and C are exactly equal and move with equal and opposite speeds, after the impact they will rebound with equal and opposite speeds.

Rule 2 is a variation on 1, where B is larger than C and all the other conditions are unchanged. The outcome is that C is reflected with the same speed, and B continues its motion as if no impact had taken place. Thus they both move with the same speed in the same direction after the impact.

Rule 3 considers the case in which B and C are equal and the speed of B is greater than that of C. After the impact they move together in the direction of B with a speed that is the average of their speeds before the impact.

The first three rules deal with bodies moving with opposite directions. The next set considers the cases in which C is initially at rest and B in motion.

Rule 4 claims that if C at rest is larger than incoming B, regardless of B's speed, after the impact C does not move and B rebounds with an equal speed.

Rule 5 considers the case in which C, initially at rest, is smaller than incoming B. After the impact they move in the same direction with a speed equal to $Bv\,/\,(B + C)$, where v is B's speed before the impact.

In rule 6, B and C are equal. After the impact, B is reflected with three-quarters of its speed, whereas C moves in B's initial direction with one-quarter of B's speed.

Finally, rule 7 deals with bodies moving in the same direction, C more slowly and B faster. Descartes considered two cases. In case 7.1, if B is larger than C and the product Bv_B is greater than the product Cv_C before the collision, then after the collision both bodies move together with speed $(Bv_B + Cv_C)/(B + C)$. In case 7.2, if B is smaller than C and the product Bv_B is less than the product Cv_C before the collision, then after the collision B is reflected with the same speed and C's speed and direction are unchanged.

I wish to draw a comparison between Marci and Descartes. They were the first, after Beeckman, to articulate a set of conditions for the study of impact involving the nature of the colliding bodies as well as their weight or size and speed. To a large extent they both conceived of impact as a contest wherein the "stronger" body defeats the "weaker." In Marci's case the key notions are impulse versus weight. For Descartes the size or bulk of the body was crucial. Moreover, within Descartes' framework, cases when the colliding bodies' speeds are equal deserve special attention, as in rules 1 and 2. Cases when the bodies have equal speeds do not figure in Marci's taxonomy of impact. Although it is not possible to draw a one-to-one comparison between Marci's porisms and Descartes' rules, in some cases their initial conditions correspond quite closely, such as in porism 1 and rule 6, or porism 2 and rule 5.

Both Marci and Descartes examined the cases in which one of the bodies is at rest. What units of measurement should be used to assess the resting body's resistance to the impact? Marci compared impulse and weight, even though speed is present in one case and absent in the other; indeed, he did so too when both bodies were in motion. By contrast, Descartes devised an elaborate accounting whereby the resistance offered by a body at rest is measured by its size times the speed of the impinging body.[43] Thus he stated that a body in motion hitting a larger one at rest rebounds without moving the larger body, despite the fact that their sizes could be almost identical. This belief led to the paradoxical conclusion that a body at rest could not be moved by an impinging body that is only marginally smaller, despite the fact that its speed could be arbitrarily large. He felt compelled to try to explain why the rule seemed not to be empirically valid, namely real bodies are not perfectly hard, and the interactions among the huge number of particles of the aether and air make it impossible to isolate a system of two colliding bodies. In this respect Descartes followed a different approach from Galileo's with regard to the imperfection of matter. Galileo aimed at a rapprochement between theory and observations by reducing or subtracting the accidents or impediments of matter. For Descartes, instead, there seems to be a gap between rules and observations that cannot be bridged in principle.

After 1644 Descartes returned to the impact rules in his correspondence, notably in the letter of 17 February 1645 to his follower Claude Clerselier and in the French edition of the *Principia*. Descartes did not provide a justification for his seven impact rules in *Principia philosophiae*, claiming them to be self-evident. Readers now as well as in his time, however, held different views. Soon after the appearance of *Principia philosophiae*, Clerselier wrote to Descartes asking for explanations. Unfortunately his letter is now lost, but Descartes' reply is extant. It appears that Clerselier had objected to rule 4. In his reply, Descartes explained: "That which is without motion has as many degrees of resistance as the other, which is moving, has of speed. The reason for this is that if it is moved by a body which moved twice as fast as another, it ought to receive twice as much motion from it"; therefore it would oppose a resistance twice as strong. Later in the letter Descartes revealed a principle on which his impact rules depend, namely "when two bodies having incompatible modes collide, there must really be some change in these modes, in order to render them compatible, but . . . this change is always the least possible."[44] It is remarkable that Descartes would hold such a general principle to be at the basis of the impact rules and refrain from publishing it. Possibly its teleological flavor was at the root of his restraint. It is plausible that in the *Principes de la philosophie* Descartes was explicitly concerned with clarifying some of his rules to some extent. In rule 3, for example, he made the connection with the principle stated in the letter to Clerselier more explicit, with the wording "it is easier" for *B* to communicate half of its excess speed to *C*. Those words, of course, suggest a teleological principle at work. Rule 4 is substantially expanded in an attempt at explaining why a body could never be set in motion by a smaller one. Descartes revisited the explanation he had given to Clerselier, arguing at length that C at rest cannot be moved by a smaller body B in motion because C's resistance is proportional to B's speed. In rule 5 Descartes tried to explain away contrary observations by invoking "the air and the other fluids which always surround hard bodies that move and which can greatly increase or decrease their speed."[45] Finally, in rule 7 Descartes added the case 7.3 where B and C have equal motion, $Bv_B = Cv_C$. B transfers a portion of its excess motion (presumably one half) to C and rebounds with the rest.

Problematic as Marci's and Descartes' analyses of impact were, they represent a major episode in the history of motion and mechanics. Different bodies and materials behave in such disparate ways that the idea of providing universal mathematical laws for their behavior may have seemed a forlorn hope. Yet Marci and Descartes attempted precisely that and sought to provide the first mathematical formulations for the *interaction* among bodies. Their work was of considerable importance, not just for its contribution to the understanding of impact but also for its applicability to other areas of inquiry as well. The studies of curvilinear motion or motion in a resisting medium, for example, could be conceptualized and investigated by means of a series of tiny impacts deflecting or retarding a body. In both cases the study of impact provided investigative tools in the form of a conceptual framework and unit measures.

5.6 The Workings of the Cartesian Universe

Descartes' *Principia philosophiae* outlined a cosmology not entirely dissimilar in scope and, occasionally, types of explanation to Plato's *Timaeus*. Descartes discussed the formation of the universe from the birth of stars and planets down to the microstructure of salt, mercury, and sulphur, the origin of earthquakes, and scores of other phenomena. Descartes dealt with several themes pertinent to motion and mechanics besides collision, notably he analyzed special cases of curvilinear motion in celestial bodies. He argued that celestial bodies are carried by vortices whirling around a central body, the sun in our system or other stars elsewhere in the universe. Their curvilinear motions result from the combination of the pressure of the vortex toward its center and the body's tendency to move away along the tangent. Thus in his account there is no real attraction; rather, what appears to be attraction actually results from the pressure of the vortex. Likewise, gravity on the Earth results from the pressure of a vortex rotating around the Earth and pushing heavy bodies down. Descartes' account relied on the conceptual framework of the rotation of the sling, discussed above. In part IV of *Principia* he

examined a broad range of phenomena loosely linked to the Earth.[46] My aim here
is to discuss those details especially relevant to the science of motion.

The Cartesian universe ultimately depends on the structure of matter and the
fluids interacting with it. In several instances Descartes had recourse to Castelli's
principle for the motion of fluids. Descartes usually did not mention his sources
by name, and in this case he did not mention Castelli, but it is possible that he was
not aware of Castelli's work and relied instead on common knowledge.[47] In the account of the formation of different types of matter, Descartes argued that smaller
particles that get separated from larger ones move faster because they have to
move through narrower and more oblique passageways. He had recourse to the
analogy of a pair of bellows, arguing that even if it is compressed slowly, the ex-
pelled air moves fast because the orifice is small. Discussing the behavior of the
matter forming the sun with respect to sunspots, Descartes invoked the analogy of
the water of a river, which runs faster where the bed is narrow than where it is
wide and deep. In both instances Descartes assumed that his fluids were incompressible. In part IV of his *Principia* Descartes tried to explain tides by means of
the compression and speed of the vortex between the Earth and the moon (fig.
5.5). In the areas where the vortex is more compressed, it would have to move
faster and the air and water would be compressed in F and H, or at low tide, and
expanded at G and E, or at high tide.[48] Thus although the ultimate components
of the Cartesian universe were collisions, Descartes relied on notions about fluid
motion too, especially river flow, to account for a range of phenomena.

Descartes' rules appear in part II of his *Principia*, but it is in parts III and IV
that he described real objects colliding. It is striking to notice that when dealing
with the real world Descartes qualified and modified his account considerably. To
begin with, he believed matter in the universe to consist of three different types,
the first two resembling fluids and the last solid bodies. The sun and fixed stars are
formed of matter of the first element, namely tiny particles of all shapes, moving
very fast and resulting from the rounding off of the particles of the second element. This second element consists of small spherical particles and forms the
heavens. The earth, planets and comets are formed of matter of the third element,
whose particles are large and of irregular shape. The three types of matter have
different optical properties, the first being luminous, the second transparent, and
the third opaque. They are formed by the rubbing together of particles and are
not in a stable state; matter is continuously being rearranged, and particles of the
first element, for example, can recombine to form larger chunks of matter of the
third element.[49]

When discussing impact in part II, Descartes had considered the size of bodies,

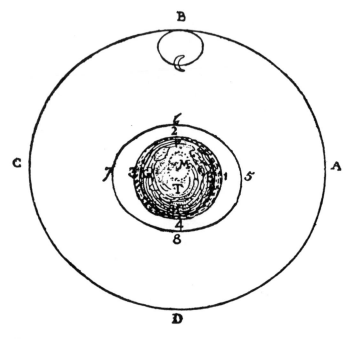

Figure 5.5. Descartes' theory of tides *(Principia philosophiae)*. Descartes tried to account for different aspects of tides on the basis of his vortex theory. The basic idea relied on the same notion underlying Castelli's proposition on the rate of water flow. The moon is carried around the Earth in a
vortex *ABCD*. The Earth does not occupy the center of the vortex *M* but is shifted by the matter of
the vortex away from the moon, so that its center is in *T*. 1234 is the surface of the sea and 5678 the
atmosphere around the Earth. Thus both directly under the moon and on the opposite side, the
matter of the vortex and the air and water on the Earth move in a narrow space and therefore are
faster and shallower in 6 and 2 and in 8 and 4 than in 7 and 3 and in 5 and 1; 2 and 4 are low tides
while 1 and 3 are high.

and size appeared to be equivalent to extension. While dealing with impacts in the
real world in parts III and IV, however, Descartes' picture became more complex,
and he introduced a new notion, that of solidity. Although Descartes' intention
here was to explain the motion of celestial bodies, he extended his remarks to bodies on the Earth as well. The concept of solidity was needed for a number of reasons. In a world of bodies penetrated by particles moving through them, sheer
geometrical extension was no longer an adequate notion. Descartes specified that
by the solidity of a body he meant only the quantity of matter of the third element, thus excluding the matter of the first and second elements, or of the fluids
flowing through it. This distinction put a wedge between quantity of matter and
sheer extension because in order to determine the latter it was no longer sufficient

to measure a body's volume; rather, it became necessary to determine its internal constitution.[50]

The notion of solidity was related to the surface and volume of a body, which for a sphere increase as the square and the cube, respectively. Solidity seemed to increase with the ratio between the body's volume and its surface. Descartes outlined a contrast in orbital motion between a tendency toward the center and an opposite one outward. The former depends on the volume of the body, because only the particles filling its place need be considered for its downward motion, whereas the latter depends on its surface, because it results from the interactions between the surfaces of the body and the fluid particles. Talking about bodies on the Earth, Descartes explained that pieces of gold or lead are more solid than other bodies because they have more matter of the third element and fewer pores. Changing a body's shape or dimensions, however, could alter its solidity. The sphere has the highest ratio between volume and surface, but size also matters, because the ratio between volume and surface grows with it.[51] Descartes was not always coherent when discussing these themes, and while one might expect a body's tendency to move toward the center to grow with the body's solidity, at times he suggested the opposite.[52]

Similar considerations apply to the discussion on gravity in part IV of *Principia*, although the term "solidity" does not appear explicitly. Descartes' heavy bodies on the Earth are pushed not by all the fluid matter surrounding them but only by the portions of that matter that can take their place. Gravity, or the weight of a body, is insufficient to determine the amount of matter of the third element in that body because the ratio between gravity and quantity of matter of the third element, or terrestrial matter, is not always the same. The variations are due to a wide array of variables, from the internal structure of the body, to the fact that fluids flowing though it can contain portions of the first as well as of the third element, and to the fact that those fluids move at different speeds depending on the size of their particles. According to Descartes, in practical terms there is no a priori way to determine gravity without actually measuring it. In the light of these comments, it would appear that Newton's pendulum experiment on the proportionality between quantity of matter and weight in book III, proposition 6, of *Principia mathematica*, was a direct response to paragraph 25 in part IV of Descartes' *Principia*.[53]

In addition to and in conjunction with solidity, Descartes often referred to the notion of agitation. Again, he was not always consistent in his account, and this notion was not clearly defined in quantitative terms. From some passages on comets agitation can be understood as the speed acquired by a body as a result of being pushed by a fluid. The agitation acquired by a body depends neither on its surface nor on its *molis*, or quantity of matter, but only on its matter of the third element. Descartes suggested that the reason for this was that the matter of the first and second elements receives no new agitation from going through a solid body, but only a change in direction or in its determination.[54] As a result of these important qualifications in parts III and IV, Descartes' notion of quantity of matter became problematic when applied to the real world. Neither extension, as we were led to believe in part II, nor weight can help us in determining a body's quantity of matter relevant to the study of its motion. In this specific respect Descartes' worldview was unsuited to a meaningful quantitative formulation.

Another important notion bearing on quantification was briefly explored by Descartes in two instances in part IV of his *Principia*. The notion is that of elasticity, and the first instance is a passage discussing air compressed in a container, where he talked of a *vis resiliendi*, a force to expand in a larger space, and then referred to several machines operating on that principle. Only in the later French edition, in which Descartes stated that the force exerted by the air is equal to that employed in enclosing it in the container, can one find an attempt at providing even a preliminary form of quantification.[55] The second instance is in a later passage, in which Descartes sought to account for the elasticity of solid bodies, including glass. He argued that elasticity, or the property of *ressort*, is found in bodies whose parts have surfaces in immediate contact. Further, the pores in those bodies are aligned so as to allow the passage of matter of the second element. When the body is bent the pores are deformed, and the fluid tends to restore them and the entire body to their original state. After some time, however, the particles carve different pores in the bodies, which lose their tendency to return to their original state. Compressed air and bent bodies both appear to be instances of the elastic properties of matter, but from the perspective of physical causation they seem heterogeneous and structurally quite different. Indeed, the resemblance we see between the two phenomena may have more to do with their mathematical description than with physical nature.[56]

These cases reveal the tension and the complex relationships among the mechanical philosophy, quantification, and the science of mechanics. This tension, at times leading to quantification, at times obstructing it, would accompany the mechanical philosophy for the rest of the century.

Generational and Institutional Changes

This intermezzo provides an outline of some significant changes of venue and major transformations in the way mechanics was practiced during the seventeenth century. I begin by listing some of the works from the 1630s to the 1650s mentioned in the previous chapters in order to consider them from a different perspective, taking into account chronological developments rather than styles of investigation and the emergence of new areas of research. The list includes a few translations and later editions that present enough new material to be considered distinct works.

—Galileo, *Dialogo,* 1632
—Cavalieri, *Lo specchio ustorio,* 1632
—Mersenne, *Les mechaniques de Galilée,* 1634
—Mersenne, *Harmonie universelle,* 1636–37
—Galileo, *Discorsi,* 1638
—Baliani, *De motu naturali,* 1638
—Mersenne, *Les nouvelles pensées de Galilée,* 1639
—Marci, *De proportione motus,* 1639
—Castelli, *Della misura dell'acque correnti,* 1639^2
—Gassendi, *De motu impresso,* 1642
—Descartes, *Principia philosophiae,* 1644
—Torricelli, *Opera geometrica,* 1644
—Mersenne, *Cogitata physico-mathematica,* 1644
—Gassendi, *De proportione,* 1646
—Cabeo, *Meteorologica,* 1646
—Fabri, *De motu locali,* 1646
—Baliani, *De motu naturali,* 1646^2
—Mersenne, *Novarum observationum tomus III,* 1647

—Descartes, *Principes de la philosophie*, 1647

—Le Tenneur, *De motu naturaliter accelerato*, 1649

—Riccioli, *Almagestum novum*, 1651

Though somewhat subjective and by no means comprehensive, this list documents a remarkable flowering of investigations into motion and mechanics. After Riccioli's *magnum opus*, however, we witness a slowdown in publications, with only a handful of relevant works appearing in the 1650s. Despite the appearance of Gassendi's imposing *Syntagma philosophicum* in his posthumous *Opera* (Lyons, 1658), his most original work on motion and mechanics dates from the 1640s. Walter Charleton's *Physiologia Epicuro-Gassendo-Charltoniana* (London, 1654) was a free paraphrase of Gassendi's work that made the new philosophy and some of Galileo's results available in English. Hobbes's *De corpore* (London, 1655) elaborated on several themes from Galilean and Cartesian sources, such as the notion of *conatus*. Charleton's and Hobbes's works were among the first contributions to the new science of motion from across the Channel.[1] Published books represent only a portion of the studies on motion and mechanics, but other forms of communication such as epistolary exchanges seem to have declined too, for reasons that will become apparent.

Galileo's *Dialogo* and *Discorsi* occupied center stage in the debates on the law of fall and the trajectory of projectiles up to the 1640s. As we have seen, Galileo's followers were by no means victorious. It does appear as if there was a deliberate attempt to attack and denigrate Galileo's results, especially from some of the most conservative quarters of the Catholic Church. According to Paolo Galluzzi, a crucial point in some of these attacks was the link between the new science of motion and Copernicanism. By the 1660s, however, the scene had changed, and some of Galileo's results on falling bodies and parabolic trajectories, for example, were tentatively accepted or at least no longer regarded as a matter of serious controversy.[2] Several factors contributed to this change, such as a major generational change, the emergence of the new scientific societies, the rise of the mechanical philosophy, and the different training received by scholars.

Midcentury saw the end of a scholarly generation, with several key figures dying within a short period of time, such as Galileo in 1642, Castelli in 1643, Cavalieri and Torricelli in 1647, Mersenne in 1648, Descartes in 1650, and Gassendi in 1655. Mersenne's death was especially significant because the French Minim had been the center of a network of correspondents engaged in debates about the intellectual issues of the day. Without him, the surviving scholars lacked a nodal point for their exchanges. But in the late 1650s and 1660s this role began to be taken on by the newly founded scientific societies, especially the Royal Society, established in 1660, the French Académie, established in 1666, and to a smaller degree the Accademia del Cimento, established in 1657. The Cimento was not primarily devoted to motion and mechanics and lasted only until 1667, but it did provide a forum for debates and publications on the trajectory of cannonballs, water flow, and the motion of the Medicean planets. The English and French academies held high status and devoted a good portion of their activities to what were perceived to be the most significant intellectual themes of the time, including the study of motion and mechanics broadly conceived. Just to mention a few examples, the Royal Society was the venue for much research on impact, motion in a resisting medium, and celestial motions, while the French Académie promoted debates on the cause of gravity, water flow, fluid resistance, and the resistance of materials. The intellectual achievements of Galileo and others, together with the perceived philosophical significance of investigations in mechanics, conferred status on the discipline at the academies and, eventually, in the universities as well.

In the first part of the century, university professors and members of religious orders were almost invariably the key figures in philosophical debates, whereas starting in the 1660s, members of scientific academies became the protagonists. In Italy the leading scholar on motion and mechanics was Borelli, a declared follower of Galileo, a professor of mathematics at Pisa, and also a member of the Cimento. In France and the Low Countries the key figures included Huygens and Mariotte, both academicians in Paris. In England the new generation of scholars included Wallis, Wren, Hooke, and Newton. Although at some point they all held teaching positions, Wallis and Newton at Oxford and Cambridge, the other two at Gresham College, they were all members of the Royal Society. Scientific societies promoted new forms of research and debate, favoring the experimental philosophy and allowing a freer handling of disciplinary boundaries and hierarchies compared with university curricula. Moreover, since many members of the new scientific societies were university professors, university teaching too was affected.[5]

Among the documents indicative of these transformations are the extensive lists of those who received dedication copies of Huygens's works. From the beginning of his career, Huygens carefully selected the recipients of copies of his works from among the most prominent scholars in his field of research. Those lists indicate that Jesuit mathematicians were among the most prominent recipients of Huygens's 1650s works on the determination of π and quadratures. Starting in the 1660s, however, matters changed. While Jesuit colleges and scholars still figured on the intellectual scene, their leading role was progressively taken over by the scientific societies in Tuscany, Paris, and London.[4]

An important effect of this change of institutional venue was the rise of mechanics and the science of motion in conjunction with the mechanical philosophy.

One of the most notable changes in status for a seventeenth-century mathematician occurred in 1610 with Galileo's transfer from the mathematics chair at Padua to the position of philosopher and mathematician to the Grand Duke of Tuscany. During the century as a whole, however, we do not see large-scale migration from the universities to the courts changing the status of scholars on motion and mechanics; therefore Galileo's case is not truly representative. His itinerary may be representative in a broader sense, however. A large number of the practitioners of mechanics were university professors of mathematics, but universities do not appear to have been key locations. Mathematics was not a major academic subject, and at most universities there were one or at most two chairs in the mathematical disciplines. Mechanics was only one among several such disciplines, and it often had to compete with the others for space in the curriculum. The emergence of scientific societies had a major effect in changing the intellectual and disciplinary landscape.[5]

The second half of the century also witnessed the emergence of a new venue for publication, the scientific journal. The *Philosophical Transactions* of the Royal Society in London, the *Mémoires* of the French Academy in Paris and Amsterdam, the *Journal des sçavans* also in Paris and Amsterdam, and the *Acta eruditorum* in Leipzig played a major role in the transformation of the intellectual climate. The new genre of the journal article provided a more visible and permanent format for brief theoretical and experimental reports than the ephemeral pamphlets and broadsheets that had been used for that purpose in previous decades, especially in astronomy. In some areas, however, such as hydraulics and water management, ephemeral publications and manuscript circulation remained common even after the emergence of scientific journals.[6]

Other important factors went hand in hand with the new institutional setting. The rise of the mechanical philosophy, the growing cosmological import of studies on motion and mechanics, and their increasing conceptual sophistication brought these subjects much closer to philosophy and refined intellectual pursuits than to the work of technicians and engineers. The more the world came to be seen as machinelike, the more the study of machines took on cosmological import. This phenomenon was enhanced by the shift in the nature of the objects themselves, from technical devices such as the lever, the wedge, the beam, and the hydraulic press to moving objects such as the pendulum and colliding, falling, projected, and orbiting bodies. Scholars studying these objects had an intellectual status different from that of the sixteenth-century theorists of pulleys and screws.

Finally, the new generation of scholars emerging in the 1660s had been trained to some extent in the new philosophy and mathematics, both inside and outside the universities. In a justly famous passage from his autobiography, Wallis mentioned informal interdisciplinary groups of scholars meeting in London and Oxford in order to study and discuss philosophical novelties ranging from Harvey and Copernicus to Galileo. Galileo's works circulated widely, and in some cases relevant portions were discussed as part of university curricula. Thus the new generation of scholars did not have to operate from the same standpoint as Galileo and the early Descartes but could build on their foundations.[7]

These remarks apply not only to mechanics and the science of motion but also to the new mathematics. Frans van Schooten the younger, professor of mathematics at Leyden, was a friend and translator of Descartes who promoted and systematized Cartesian methods and results. The progressively larger editions of his translation of Descartes' *Géométrie* were studied by a new generation of mathematicians, such as Huygens and Newton. Wallis and Hooke accepted and applied the new method of indivisibles put forward by Cavalieri and Torricelli.[8] The new mathematics opened new vistas in the study of motion. Some of the best mathematicians of the second half of the century, such as Huygens, Leibniz, and Newton, were able to apply the new mathematics to areas that had once seemed too complex for mathematization, such as the motions of bodies in resisting media or the elastic properties of solid bodies. The scene awaiting us in the following chapters is one of exciting novelties explored from a variety of perspectives.

CHAPTER SIX

The Equilibrium and Motion of Liquids

The speed of water flowing through a certain cross section of an inclined canal is the same as if the water flowed out of a vessel through an orifice similar and equal to that cross section, as distant from the surface of the water [in the vessel] as the cross section of the canal is from the horizontal line through its origin.

Domenico Guglielmini

6.1 A Characterization of a Research Tradition

In the second half of the seventeenth century the mathematical study of liquids, notably water and other incompressible fluids, was characterized by two types of approach, one based on motion and the other on statics. The former was pioneered in Italy, where scholars framed their views largely in a Galilean fashion, treating fluids very much like solid bodies. For example, falling bodies were invoked when investigating the flow of water spurting out from a pierced cistern, river flow was considered analogous to a body or system of bodies rolling down an inclined plane, and in one instance a fluid was explicitly compared to a crystal, thus exerting no lateral pressure.[1]

North of the Alps the Italian tradition based on motion was also quite influential, but scholars such as Pascal and Boyle investigated the behavior of water and fluids in general on the basis of static principles, having recourse to the notion of pressure. Their approach was not based on the behavior of solid bodies but rather tried to develop an autonomous understanding of fluids *juxta propria principia*. The emphasis on paradoxes in Pascal and Boyle, for example, is indicative of their focus. The behavior of fluids is paradoxical precisely because it is so unlikely that of solid bodies. One example that we encountered earlier is that of a column of fluid, whose height seemed to be a more significant variable than its weight. In

Italy, mathematics professors Stefano degli Angeli at Padua and Geminiano Montanari at Bologna followed this approach to some extent.[2]

As we have seen, water management was not an abstract area of inquiry pursued for its own sake, but rather a pressing and expensive practical problem affecting entire regions. Therefore the study of the equilibrium and motion of fluids presents some peculiar features associated with the sources, the practitioners, the objects, and methods of inquiry. Much of the literature on water management consists of small pamphlets devoted to opinions, controversies, or arbitrations. One midcentury engineering treatise on water management advertised on the title page its utility not only to engineers and technicians but also to those involved in legal litigations.[3] A great deal of the literature circulated in manuscript form. Classic themes include the preservation of the Venetian lagoon and the regulation of the Reno and Po rivers. As to the latter, for example, visitations to the crucial area around Ferrara by cardinals sent from Rome in 1625, 1660, and 1692–93 were accompanied by a flood of reports and memoranda trying to sway action one way or the other. Several works by Castelli (as we have seen), Giandomenico Cassini, and Domenico Guglielmini stem from those visitations.[4] Among the practitioners we find engineers and technicians as well as university professors of mathematics, the former often basing their views on practice, the latter on theory. The mathematical science of waters turned out to have cosmological implications too, since Castelli's fundamental theorem was used by Descartes in several passages of *Principia philosophiae*, as we saw in the previous chapter, and by Newton in book 2 of the *Principia mathematica* in the attempt to refute Cartesian vortices, as we shall see in chapter 9. This connection between practical and technical applications, on the one hand, and broader philosophical and cosmological concerns, on the other, represents one of the characteristic features of mechanics and the mechanical philosophy of this period.

This chapter presents a selection of themes and debates related to the science of waters. Whereas problems involving equilibrium were comparatively manageable, the flow of water was remarkably complex for the conceptual, mathematical, and experimental tools available in the seventeenth century. This intractability and lack of conceptual systematization is probably why this is an unusual topic for a study of seventeenth-century mechanics. Many problems were investigated, from whether water in a river accelerates to the relation between the height and speed of water. According to one tradition linked to Castelli, the speed of water in a given river was thought to be proportional to its height (meaning the height of its surface above the riverbed, or in other words its depth), whereas according to a rival tradition linked to Torricelli, the speed was as the square root of the height. Torricelli relied on the pierced cistern, following the analogy between water spurting out of a hole at the base of a container and water in a river being pushed along by the water behind. Both traditions relied on the theory of proportions.

Section 6.2 examines the influential second book of Castelli's pioneering *Della misura dell'acque correnti* and subsequent treatises emerging around the time of the Cimento Academy. Sections 6.3 and 6.4 deal with the situation across the Alps in the period from the posthumous publication of Pascal's works to Mariotte. Equilibrium is the subject of section 6.3, devoted to Pascal and Boyle, whereas motion is at the center of section 6.4. In section 6.5 I return to Italy, this time to Bologna, where in the 1690s Domenico Guglielmini was gaining a prominent position as the leading expert on hydraulics and water management. Finally I outline some mathematical problems associated with the combination of notions derived from the equilibrium and motion of water, notably pressure and speed. Although a new mathematical approach allowing the combination of height of fall, pressure, and speed emerged only in the eighteenth century, some experimental work along those lines was undertaken in the 1690s.

Experiments and debates on the management of rivers, canals, fountains, and water conduits occur throughout the seventeenth century and extend well into the next. These studies highlight the existence and importance of another area of study, besides falling bodies, intersecting motion and mechanics and dealing with issues such as the relation between mathematics and nature, theory and experiment, the nature and behavior of fluids. Although the science of waters did not lead to spectacular results comparable to those achieved in other areas, it is revealing of how mechanics operated at the time and of its reliance on objects. The inclined plane, the pierced cistern, and the hydraulic press were among the key devices mathematicians relied on.

6.2 Studies around the Time of the Cimento Academy

Toward the end of his life Castelli composed a second book to his treatise on the measurement of flowing waters and sent it to his friends in Tuscany, especially Torricelli, who was supposed to prepare it for publication. Torricelli did nothing, both because he was completing his own *Opera geometrica* and probably also because he disagreed with Castelli.[5] Castelli's manuscripts circulated quite widely, and a couple of decades later the Bologna publisher Manolessi embarked on putting out an edition. The patron of the Cimento Academy Leopoldo de' Medici, probably acting at Borelli's instigation, intervened to stop the publication in order to allow his mathematicians to correct the proof of the second proposition in book 2 of Castelli's work, namely the fundamental proposition of the new edition (fig. 6.1). The statement of the proposition seemed less problematic than the proof.

82   D E L L A   M I S V R A

PROPOSIZIONE II. PROBLEMA II.

SE vn Fiume muouendofi con vna tal velocità per vn fuo Regolatore hauerà vna data altezza viua, e poi per nuoua acqua crefcerà il doppio, crefcerà ancora il doppio di velocità.

Sia l'altezza viua d'vn Fiume nel Regolatore A B C D, la perpendicolare F B, e poi per nuoua acqua fopraue-nuta al Fiume, fi fia alzata l'acqua fino in G, fiche G B, fia doppio di E B, di-co, che tutta l'acqua G C farà il doppio di velocità di quella, che era E C; Im-peroche hauendo l'acqua G F per fuo letto il fondo E F egualmente inclinato come

il letto B C, & effendo la fua altezza viua G E, eguale al-l'altezza viua E C, & hauendo la medefima larghezza, B C, hauerà per fe ftetta vna velocità eguale alla velocità della prima acqua F C, ma perche oltre il proprio moto, che viene portato dal moto dell'acqua E C, haueua an-cora oltre al proprio moto, il moto della E C, e perche le due acque G C, & E C fono fimili di velocità per la ter-za fuppofizione, però tutta l'acqua G C farà doppia di ve-locità di quella, che haueua l'acqua E C, che era quello, che fi doueua dimoftrare.

82   D E L L A M I S V R A

PROPOS. II. PROBLEMA II.

SE vn Fiume muouendofi con vna tal velocità per vn fuo Regolatore hauerà vna data altezza viua, e poi per nuoua acqua crefcerà il doppio, crefcerà ancora il doppio di velocità.

Sia l'altezza viua d'vn Fiu-me nel Regolatore A B C D, la pédicolare per E B, e poi per nuoua acqua foprauenuta al Fiume, fi fia alzata l'acqua fi-no in G, fiche G B fia doppio di E B, dico, che tutta l'acqua G C farà il doppio di velocità di quella, che era E C.

Non fi mette la dimoftrazione della propofta perche da let-tere fcritte dall'Autore ad Amici, cofta non efferfi fodisfatto, e che non intendeua di publicarla fenza vna più falda dimo-ftrazione, la quale fperaua di confeguire. Ma preuenuto dal-la Morte non potè dare, ne à quefta, ne al rimanente del fe-condo Libro l'eftima mano. Onde fi è ftimato più opportu-no il tralafciarla, che il contrauenire alla mente dell'Autore; E ciò ferua ancora d'auifo à coloro, che fi trouaffero hauer co-pia manufcritta di quefto libro con la detta dimoftratione. Per hora fi contenti il Lettore della notizia di così bella, & vtile conclufione, della verità della quale egli può con poca fpefa, e con molto diletto venire in ficurezza per mezzo dell'-efperienza da farfi in modo fimile à quello, che viene fpiega-to nel fecondo Corollario della quarta Propofizione di quefto, con la fua Tauola, & apprefio con l'vfo di effa.

Figure 6.1. Two versions of page 82 of Castelli's *Della misura*. Compare Castelli's defective proof with the corrected version instigated by Borelli, in which Castelli's original text was replaced with the explanation that the author had been uncertain about this point. The proposition states that the speed of water through a *regolatore* is proportional to the water's height. By *regolatore* Castelli meant a flat riverbed with two perpendicular banks, providing a rectangular cross-section that en-abled him to rely on simple geometry. If the width is fixed, the quantity of water flowing through a *regolatore* in a given time is as the height and speed of the water. If the speed is as the height, the quantity of water is proportional to the square of its height.

Manolessi said he would comply, replacing the defective section with an explana-tion by Borelli that the proof was lacking because the author had not found a sat-isfactory one. The third edition of Castelli's work appeared in 1660, but some copies circulated with Castelli's original proof.[6] The edition is indicative of the range of manuscripts in circulation that were available to Manolessi. It includes *Considerazioni* on the Venetian lagoon, on the draining of swampy areas near Rome (the "Paludi Pontine"), and on the draining of the area between Bologna and Ferrara, as well as letters and memoranda to the Piarist mathematician Fami-ano Michelini, to the engineer Alessandro Bartolotti on the Fiume Morto and Ser-chio, and to others.

Castelli tried unsuccessfully to prove a relation between the height of the water of a river or canal and its speed, namely that the speed is proportional to the height. If the width is constant, the rate of flow is proportional to the square of the height. One of the obvious defects in his approach is that he considered the entire cross-section to be moving at the same speed, as Cabeo had already objected in 1646.[7] Castelli even claimed that he had performed trials involving one hundred siphons so as to prove his proposition. One would like to know more about this alleged experiment and, especially, whether Castelli actually performed it. He talked of a container with one hundred siphons or bent pipes connected to an outside source, pouring water at the same level below the water surface. From the container the water overflowed into a receptacle below, and from there it went into a canal whose height was divided into ten equal parts. Upon shutting off 19 siphons (and leaving $81 = 9^2$), the water level in the canal went down by a tenth. Shutting off 17, 15, etc., additional siphons (leaving $64 = 8^2$, $49 = 7^2$, etc.), the water level decreased by additional tenths of the overall height. In conclusion, the experiment proved that the quantity of water flowing in a river or canal is proportional to the square of the height.

In the first edition of *Della misura dell'acque correnti* Castelli had thanked Giovanni Ciampoli for his help and financial support in performing an experiment using a barrel full of water with a mere two pipes attached to it. It is thus remarkable that no acknowledgment or descriptive detail accompanied his report of an experiment using no fewer than one hundred siphons. In a letter of January 1642 to Cavalieri, Castelli reported a thought experiment involving a river similar to the trial with one hundred siphons. Castelli made no pretense of having performed it but rather presented it as a "scherzo della natura," or curiosity.[8] These observations suggest to me that Castelli's hundred-siphon experiment too was imaginary rather than real and that he was well aware of the rhetorical significance of experiments and real models. This is not to suggest that he was deliberately lying. Presumably he had intended to try the experiment and wrote about it as if it had been performed, but after his death in April 1643 this part of the manuscript was published without emendations.

In his treatise Castelli described another experiment, similar to the preceding one, in which a dam with twenty-five holes or siphons releases water into a lower basin, whence the water flows out into a canal (fig. 6.2). He made clear that this experiment had not been performed and that the accompanying figure was provided purely for clarity's sake. The siphons are arranged in groups of 1, 3, 5, 7, and 9, and once again, according to Castelli, the square of the height of the water flowing out of the canal is as the quantity of water. Castelli also pointed out that the water flowing down the canal is lower at the end, because of its greater speed. Thus water flowing down an inclined plane does accelerate.[9]

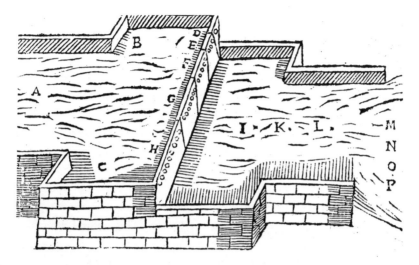

Figure 6.2. Castelli's thought experiment *(Della misura)*. Castelli's thought experiment intended to show that the quantity of water flowing in a canal is as the square of the water's height.

In 1656 the engineer Giovanni Battista Barattieri published the first part of a large folio treatise on water management, *Architettura d'acque*. The second part followed in 1663. Barattieri had access to Castelli's manuscripts and in 1656 had suggested that the speed of water is proportional to its height. Barattieri also claimed that Castelli's proposition was already known to an engineer earlier in the century, Alessandro Betinzoli from Crema. Betinzoli claimed that the square of the height was as the quantity of water, which would grow as the cross-section and "quantità della gravezza" combined. A doubling of the height doubles the cross-section, but it also doubles the speed. Betinzoli seemed to be relying on the dubious notion that the height is proportional to the weight and the weight to the speed.[10]

In 1663 Barattieri reprinted many of the texts in the 1660 edition of Castelli's work from his own manuscript sources. He agreed with Castelli's second proposition, but as we have seen he believed in the proportionality between gravity and speed. Barattieri also tried to confirm it independently with an experiment conducted with the Codogna aqueduct, one of the largest south of Milan. The aqueduct was divided into four unequal branches, all with the same slope. Barattieri considered only two branches with widths 67 and 86 units, through which water flows in proportion to the width. The height of the water was 8$\frac{3}{12}\frac{1}{6}$ linear ounces, whose square is 67. Closing the wider branch forces all the water to flow through the narrower one, whose new height was 12$\frac{4}{12}\frac{1}{2}$; its square is 153, namely the sum of 67 and 86. In this way, by measuring only widths and heights, Barattieri could claim that the quantity of water is proportional to the square of the height. In his figure Barattieri had given values for the water flowing through the two branches as 37⅓ and 48⅔ ounces of water, whereas in the text he wrote 37⅔ and 48⅓. He acknowledged the discrepancy, arguing that the values in the figure were experimental whereas those in the text were based on theory. The discrepancy "is not sensible and should not be seen as a defect" because measures cannot be absolutely precise. The reader is left to wonder how Barattieri managed to measure those values to such a degree of precision, and indeed whether he performed the experiment at all. I suspect that Barattieri himself was concerned with the perfect agreement between theory and alleged experimental results and introduced a small error as an afterthought so as to make his alleged experimental report appear more realistic. My skepticism was not shared by Domenico Guglielmini, who disagreed with Castelli's proposition about the proportion between the speed of a river and its height, but in a letter to the Bologna professor Geminiano Montanari expressed concern that the results of Castelli's and Barattieri's experiments seemed to confirm it.[11]

Castelli and Barattieri were not alone in claiming to have performed experiments on the proportion between rate of flow and square of the height, or speed and height of a canal. In the 1650s and 1660s the astronomer Giandomenico Cassini, professor of mathematics at Bologna, was involved in matters of water management in the Papal States and published a series of pamphlets and documents on the Reno and Po rivers. Cassini too agreed with the proportion we have seen and argued that deviating the Reno into the Po would not cause the water to rise as much as was feared. Indeed, this was the main concern behind the attempts to determine a relation among speed, height, and rate of flow. He claimed to have performed an experiment with nine canals, where one canal produced a given height, four canals twice that height, and nine canals three times that height. Cassini's claim to have performed the experiment "questo stesso giorno" in Rome, probably in 1657, appears to be the most reliable, despite his shaky theoretical explanation. Much like Barattieri the year before and some engineers earlier in the century, Cassini too believed in the proportion between weight and speed.[12]

Castelli's 1660 edition is not the only work with an interesting publication history. In 1664 the former holder of the Pisa chair of mathematics, Famiano Michelini, put forward *Trattato della direzione de' fiumi*, a handsome volume with a title page on which the author was named as Philosopher and Mathematician to Leopold, who had paid for its publication. The book dealt mainly with the behavior of rivers and ways to prevent damage from them, but Michelini argued that theory is the foundation of practice. Unfortunately, his theory was highly problematic. Since Michelini was in poor health, Leopold asked Borelli to go over the manuscript before printing it. Michelini and Borelli disagreed on a crucial point of the treatise, namely the author's belief that still water exerted no lateral pressure, only pressure on the bottom, much like a solid crystal. In preparing the manuscript for the press Borelli appears to have mitigated Michelini's stance by claiming that water does exert pressure against the sides of its container to a small degree. Even so, Michelini's views could not be hidden and came under the scrutiny and criticism of other mathematicians, such as degli Angeli, the Jesuat former student of Cavalieri and professor of mathematics at Padua. Michelini's views represent the most extreme attempt to understand the behavior of water in terms of solid bodies.[13]

The Pisa professor Giovanni Alfonso Borelli devoted portions of his treatise *De motionibus* to the flow of liquids, which he considered to consist of small spheres with interspersed *vacua*. In order to account for phenomena we associate with viscosity, Borelli proposed, in terms typical of the corpuscular tradition, that water particles are covered by a woolly coating, or "tenuissima lanugine." Unlike Castelli and others, however, Borelli studied primarily flow in closed conduits rather than open beds. Since he conceived his work as a prelude to a treatise on the motion of animals, this choice probably reflected his anatomical interests.[14]

Overall, the Ferrara Jesuit Riccioli, in his gigantic *Geographia et hydrographia reformatae*, agreed with Castelli's proportions. On another matter, however, he agreed with Cabeo and Barattieri against Castelli in denying that water accelerates in rivers and canals.[15]

6.3 Pressure and Equilibrium in Pascal and Boyle

The barometric experiment provided a link between the behavior of incompressible and compressible fluids, such as water and mercury versus air. This connection is especially evident in the publications of Pascal and Boyle, who worked extensively on the equilibrium of fluids. In this section I focus on incompressible fluids, while air and its elastic properties will be briefly discussed in chapter 8.[16]

In 1663 Florin Périer published a work containing a range of materials he had put together from the papers of his brother-in-law, Blaise Pascal, who had died the year before. According to the editors of Pascal's works, his papers date from the first half of the 1650s and rely on several Italian sources, such as Benedetti, Galileo, and Torricelli, often transmitted via Mersenne. Those papers include a treatise on the equilibrium of liquids, one on the weight of the air, fragments from a longer work on the same subject, the reprint of the report of the Puy-de-Dôme experiment showing that air pressure decreased with altitude, epistolary exchanges, records of weather reports by Périer, and a brief examination of Boyle's

HUITIÈME TABLE

Pour assigner la hauteur à laquelle l'eau s'éleve et demeure sus-
penduë en l'experience ordinaire.

Quand l'Air est chargé

	LE PLUS.		MEDIOCR.		LE MOINS.		DIFFER.	
	pd.	pc.	pd.	pc.	pd.	pc.	pd.	pc.
A Paris...............	32.		31.	2.	30.	4.	1.	8.
A Clerm...............	29.	8.	28.	10.	28.		1.	8.
A Lafon...............	28.		27.	2.	26.	4.	1.	8.
Au Puy...............	26.	3.	25.	6.	24.	7.	1.	8.

DIFFERENCES D'UN LIEU A L'AUTRE

Quand l'air est chargé.

De	LE PLUS.		MEDIOCR.		LE MOINS.	
	pd.	pc.	pd.	pc.	pd.	pc
Par a Cler....................	2.	4.	2.	4.	2.	4.
Cler. a Lafo....................	1.	8.	1.	8.	1.	8.
Laf. au Puy....................	1.	8.	1.	8.	1.	8.
Cler. au Pu....................	3.	4.	3.	4.	3.	4.
Par. au Puy....................	5.	8.	5.	8.	5.	8.

Figure 6.3. Pascal's tables *(Physical Treatises).* These tables were presented by Périer as fragments
from Pascal's manuscripts. Despite their appearance of being empirical reports, they contain data
calculated from some reference experimental values by means of simplifying assumptions and
interpolations.

air pump experiments. Périer also included some numerical tables including not
direct experimental data but rather data calculated from some reference experimental values by means of simplifying assumptions and interpolations (fig. 6.3).[17]

I focus here primarily on the treatise on the equilibrium of fluids, in which
Pascal discussed a number of experiments and paradoxes related to liquids. For example, Pascal arranged a tube of water with a plunger at the bottom to be held in
position by a thread to a counterbalancing weight attached to a balance (fig. 6.4).
The counterbalancing weight depends only on the height of the water in the tube,
not its weight; therefore if the tube is thin and tall, a large weight may be required
to counterbalance a tiny amount of water. Note that Pascal used the balance here
for a purpose quite different from Galileo's in his *Discorso* on bodies in water: Pascal used it not to show that hydrostatics works like statics but, on the contrary, to

highlight the differences between them and the paradoxes resulting from attempts to treat them as comparable.

In the same vein, Pascal introduced a new machine for multiplying forces, the
hydraulic press, also illustrated in figure 6.4. He had recourse to two principles of
mechanics to explain the machine's operation, one based on the relation between
distance and force, the other on the center of gravity of the system. The former
resembles Galileo's treatment of the problem in *Discorso* on bodies in water: "It
is remarkable that this new machine exhibits the same constant relation that is
characteristic of all the old machines, such as the lever, the wheel and axle, the
endless screw, and others, which is that the distance traversed increases in the
same proportion as the force."[18] The latter was inspired by Torricelli's principle
and states "that a body never moves by its own weight without lowering its center
of gravity."[19] Thus Pascal emphasized both the paradoxical behavior of liquids by
means of the balance and the analogies of the hydraulic press to simple machines
and mechanical principles. His hydraulic press represents a paradoxical twist:
Simple machines were initially thought to act by multiplying force, though they
became more and more associated with natural operations and natural philosophy.
Hydrostatics, by contrast, was traditionally intended to describe nature, but it led
to a machine for multiplying forces.

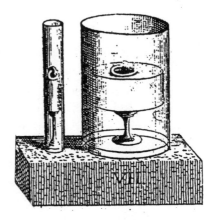

Figure 6.4. Pascal's hydraulic paradox *(Physical Treatises)*. In figure V Pascal used the balance to
highlight a hydraulic paradox, whereby a large weight holding the bottom of the tube in place by
means of a thread is needed to counterbalance a tiny amount of water in the narrow tube. In figure
VII a small weight keeps a large one in equilibrium through the hydraulic press.

Pascal's work relied on the notion of pressure as the key tool for investigating hydrostatics and for carrying out experiments intended to highlight the surprising and paradoxical features of the behavior of fluids. By "pressure," Pascal understood force or weight over surface. He believed that pressure was transmitted instantaneously through the fluid and that it was exerted in all directions, hence his concern with the strength of the sides of the press.

In his posthumously published work Pascal commented on Boyle's *Nova experimenta physico-mechanica de vi aeris elastica* (Oxford, 1661), while a few years later Boyle reciprocated with an extended commentary on Pascal's own work, the *Hydrostatical Paradoxes*. Boyle treated hydrostatics as part of experimental philosophy, and his narrative differs greatly from Pascal's. Whereas Pascal relied on extrapolations and adopted an impersonal style aiming at teaching the general principles underlying his experiments, Boyle painstakingly reported where, when, and how his experiments were performed.[20] Despite these differences, Boyle too relied on the notion of pressure in his work on hydrostatics and, much like Pascal, did not deal with the motion of water. Castelli is never mentioned in his publications.

6.4 Studying the Motion of Waters North of the Alps

It is not clear how well known the chief works of Italian scholars were outside Italy. The Jesuit Gaspar Schott provides a useful list of authors on the subject including Castelli, although his work is not discussed. Schott discussed efflux from a small hole in a tube, claiming that the quantity of water is as the square root of the height. His source was Mersenne rather than Torricelli, who is not mentioned there, although his name appears in the initial list.[21] Here I provide a preliminary investigation centered on France and England.

In 1664 a French translation by Pierre Saporta of the first book of Castelli's treatise along with Torricelli's essay on the motion of waters from *De motu* appeared at Castres, in the South of France. Saporta had wide-ranging intellectual interests and was a correspondent of Henry Oldenburg's, whom he had met at Castres in the winter of 1658–59.[22] Both the venue and the motivation for his double translation are of interest. Castres is situated about forty miles east of Toulouse, both locations being closely linked to key events of Fermat's life, who was a lawyer and *parlamentaire* at Toulouse and president of the Chambre de l'Edict at Castres, where he died early in 1665.

Saporta's translation of Castelli's work is dedicated to the Royal Committee for the Construction of the Canal du Midi, joining the Mediterranean to the river Garonne near Toulouse and thus to the Atlantic Ocean. This extraordinary canal was built between the 1660s and 1680s at the instigation of the tax collector Pierre-Paul Riquet and the engineer François Andreossy. In order to demonstrate the project's feasibility during the initial planning stages, Riquet, Andreossy, and the *fontainier* Pierre Campmas had a huge model of the canal built at Riquet's castle of Bonrepos, using the moat fed by two small lakes. The model, apparently including tunnels, locks, and supply channels, turned out to be of crucial importance, for the 1666 edict by Louis XIV approving the construction of the canal mentioned specifically that the model was "so cleverly contrived and satisfactorily achieved" that there was every reason to foresee a happy outcome. The construction of this elaborate model shows interesting analogies with the attempted construction of the fountain at the Boboli gardens discussed above in section 3.3. In both cases the behavior of water proved hard to envisage and prompted the construction of extensive models to assess the project's feasibility.

The translation of Torricelli's work, meanwhile, was dedicated to Fermat, who had prompted Saporta to include it in his volume, which was published under Fermat's auspices. At the end of this little volume Saporta also included Fermat's interpretation of a letter by Synesius to Hypatia on an instrument to measure the specific weight of liquids for medical purposes. This episode represents a connection via Saporta and Fermat between the Italian hydraulics tradition and a major engineering project in southern France. The visits of French engineers such as Andreossy to Italy, as well as the Low Countries, in order to study water works further emphasize this connection.[23]

In his *Cursus seu mundus mathematicus* (Lyons, 1674), an encyclopedic work in three gigantic volumes, the French Jesuit Claude Dechales dealt with virtually the whole range of the mathematical disciplines, broadly conceived, including hydrostatics, rivers, and hydraulic machines. He taught mathematics at several French cities before moving to the University of Turin, where he died in 1678. His sources on river flow were largely Italian and included the engineer Gianbattista Aleotti, Castelli, Barattieri, Cabeo, and Baliani, who had discussed relevant issues in the second edition of *De motu*. Dechales was apparently the first to establish in print a link between river flow and the efflux of water from a pierced container. Thus he denied Castelli's proportionality between speed and height, arguing instead for a proportionality between speed and the square root of the river's height. Not surprisingly, Dechales devoted some space to the law of efflux and reported the experimental data in a table showing that the square root of the height of water in a cistern suspended above the ground is as the horizontal distance traveled by the water spurting horizontally from the cistern in the time from when it leaves the cistern to when it reaches the ground. Since the time of fall for a body shot horizontally is constant, the speed is as the distance traveled; therefore Dechales could retrieve the standard form of Torricelli's proposition whereby the square root of the height is proportional to the speed. The arrangement of Dechales' experiment resembles some of Galileo's trials (described in section 3.2). Dechales proceeded to use his result by comparing, on the one hand, the depth of a river to the height of a tube out of which water flows and, on the other, the river's cross-section to the size of the hole in the tube. Much like Castelli, Dechales did not consider that the speed of water varied in a cross-section, and in this respect his account looks quite similar to that of the Italian mathematician. Dechales devoted his attention to the problem of accurate leveling, arguing that the Earth's curvature is a significant factor in determining slope because even a height as small as a foot over a distance of one mile does make a difference.[24]

The study of the equilibrium and motion of water was the subject of studies at the Paris Académie in the late 1660s. The academicians focused on three themes: the pierced cistern, Castelli's proportion between speed of a river and its height, and the force of water jets. Several academicians were involved at different stages of the experimental investigations, including the astronomer Jean Picard, Huygens, Roberval, and Mariotte. The experiments tended to shed doubt on Castelli's results on river flow and to confirm Torricelli's. In the case of the pierced cistern, they showed that the same amount of water comes out regardless of the containers' cross-sections, provided that the size of the orifice was much smaller than the containers and that the heights were the same. As to water jets, the investigations were linked to utilitarian aims, such as measuring the pressing and moving forces exerted by running waters.[25]

Some of the experiments share similarities with Galileo's attempts to measure the force of percussion. For example, the academicians tried to measure the force of a water jet falling on one side of a balance by trying the keep the balance in equilibrium with a weight on the other side. They believed the force of a water jet emerging from a small orifice at the bottom of a large container to be proportional to the jet's cross- section and to the height of the water in the container, which was proportional to the square of the speed. Or, to put it a different way, a jet is balanced by a weight equal to that of a cylinder of water with a base equal to that of the orifice and a height equal to the height of the water in the container.[26] Another experiment involved measuring the force of running waters by determining the force necessary to keep in place against the current a body floating in a canal constructed for the purpose as well as one floating in the river Seine. The body was attached to a weight by means of a rope passing through a pulley. These experiments were apparently conceived by Huygens, who also added important commentaries. First, in line with his views about relativity of motion in the case of impact, he argued that the two cases of a moving weight in still water or a still weight in moving water are symmetrical.[27] Moreover, in the case of the Seine, Huygens argued that in addition to the simple resistance of water proportional to the speed squared, there was another form of resistance due to the friction against the sides of the body.[28] Huygens also devised rather elaborate instruments for measuring air resistance and performed experiments leading to tables of numerical results. In all cases the moving force of water or air appeared to be proportional to the square of the speed and, conversely, its resistance to motion followed the same proportion. We shall see in section 7.4 how this discovery affected Huygens's research.[29]

The academician Mariotte became progressively more involved in these investigations and composed a treatise relying to a large extent on the Académie's work. By the time of his death in 1684 Mariotte had not quite completed his treatise on the motion of waters, which was seen through the press by Philippe de la Hire in 1686. Mariotte's posthumous treatise addresses a rather heterogeneous mixture of themes at the margins of its subject matter, such as observations on the variations of the seconds pendulum near the equator, air resistance encountered by water jets, and the resistance of solid bodies to rupture. Such topics were introduced because Mariotte used the pendulum to measure time and because he wanted to ascertain the resistance of water pipes.[30] Even with respect to the motion of fluids, Mariotte's treatise is rather heterogeneous and ranges from discussions on the origins of springs and winds to hydrostatics and the behavior of jets.

The third section of Mariotte's treatise is devoted to the measurement of flowing waters and jets. Mariotte measured the quantity of water escaping from orifices in a given time. He argued that, if the height is fixed, the quantity of water is proportional to the size of the orifice. If the size of the orifice is fixed, the quantity of water is as the square root of the height. Rather than focusing on the results themselves, I wish to discuss briefly the way in which they were presented and argued for. Mariotte carefully described how he performed the experiments, and he noted some discrepancies with predictions. For example, he argued that larger orifices give slightly less water than expected, and great heights of about 35 feet give 1/17 or 1/18 less water than expected, whereas small heights of about 6 or 7 feet give more. The discrepancies were attributed to different forms of friction. Mariotte reported both sets of results in tables. For example, in the case of variable heights he chose as his standard a height of 13 feet giving 14 pints of water per minute (fig. 6.5). It is clear that these tables report not actual experimental data but values that Mariotte had adjusted to serve as a guide while instantiating in numerical form the rules he had discovered. The same applies to Mariotte's table pertaining to variable orifices.[31]

Mariotte was also interested in water jets exiting from orifices at the bottom of

Hauteurs des Reservoirs	Depenfe d'eau,
6 pieds	9 pintes $\frac{1}{2}$
9 pieds	11 pintes $\frac{2}{3}$
13 pieds	14 pintes
18 pieds	16 pintes $\frac{1}{2}$
25 pieds	19 pintes $\frac{1}{3}$
30 pieds	21 pintes $\frac{1}{3}$
40 pieds	24 pintes $\frac{1}{2}$
52 pieds	28 pintes.

Figure 6.5. Mariotte's table *(Traité du mouvement des eaux).* This table gives the quantity of water spurting out of a hole of three lines at the bottom of a container in relation to the height of the water in the container, given in the first column. The second column gives the amount of water. Mariotte did not report here his experimental values, given in the text, but rather provided extrapolations from the case of a height of 13 feet, giving 14 pints of water in one minute. His last value is 52 feet, giving 28 pints of water. The ratio between the most unlikely height of 52 feet and 13 feet is 4, whose square root is 2. Hence if the height is quadrupled, the quantity of water doubles from 14 to 28.

a container and in comparing the jet's height to the height of the water in the container. He argued that the jet's height is lower than the height of the water in the container, and that the difference is proportional to the square of the jet's height. Once again, a table of theoretical data accompanies the rule. The two methods of determining the speed of the water—by measuring the quantity of water escaping from the orifice in a given time, and by measuring the height reached by the jet—seem to be equivalent; however, there are some subtle differences that were later investigated by Newton, involving the contraction of the vein, that is, the narrowing of the jet as it exits from the orifice.[32]

In 1661 Thomas Salusbury published an English translation of the third edition of Castelli's treatise, which had appeared just the year before. On the title page Salusbury emphasized the utility of Castelli's work for draining fens and for water management, major concerns in England at the time in connection with the project of draining the fens north of Cambridge.[33] Salusbury's translation was available to Newton, who implicitly referred to Castelli's fundamental proposition about cross-section and speed in the scholium to proposition 53, book 2, of his *Principia mathematica*. In the same book, proposition 37, Newton also provided a faulty analysis of the problem of efflux from a pierced cistern.[34]

Also in England, the treatise on *Mechanica* by John Wallis contains a section on *Hydrostatica* dealing with Torricelli's barometer and the equilibrium of fluids, but Wallis did not discuss the pierced cistern. In 1675 he published a short pamphlet on gravity largely devoted to fluids, in which he discussed the outflow of water from a pierced cistern, arguing that pressure of the upper parts of water over the lower ones makes water escape with greater speed from a lower orifice. Wallis did not provide a relation between height and speed, nor did he refer to Torricelli, despite his familiarity with Torricelli's *Opera geometrica*. He also discussed the issue of lateral pressure, reaching opposite conclusions to Michelini. Wallis argued that lateral pressure exists both in solids and in fluids, and is overcome by cohesion in the former and by the strength of the container's sides in the latter.[35]

6.5 Guglielmini and the Bologna Scene

Domenico Guglielmini was the leading Italian scholar on river hydraulics toward the close of the seventeenth century. He was a student of Marcello Malpighi in medicine and Geminiano Montanari in mathematics. Soon after gaining his medical degree in 1678, Guglielmini was appointed to a mathematics chair, following Montanari's departure for Padua. Montanari had already dealt with problems of hydrostatics at debates at the Accademia della Traccia in Bologna, choosing a representation of water consisting of small spheres that was to be adopted by Guglielmini in 1697. In a 1680 letter to Montanari, Guglielmini outlined some of his views on hydraulics and mentioned some of his sources, such as Baliani, Mersenne, and Dechales. To them we should add the classic works by Castelli and Torricelli and, by the time his book was published, Mariotte.[36]

In the 1680s Guglielmini was repeatedly consulted by the Bologna Senate on matters of water management, and in 1686 he was made Soprintendente alle Acque, thus joining practical tasks with the theoretical ones associated with his mathematics chair. In 1694 a new chair of hydrometry was established at Bologna University, and immediately afterward Guglielmini was called to the new position, which he held along with his mathematics chair. Although several scholars had taught matters pertaining to hydraulics as part of their duties as professors of mathematics, this was the first chair at a European university specifically devoted to the study of waters.[37] Guglielmini's interest in the motion of fluids was linked to his medical interests, and in his address to the reader in *Aquarum fluentium mensura* (1690–91) he expressed his wish to extend his hydraulics studies to medicine, since the motion of fluids was of paramount importance to the study of the human body. Guglielmini saw medicine, both theoretical and practical, in terms of the mechanistic tradition advocated by Borelli and Malpighi.[38]

In the address to the reader Guglielmini outlined two chief problems with Castelli's work. With regard to proposition 1, book 1, stating the inverse proportion between cross-sections and speeds, the problem is that the speeds in the cross-sections vary with the depth. And proposition 2, book 2, stating that the speed of water through the cross-section of a river is proportional to the water's height, contradicts the law of efflux. Both criticisms rely on the law of efflux, which is Guglielmini's beacon in all matters relating to river hydraulics. Following a tradition that we have seen instantiated many times with many devices, Guglielmini conceptualized a complex object in nature like a river in terms of water flowing out of a pierced cistern. According to Guglielmini, the speed of water through a given cross-section in a river is proportional to the square root of the height from that cross-section to a horizontal from the source (figs. 6.6–6.8). Moreover, the profile of speeds in a cross-section of the river is proportional to the square root of the height to the horizontal from the source. He reformulated Castelli's inverse proportion between cross-sections and speeds in terms of average speeds. Guglielmini considered the steady flow of a river without taking friction and other hindrances into account, and this assumption led him to the strange conclusion that water near the bottom of the river moves faster than that closer to the surface. He thought of providing experimental evidence for this unlikely claim by means of a pendulum attached to a graduated arc. When the bob is immersed at different depths in the water, the pendulum is deflected and the degree of deflection measures the speed of the water. In conclusion, he used the law of efflux twice, once for the average speed of the river with respect to the height of its source and again for the profile of speeds in a given cross-section. Both usages differ from Dechales' account, which took speed through a cross-section to be uniform and considered not the height from the source of the river but rather the actual depth of the river.[39]

Given the enormous importance he attached to the law of efflux, it is not surprising that Guglielmini devoted attention to confirming it by means of a careful experiment performed on 14 October 1683 with the help of the Olivetan abbot Taddeo Pepoli and the physician Giovanni Lodovico Donnelli. The experiment's results were presented in a table and compared with the results predicted by theory. The discrepancy was so small that Guglielmini could conclude that the quantity of water flowing from the orifice was indeed proportional to the square root of the height of the vessel.[40] This was not the only experiment described by Guglielmini. In the scholium to proposition 2, book 4, he tried to test the relationship between speed and height of flowing water by means of a device consisting of an iron container attached to a sloping channel also made of iron. Water

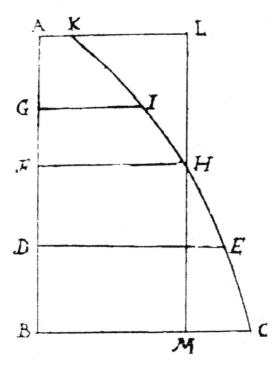

Figure 6.6. Guglielmini's profile of speeds in a river *(Aquarum fluentium mensura)*. The river is of depth *AB*, and *AK*, *GI*, *FH*, etc., are the speeds at corresponding heights *A* (the surface), *G*, *D*, etc. *FH* is the average speed. This illustration should be considered together with the follow-ing two, which explain why the profile of speeds is parabolic and how the surface speed *AK* is determined.

flowed from the container into the channel. An obstruction was lowered into the channel, not all the way to the floor but far enough down to cause water to build up behind it. Despite the higher level of water at the obstruction, the speed of the water flowing under it did not increase, but rather decreased because of its verti-cal motion. Despite the inconclusive result, this experiment is an intriguing in-stance of experimental activity related to water flow.[41]

In the first half of the 1690s Guglielmini was involved in a controversy with the Huguenot Denis Papin about the flow of water in siphons. Papin argued that water in a siphon inclined on the horizontal does not accelerate; thus it did not fol-low Galileo's law of falling bodies, a claim rejected by Guglielmini. In this case too he constructed a system of pipes to prove his point experimentally.[42]

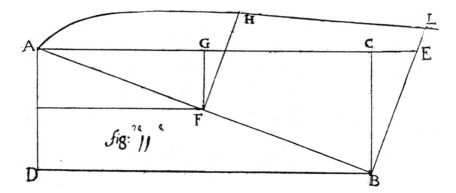

Figure 6.7. Guglielmini's law of efflux (*Aquarum fluentium mensura*). According to Guglielmini, the law of efflux rules water flow and determines also the speed of the water of a river. At *F*, for example, the speed is proportional to the square root of the segment *FG*, representing the distance to the horizontal *AE* from the source *A*.

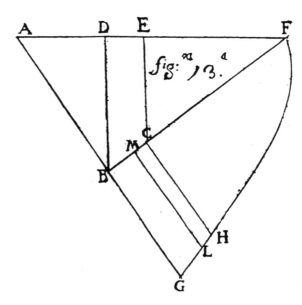

Figure 6.8. Guglielmini's profile of speeds in a river (*Aquarum fluentium mensura*). The profile of speeds *FHLG* along river *AB* resembles that in figure 6.6, which appears rotated by about 45 degrees. Notice that the surface speed at *F* is nil, because *AF* is the parallel to the horizon from the source *A*.

In 1692–93 Guglielmini produced a large number of reports and opinions pertaining to the visitation to the Po and Reno areas by Cardinals Ferdinando D'Adda and Francesco Barberini.[43] In 1697 he published a new treatise on river hydraulics, *Della natura de' fiumi*, alternating theoretical and applied chapters. Unlike *Aquarum fluentium mensura*, his new work provided a microanalysis of water as consisting of minute smooth spheres, arranged in a form suggestive of crystallography (fig. 6.9). Guglielmini discussed at length the stacking and arrangement of the spheres, the angles among their centers, and their spatial configurations, arguing that the empty spaces between the spheres are what explains water's transparency to light. Nor was water the only fluid subjected to such analyses; the microstructure of mercury and air too was discussed in connection with their macroscopic properties. Mercury was believed to consist of oblong particles having a much smaller proportion of empty spaces among them, explaining both its higher density and its opacity. Air was thought to consist of spirals wound around spheres, thus explaining its elasticity and lower density.[44] The analogy between spheres and water was first established for hydrostatics and Torricelli's pierced cistern, and was then extended to the cases of waters flowing down a slope and resistance to motion. Thus the application of the rules about bodies descending along inclined planes was justified at the microscopic level, and the assumptions could be tested in the established case of hydrostatics before extending it to the more complex case of motion. For example, Guglielmini was able to explain the hydrostatic paradox and the existence of lateral pressure by arguing that in a heap of spheres, pressure depends on their height and was exerted also toward the sides.

Although Guglielmini's microanalysis does not explicitly mention Papin, it was nevertheless probably designed as a response to his criticisms.[45] If a fluid is a collection of solid bodies, then clearly the same laws apply to both, at least in principle. In chapter 4 of the work Guglielmini turned to the study of flowing water, arguing that although the same laws apply to solid bodies and fluids, in practice friction acts in different ways in the two cases. This time the profile of speeds is inverted with respect to the *Aquarum fluentium mensura*, since the speeds at the bottom of the river are smaller. Few illustrations could better capture the spirit of Guglielmini's endeavor than his representation of a river as a collection of spheres going down an inclined plane (fig. 6.10). This representation can explain why the water tends to be shallower the further it descends, because the spheres are more spaced.[46] The speed of rivers depends on two principles acting independently, slope of the riverbed and height of the water.

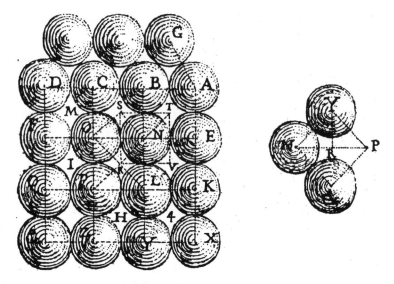

Figure 6.9. Guglielmini's layers of spheres *(Della natura de' fiumi).* The drawing on the left depicts a layer of spheres, whereas the one on the right shows their vertical arrangement. It is easy to see from the second drawing that the lowest sphere is pushed both downward and sideways.

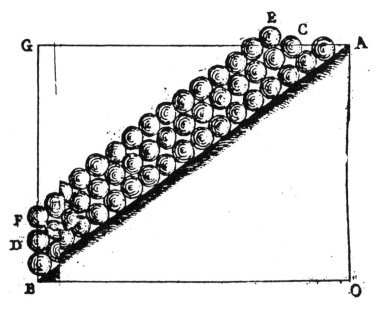

Figure 6.10. Guglielmini's river as a collection of spheres *(Della natura de' fiumi).* Guglielmini's crude representation of a river allowed him to draw some conclusions about water flow from the theory of motion along inclined planes.

6.6 Experiments Combining Pressure and Speed

I conclude this chapter with some reflections inspired by the collaboration between the distinguished Modena physician Bernardino Ramazzini and the lawyer and mathematician Giovanni Battista Boccabadati, who lectured on hydraulics from 1684 to 1689, also at Modena.

The science of waters falls into two traditions, one based on equilibrium and pressure, the other on motion and speed. Both shared a common mathematical language based on the theory of proportions. In the one case the point was to determine the proportionality between height and pressure; in the other it was to determine the proportionality between height and speed. But for both traditions, and in all the examples we have seen, the problem amounted to finding the correct relation between two variables. The theory of proportions, however, was ill suited to account for problems involving more than two variables, such as height, pressure, and speed taken together. What was needed was not a new way of combining proportions but rather an equation that could link variables expressed by means of appropriate parameters. At the turn of the century we begin to see equations replacing proportions in works on the equilibrium and motion of water, by such scholars as Jakob Hermann and Bernardino Zendrini, up to the classic work by Daniel Bernoulli.[47]

Although conceptual issues too were significant, part of the problem was mathematical, due to the correlation between the theory of proportions and the relations allowed by that theory. At the experimental level, however, the situation was different, and Ramazzini and Boccabadati found a way to link the disparate variables by means of an experiment using a container attached to a system of pipes (fig. 6.11). Thus a modified form of the pierced cistern was used to investigate a complex phenomenon. The experiment was spurred not by mathematical or strictly theoretical concerns but rather by the desire to discover why in the area around Modena water gushed with such force from relatively shallow wells. Ramazzini was aided in his field investigations by several unnamed *putearij,* or well-diggers. According to Ramazzini, the source of the water was a huge underground reservoir situated in the nearby hills, a physical situation that could be represented with a water container attached to pipes. The problem was to determine the pressure along an underground conduit filled with running water, allowing the water to spurt forth from the ground. Ramazzini claimed that the problem of determining the variations of pressure along a streamline had not been previously investigated. Although he had some working hypotheses, his experiments were clearly heuristic and provided him with surprising results. The experiments may

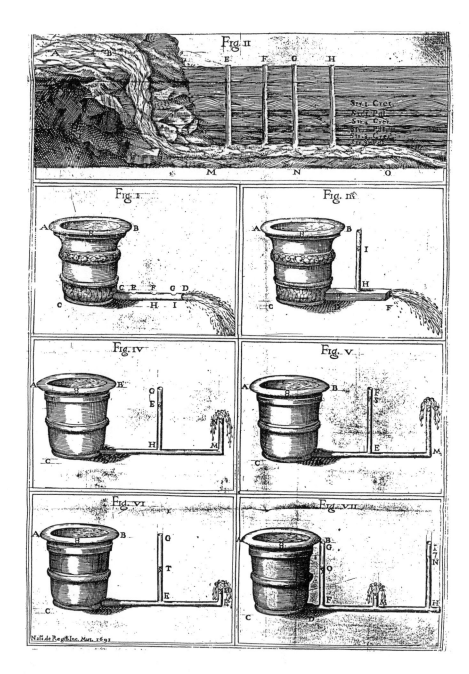

have potentially linked height of water in the container, speed in the horizontal pipe, and elevation in the vertical pipes, but Ramazzini and Boccabadati ended up establishing a simple proportion between the elevations or pressures of water at different points along the horizontal pipe.[48]

Ramazzini's work had little impact in the literature on the science of waters and was not, for example, included in the second edition of the *Raccolta d'autori che trattano del moto dell'acque* in the second half of the eighteenth century. His sources in hydraulics, such as Galileo, Stevin, Boyle, Borelli, and Gugliel-mini, were mentioned in passing rather than used in the research. Parts of his work were translated into English as relevant to debates on the history of the Earth and were reprinted at Padua in 1713 together with a treatise by Francesco Ariosto on oil found around Modena and its medical uses. These editions suggest that Ramazzini's work was seen as less relevant to the science of waters than to the history of the Earth and the local history of the Modena area. Nonetheless, his use of the pierced cistern to account for the wells around Modena offers another example of the versatility of mechanical devices.[49]

Figure 6.11. (opposite) Ramazzini on wells *(De fontium).* Note the analogy between Ramazzini's representation of the Modena wells in figure II at the top and the cistern attached to a system of pipes in the six figures at the bottom. In figure I, Ramazzini attached to a vessel *ABC* an open tube with holes *E, F,* and *G* in the upper part and a thin slit *HI* underneath. Water gushed forth from the orifice in *D,* though not from *E, F,* and *G.* As to the slit *HI,* water started coming out only when the vessel was almost empty. When the orifice in *D* was partially occluded with a finger, however, water spurted out from *E, F, G,* and *HI.* In order to study the behavior of water, Ramazzini introduced a glass pipe *HI* (figure III), which allowed him to see how much the water rose depending on how much the orifice in *F* is obstructed. Following Boccabadati's advice, Ramazzini attached vertical pipes of different lengths to study further the pressure in the pipes. The remaining figures indicate different pressures at different points in the horizontal pipe, establishing that the level of the water in the glass pipe is proportional to the average between the levels in the vessel and in the vertical end-pipe, such as *NM* in figure IV.

CHAPTER SEVEN

Projected, Oscillating, and Orbiting Bodies

> This inflection of a direct motion into a curve by a supervening
> attractive principle I shall endeavour to explicate from some
> experiments with a pendulous body.
>
> *Robert Hooke*

7.1 The Tools of Investigation

This chapter deals with the study of projected, oscillating, and orbiting bodies from the 1660s to the early 1680s. This area constituted the backbone of traditional histories of seventeenth-century mechanics, but now it has been recognized that it is unhelpful to give it exclusive attention or separate it from other fields. Issues such as the conceptualization and mathematical representation of speed, acceleration, and force are common ground among many domains. As in previous chapters, here too we witness in the behavior of objects on the Earth the interplay between cosmological and technological implications. Often the tools used to study celestial motions were objects of common experience, including rotating levers, vertical and conical pendulums, floating bodies oscillating on the surface of a fluid, and projectiles. The trajectory of falling bodies on the Earth had implications for the world-system, for example, and the conical pendulum was used both to construct more accurate clocks and to conceptualize orbital motion.

I am especially interested in exploring how projected and oscillating bodies were used in different fashions in the study of orbital motion: the interplay among these objects is at the center of this chapter. I focus on the problem of relative motion, especially on a moving Earth, the components of curvilinear motion, the actions or forces responsible for them, and the analysis of orbital trajectories. While discussing the problem of fall, scholars—from Riccioli and Borelli in the 1650s and 1660s to Hooke and Newton in 1679–80—debated the issue of the components of curvilinear motion. What was at stake was the primacy of rectilinear over curvilinear motion in the new worldview and the nature and location of an outward, or centrifugal, tendency or force. On the latter issue I identify two investigative traditions, one prevalent on the Continent and the other in England.

I first address the analyses of the motion of the satellites of Jupiter put forward by Fabri and Borelli in the mid-1660s. Whereas Fabri relied on projectile motion, Borelli used a wide set of tools ranging from the lever to the pendulum. Section 7.3 investigates a debate on the trajectory of falling bodies on a moving Earth, focusing on the views of Borelli and degli Angeli. Both examined the case whereby the falling body is projected in the direction of rotation and relied on the conical pendulum. We have seen in chapter 4 that this problem had been debated in the aftermath of Galileo's *Dialogo* and continued to attract scholarly attention well after the time covered here. Section 7.4 examines the debates on the trajectory of projectiles in a resisting medium in Italy, England, and France, from the Cimento Academy to James Gregory and Blondel. Section 7.5 deals with the main work on mechanics before Newton's *Principia*, Huygens's *Horologium oscillatorium*, which put forward a sophisticated new theory of the vertical and the conical pendulum. Section 6 explores the studies of orbital motion in England by Hooke, Wren, Wallis, Newton, and Flamsteed. Initially Wallis, Hooke, and Wren explored the motions of planets and satellites using an analogy with the conical pendulum.

In this chapter mathematics takes on a significant role, in two respects. Some of the changes occurring in mathematics were becoming increasingly relevant to mechanics. In the case of Huygens's *Horologium oscillatorium*, a new curve—the cycloid—entered the scene; but this was not an isolated case, since many areas of study, from centers of oscillation to the motion of projectiles in resisting media, required increasingly sophisticated tools. Moreover, mathematics could be used in different ways with regard to physical causes. For example, gravity could be explored having recourse to its causes, as Huygens did at a debate at the Paris Académie, or purely in mathematical terms, as he did in *Horologium oscillatorium*. The same applies to several scholars, such as Newton up to the mid-1680s, who considered gravity mathematically and at the same time firmly believed in a physical mechanism such as a fluid medium generating it. These different approaches were to emerge as a major theme later in the century.

7.2. The Analyses of Orbital Motion by Fabri and Borelli

It should come as no surprise to find astronomy closely tied to mechanics and the science of motion, since this link was already present in Galileo and the following generation, notably Gassendi. The study of the trajectory of planets, satellites, and comets required a careful conceptualization of the forces at play. In his study of terrestrial motions, Galileo had argued that the trajectory of projectiles is parabolic and results from the composition of two independent motions, one uniformly accelerated downward and the other horizontal rectilinear uniform. We should not underestimate or take for granted Galileo's originality, given that around the middle of the century Riccioli, for example, had problems accepting that different motions could be combined as if they were independent. We have seen that Galileo considered the rectilinear component to be an approximation over a short distance of a circle arc and did not see planetary trajectories as resulting from the combination of a rectilinear uniform motion with one due to a central attraction. Rather, he considered those trajectories to be natural. Conceiving the motion of an unimpeded body to be uniform and rectilinear, as Descartes and possibly Gassendi had done, the curvilinear motion of celestial bodies could be considered in a new way requiring something preventing them from flying away along the tangent.

A different tradition too played an important role in the conceptualization of celestial motions, namely Kepler's celestial physics. With the dissolution of the solid celestial orbs that for many centuries had been thought to carry the planets in circles, Kepler argued that the heavens were fluid, talked of a vortex carrying the planets, and introduced a central force, which he thought was magnetic. His *Astronomia nova* (Heidelberg, 1609) and *Epitome astronomiae Copernicanae* (Linz, 1618−22) stimulated reflections on these matters. From the middle of the century onward, vortex theories became the main type of explanation for celestial motions. Within its framework, however, the vortex often replaced any type of attraction in the attempt to explain all phenomena in terms of contact actions among particles.[1]

I am especially interested in the interplay between astronomy and the science of motion rather than in purely astronomical issues, very much in the Keplerian tradition of combining astronomy and physics. Useful starting points are the near-contemporary works on the satellites of Jupiter by Fabri and Borelli. Although both were hastily drafted in response to external concerns and stimuli, they reveal much about the state of the field in the mid-1660s, and curiously, despite their differences, both reveal in different ways the vitality of the Galilean legacy.

The context for both works was the observation of the shadows of the satellites of Jupiter on the body of the planet by the astronomer Giandomenico Cassini and the instrument maker Giuseppe Campani in 1664. Around 1660 Fabri had put forward the idea that the appearance of Saturn could be accounted for in terms of dark and light satellites oscillating behind the planet. Fabri thought that the satellites of Jupiter moved in the same way because they are never seen in front of

Jupiter, even though they are very bright, and because their shadows on the body of Jupiter had not been observed.[2] Both observations, however, had just become available, and Fabri found himself in an awkward position. At the end of his *Dialogi physici in quibus de motu terrae disputatur* (Lyon, 1665), Fabri added two letters to his Lyon friend Claude Basset outlining his revised views. Fabri believed the orbital trajectory to be composed of two parabolic arcs with a common basis. He ruled out circular or elliptic trajectories on the grounds that the motion of the satellites, as seen from the Earth, is uniform—a claim disputed by most of his contemporaries. Despite its shortcomings, it was a revealing attempt to combine the science of motion with astronomy. Fabri took the circular motion of the satellites around the center of the universe as natural and then attempted to construct a trajectory satisfying his data, whereas we would expect a central symmetry around Jupiter. His efforts resulted in a bizarre transposition in heaven of Galileo's parabolic trajectories, with two unlikely cusps at F and G (fig. 7.1). Although his account in *Dialogi physici* does not appear to have been widely known, few accounts could provide a more direct illustration of the links between projected and orbiting bodies.[3]

Borelli's *Theoricae Mediceorum planetarum* (Florence, 1666) sought to provide a purely theoretical analysis of the motions of the Medicean planets as part of the European debate arising in the wake of the recently observed shadows of the satellites of Jupiter on the body of the planet—a required topic for a Medici mathematician. The dispute on the Medicean planets allowed forays into the solar system while avoiding dealing directly with Copernicanism, a dangerous topic in Italy in those years. Despite this precaution, Borelli established an analogy with planets that left little doubt as to his cosmological beliefs. *Theoricae* was published by the Medicean press and circulated among a select group of intellectuals and patrons. Borelli's work is of interest in several respects, such as his integration of celestial and terrestrial physico-mathematics; his attempt in a Keplerian tradition to conceptualize curvilinear motion as resulting from an imbalance between an outward and an inward tendency; his effort to provide a quantitative analysis of those two components; and his experiments in the tradition of the Cimento. Borelli's method of investigation was heavily dependent on analogies among different phenomena and fields, on the assumption that nature's method of operation is both uniform and simple. He often extended his analogies from physico-mathematics to medicine and anatomy, and even his work on the Medicean planets starts with analogies between the solar system and the animal body.[4] At times one finds so many analogies in Borelli's work that one wonders which one he adhered to, since those presented early on were later modified or cast aside.

It is not difficult to detect the strong Keplerian flavor of Borelli's account, not

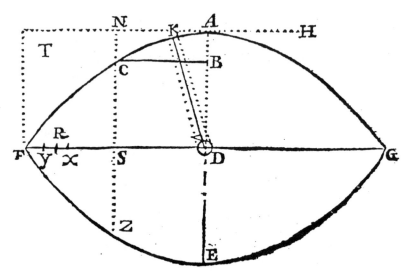

Figure 7.1. Fabri's orbits for Jupiter's satellites *(Dialogi physici)*. The trajectory of the satellites of Jupiter, placed at *D*, is depicted as two parabolas with a common basis *FG*. Fabri believed that Jupiter and its satellites moved along circular arcs so large that straight lines could approximate a small portion of them. The orbital motion of the satellites consisted of a nearly rectilinear motion, *AH*, and a uniformly accelerated one toward the center of Jupiter in *D*. Fabri also believed that Jupiter and its satellites moved with the same absolute speed around the center of the universe, namely the Earth. The key idea was that a satellite beyond Jupiter or, according to Fabri, further away from the Earth, had a smaller angular speed than Jupiter and therefore remained behind the planet, whereas a satellite in front of Jupiter had a larger angular speed and therefore moved ahead. This explanation accounted for the nearly rectilinear component in terms of a difference between absolute and angular speed. Thus a satellite in *A*, for example, remains slightly behind Jupiter and falls with a uniformly accelerated motion toward *D* until it reaches *F*. There it starts moving ahead of Jupiter, and it also moves with a uniformly decelerated motion away from *D* along the parabola *FEG*. While at *F* the satellite has enough speed to proceed further in the direction *DE*, slowing down until *E*, where it starts falling back toward *D*, reaching *G*.

just in his attempt to bring together physics and astronomy but also in his claim that the satellites' trajectories are elliptical and that the sun pushes the planets with a lever, as it were. Borelli conceived orbital motion as if it were taking place on a rotating lever moved by the rotating sun. The inward tendency was due to an appetite of the satellites, or planets, to move toward the central body, whereas the outward tendency was due to their circular motion generated by the light emitted by the central body. Throughout his account Borelli provided examples and experiments with pendulums, magnets, and rotating bowls, thus treating celestial and terrestrial phenomena in a similar fashion. For example, Borelli described a device meant to illustrate the motion of celestial bodies by means of rotating

wooden sticks and magnets (fig. 7.2). Without mentioning his name, Borelli examined Roberval's hypothesis whereby planets float on a fluid whose density increases away from the center, occupied by the sun. The fluid's density depended on
the sun's heat, and the planets' distance from the sun depended on their specific
gravity, following the laws of hydrostatics.[5]

Borelli believed the speed of a body moving in a circular motion, due to the
same motive power, whether from an internal and natural or external and violent
action, to be constant. He tried to support this claim by means of an experiment
with a strong Galilean flavor, allegedly showing that the speed of a pendulum bob
is not affected by a nail on the vertical hindering its oscillations (fig. 7.3). The
choice of the vertical pendulum was unfortunate, because the bob's speed varies at
each point.[6]

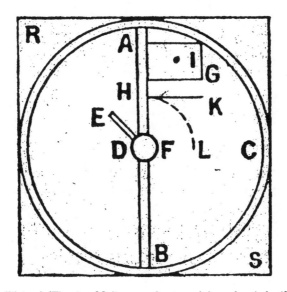

Figure 7.2. Borelli's bowl *(Theoricae Mediceorum planetarum).* A wooden circle *ABC* with a
wooden diameter *AB* is in a container full of water *RS.* A magnet *F* is in the center *D. I* is an iron
ball floating on a small piece of cork *G* capable of moving freely on the water. When one moves the
cork with the ball sufficiently close to the magnet *F*, the ball starts being attracted. If at this point
one rotates the whole wooden frame *ABC*, the diameter pushes the ball and makes it rotate. If the
magnetic attraction is compensated by the tendency *HK* generated by the circular motion to move
away from the center along the tangent, then the ball *I* will be in equilibrium, just like a planet
with respect to the sun or a satellite with respect to a planet. In his typical style, Borelli also provided an alternative account without the magnet and the floating cork. This time the diameter *AB*
is not flat but raised in *A* and *B*, so that a ball placed on it tends to slide toward *D* along a groove cut
in *AB.* By rotating *AB*, so as to have two rotating inclined planes, it is possible to generate a force
opposed to the weight of the ball so that the ball stays in equilibrium. In some respects his approach
is not entirely dissimilar to Huygens's, as we shall see below.

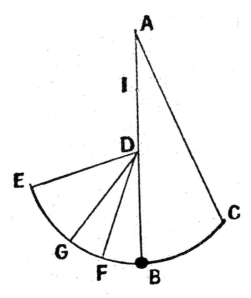

Figure 7.3. Borelli's pendulum *(Theoricae Mediceorum planetarum).* Borelli claimed that the speed of a body moving along the arcs *BC* or *BE* is the same, although the angular distance covered in the same time is greater over *BE* than *BC.* He tried to prove this statement by arguing that the speed of a bob would be the same whether *AB* is free or hindered by a nail in *D.* The idea of using a nail in *D* to shorten the pendulum is clearly borrowed from the third day of the *Discorsi* (fig. 3.16). Borelli claimed that the bob would reach point *G.* Surprisingly, he did not state that *C* and *G* should be at the same height, but rather claimed on the basis of experience that the angles *GD̂B* and *BÂC* are inversely as the square roots of the lengths of the pendulums, $\sqrt{AB}:\sqrt{DB}$. From here Borelli argued that the speeds of the pendulum are equal and also that without external influences, the speed of an orbiting body would be constant and the body would traverse equal distances in equal times at different distances from the center. The analogy between the vertical pendulum and an orbiting body is problematic, however, because an orbiting body at a constant distance from the center has a constant speed, whereas the speed of the pendulum varies at each point.

In the case of orbiting bodies, however, we see that their speeds are not constant but rather increase closer to the central body, whether the sun or a planet. Thus Borelli argued that the light of the rotating sun pushes the planets with a force inversely as the distance, much as Jupiter does with its satellites. Here Borelli seemed unconcerned that Jupiter is not a luminous body like the sun. At this point he had recourse to the impact laws and implicitly criticized Descartes for refusing to accept that a large body cannot be set in motion by a smaller one. The action of the central body is assimilated to a lever, increasing further the barrage of mechanical analogies. Borelli argued that the rotation of the central body, either the sun or Jupiter, moves orbiting bodies with an equal force, opposed by the body's resistance. Since the resistance increases with the distance from the center, the circular component of the speed will be inversely as the distance from the center. As to radial motion, Borelli considered the tendency of orbiting bodies toward the center to be constant. The outward tendency, however, was variable and produced an oscillating motion that, composed with the circular one, generated ellipses. Borelli compared the oscillating motion with that of a cylinder floating vertically in a bucket of water removed from its equilibrium position. Removing accidental perturbations, the oscillations would not stop, a situation much more likely to occur in the celestial aether.[7]

Borelli's aim was to find analogies between planets and satellites, but like several of his contemporaries he failed to mention Kepler's harmonic or third law, originally formulated in *Harmonice mundi* (Linz, 1619).[8] Moreover, again like several of his contemporaries, Borelli did not distinguish between a planet's orbital speed and its circular component. Without this distinction, his claim that the motion of the planet (or satellite) is inversely as its distance from the center is incorrect and implies an inaccurate understanding of Kepler's second or area law.[9] In his treatment of the outward tendency, which Huygens called centrifugal, Borelli seems to have believed that it varied inversely as the distance from the center; therefore the outward tendency was proportional to the orbital speed. Borelli did not carry out an investigation of how the outward tendency varied and simply adopted an intuitively simple relation. In conclusion, Borelli's ingenious work contains valuable insights but betrays the haste with which it was composed. Borelli introduced outward tendencies in the account of orbital motion, but Koyré's claim that this would represent his great innovation seems grossly overstated, given that Descartes' *Principia* had unequivocally had recourse to them in 1644. Borelli's mathematics appears remarkably simple, and nowhere did he *prove* that the orbital motion resulting from the conditions he had stated was indeed an ellipse. He often tried to follow a Galilean style and relied extensively on geometry and proportions, as well as some theorems on conic sections from Apollonius.[10] Unlike Galileo, however, Borelli did not attempt to use indivisibles or infinitesimal procedures.

7.3 Falling Bodies on a Moving Earth

In *Astronomia reformata* (1665) Riccioli reiterated the views he had put forward in *Almagestum novum* (1651) about falling bodies on a moving Earth as part of the debates dating back to Galileo's *bizzarria*. In the mid-1660s Riccioli's claims were challenged by a number of mathematicians who disagreed on the trajectory of falling bodies but agreed on composition of motion. In the 1660s Riccioli's main antagonists were Borelli and the Padua professor degli Angeli. In the new phase of the debate the Cimento Academy, under Leopold's protection, played an important role both publicly through Borelli and privately through Medicean diplomatic channels. A contemporary account was provided by the Scottish mathematician James Gregory, a student of degli Angeli's at Padua.[11] This controversy highlights that Gassendi's and Descartes' notion that the motion of a body left to itself is uniform and rectilinear was rarely challenged but not always applied.

In *Astronomia reformata* Riccioli provided a list of seventy-seven arguments against Copernicanism. The Jesuit had great difficulty accepting that different motions could be combined without one of them overpowering and annulling the effects of the other. A key feature of Riccioli's position was that there were observable differences in the behavior of bodies depending on whether the Earth was in motion or not. Since observations showed the effects he expected on a stationary Earth, he concluded that the Earth does not move. In this extreme form, his views about composition of motion were rather isolated, and Riccioli admitted that he had found few supporters even among colleagues, members of his order such as Honoré Fabri and André Tacquet, and friends such as Giandomenico Cassini.[12] The issue, however, involved not only the composition of motion but also the motions being composed, and this was an area of considerable uncertainty.

Following a preliminary essay sent anonymously to Riccioli in 1666, Borelli put forward his argument in *De vi percussionis* (Bologna, 1667). There he argued that the trajectory of a body dropped from a tower in the hypothetical situation of a rotating Earth would be an irregular curve resulting from a uniform curvilinear motion acquired from the Earth's rotation and a uniformly accelerated one (fig. 7.4).[13] Borelli's position implied that the behavior of falling bodies differed depending on whether the Earth moved or not, thus making potential tests possible.

Stefano degli Angeli entered the controversy against both Riccioli and Borelli. His attacks created considerable embarrassment because degli Angeli was a closet Copernican and a student of Cavalieri's, just as Borelli had been a student of Castelli's. As Koyré rightly put it, degli Angeli "grasped the meaning of the Galilean relativity of motion." According to degli Angeli, however, the circular component of the velocity of the body falling from the tower was not conserved but diminished as the body fell. More precisely, the circular motion is proportional to the radius and thus diminishes exactly in the amount required to make the body fall parallel to the tower, with no deviation. Thus degli Angeli adopted neither the notion that motion is uniform nor that it is rectilinear. His justification, referring to Borelli's figure (fig. 7.4), was that "the moving body is not on the arc *AB*; except in one point at the beginning of the motion, in the progress of which it is always elsewhere. And thus it does not move on this [arc], as it is assumed."[14] This claim may appear not wholly implausible after the first instant, but it is hard to grasp

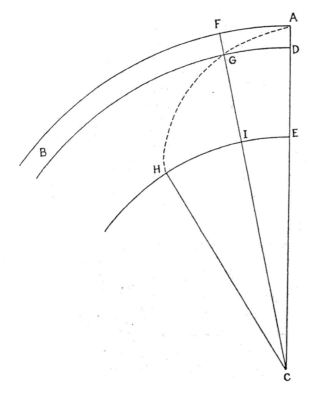

Figure 7.4. Borelli's trajectory *(De vi percussionis).* The figure shows the trajectory of a body falling on a moving Earth. *C* is the center of the Earth, *EIH* its surface, and *EA* the height of the tower from which the body falls. *AGH* is the trajectory of the falling body, composed of a circular motion *AFB* and an accelerated one toward *C*. Borelli argued that since the circular motion in *A* is conserved, the angular displacement along *EIH* is going to be greater than along *AFB*, and the angle *ICH* is greater than the angle *ACF*. In conclusion, the body does not fall at the foot of the tower.

why degli Angeli did not consider that the body in *A* has been moving along the arc *AB* while it was on the tower. Borelli's figure, showing only the trajectory of the falling body starting from *A* and excluding—not without reason—the previous motion approaching *A* from the right, may have led the Padua professor astray.

There was another issue raised by degli Angeli. He pointed out that according to Borelli's views the two different cases of a moving versus a stationary Earth would lead to different trajectories for falling bodies. In the latter case the trajectory would always be parallel to the tower, whereas in the former it would deviate toward the east. Since no eastward deviation had ever been observed, Borelli would have provided Riccioli with the argument he was looking for to prove the immobility of the Earth.[15]

In his reply, Borelli introduced a novel element in the form of experiments designed to show that transverse velocity is conserved, an estimation of the order of magnitude of the eastward deviation, and a correction to his previous understanding of the body's trajectory. In the earliest stages of the controversy, Borelli had tried to use pendulum experiments to convince Riccioli of the independence of different components of motion.[16] In addressing degli Angeli, Borelli tried to use pendulums again, but this time he had recourse to a conical pendulum whose thread is shortened to a quarter of its original length while the pendulum is rotating. Although Borelli's intuition that the pendulum would have a shorter period was sound, he was not coherent in stating that the period would be half the original one. He was also incorrect in claiming that in this way the velocity of the bob did not change. As we saw in the previous section, Borelli had tried to use a similar argument in the nearly contemporary *Theoricae*. Since the shorter pendulum moves along a path whose length is a quarter of the original, following his assumptions the period of the shorter pendulum should also be a quarter. Moreover, as degli Angeli was to object, if the length of the pendulum is shortened, its inclination would change, and the two paths would not be proportional to the lengths of their threads. Borelli presented another experiment in which a ball rotates inside a hollow cone pointed downward and moves faster the closer it gets to the cone's vertex.[17] Huygens was to produce a much more sophisticated analysis along similar lines (see sec. 7.5).

Addressing now the deviation from the perpendicular, Borelli admitted that he was troubled by this phenomenon, and he tried to counter it with a two-pronged approach (fig. 7.5). On the one hand, he estimated the actual deviation following his initial hypothesis, showing that for a fall of 240 feet the deviation was eight-tenths of an inch and therefore, in his opinion, unobservable.[18] The height was covered in precisely four seconds, relying on the odd-number rule and the value of fifteen feet for the first second of fall. Following his assumption, Borelli compared the circular path traversed by the body released from the tower, imagining gravity to be absent, with the path traversed by the bottom of the tower, finding the correct result. This noteworthy estimation of the order of magnitude of the eastward deviation was produced not as a routine procedure but only as a result of Borelli's worry that he was involuntarily providing Riccioli with a proof of the immobility of the Earth. Borelli's concerns produced another important correction. Neglecting the action of gravity, he had initially assumed the motion of the body to be circular, but now he realized that its motion must be rectilinear. This correction came conveniently to hand because it enabled him to reduce further the amount of the eastward deviation.[19] In his estimation, however, he first argued qualitatively that the deviation over short distances would not be noticeable. A fall

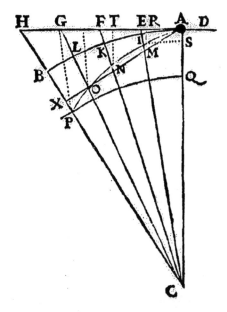

Figure 7.5. Borelli's revised trajectory *(Risposta alle considerazioni)*. In Borelli's revised trajectory
one component of motion is no longer circular but rectilinear, as *AEFGH*. Borelli first argued qualitatively that the line of fall *AMNOP* does not deviate from the tower or the perpendicular, at least
over short distances. This was due partly to the fact that gravity does not act along parallel lines but
has growing components *ER, FT,* etc., pushing the falling body back toward the tower. When it
came to making the calculation, he argued that the deviation would be the difference between *AH*
and the arc *AB*, measuring by how much the rectilinear segment exceeds in length the circular path
described by the top of the tower. This difference, which can be determined from the known dimensions of the Earth and the time of fall, was found to be too small a fraction of an inch to be
measured. The correction due to the introduction of the rectilinear component *AH* was conceptually significant, but quantitatively it was negligible; the eastward deviation in both cases is virtually
identical. Borelli, however, was not right in thinking that the convergence of the lines of gravity
would virtually nullify the eastward deviation (see note 19). Then Borelli proceeded to consider that
on a moving Earth a body falls along a path slightly longer than the height of the tower, and to estimate the additional time required to fall that distance, which turned out to be utterly negligible.
In conclusion, he found the deviation from the foot of the tower to be negligible; therefore no experiment could determine whether the Earth rotates or is still.

over longer distances would produce a noticeable deviation, but at the end the ball
would reach the center of the Earth. This conclusion was standard at the time and
shows that scholars did not see an analogy between falling and orbiting bodies.

Borelli's attempts toward a quantitative estimation of the eastward deviation
were significant, especially if compared with degli Angeli's work. In his reply,
degli Angeli insisted that there was no eastward deviation and invoked air resis-

tance as a possible cause counterbalancing exactly the effect treated by Borelli.[20] The invocation of air resistance appears entirely arbitrary in this context.

References to falling bodies and Galileo's *bizzarria* can be found elsewhere in the literature. For example, Nicholas François Blondel argued in a treatise on ballistics, *L'art de jetter les bombes,* presented to the Académie in the 1670s but for political reasons published in Paris only in 1683, that Galileo's fancy appeared very plausible and led to results indistinguishably close to the odd-number rule. Blondel was a military man as well as a man of learning. He had been involved in several military campaigns and geographical expeditions. He succeeded Gassendi upon the latter's death in 1655 in the mathematics chair at the College Royal, in 1669 he was elected to the Académie des Sciences and two years later to its counterpart for architecture.[21]

7.4 Projectiles and Air Resistance

We have seen that following the publication of the *Discorsi* Galileo's views about parabolic trajectories were far from meeting with universal approval, objections ranging from the form of the law of fall to the role of air resistance. Galileo had despaired that air resistance could ever come under the compass of mathematical treatment, but in the second half of the century mathematicians and experimentalists tackled the problem from different angles. Air resistance is often called a second-order effect, namely one that somewhat alters parabolic trajectories by producing perturbations smaller than the initial impulse or gravity. Air resistance is difficult to measure and calculate, since it does not lead to simple mathematical solutions. Some early attempts simply aimed at assessing how accurate Galileo's claims were, but more refined experiments tried to determine a mathematical relation for air resistance, whether it was proportional to the speed or to the square of the speed, for example. In a celebrated experiment, Boyle showed that light bodies like feathers fall in an evacuated tube much like heavy bodies or dead weights.[22]

The issue was of considerable practical as well as theoretical interest and was addressed by the academies at Florence, London, and Paris. The Saggi of the Cimento Academy reported experiments performed from the Leghorn fortress to test whether the time of fall of iron balls was the same whether they were fired point blank or simply dropped. They argued that this would be rigorously the case neglecting air resistance, but they still proceeded with the test. Time was measured by the oscillations of a half-second pendulum, and the fired shot was found to be in the air half an oscillation longer than the dropped one. A similar trial was performed by the Royal Society in January 1671 without a clear outcome. The Cimento tested other propositions by Galileo. For example, he had claimed that when shooting vertically downward, the effect would be greater from a short distance than from a long one, because air resistance would slow the projectile down. The Cimento allegedly confirmed this result. Moreover, Galileo had claimed "that the power impressed on projectiles is not destroyed by a new direction of motion." The academicians tested the claim, following Galileo's statement in the *Dialogo*, by shooting with cannons and crossbows from a moving wagon pulled by horses along a flat plain. In all cases the balls allegedly fell close to the shooting device. In *De vi percussionis* Borelli defended Galileo's claims about composition of motion.[23]

Beginning in the 1660s the Royal Society was the venue for several experiments and discussions on falling and projected bodies in air and water, Hooke being a chief figure. In 1661, for example, experiments on recoil were carried out in the courtyard of Gresham College. In January 1669 the Society tried to investigate the proportion between air resistance and speed of a body by means of pendular oscillations. The investigators found that there was no direct proportionality, since resistance decreased more rapidly than speed, though it was not clear what the exact proportion was. In January 1671 Hooke performed the experiment of simultaneously dropping and projecting two balls from a window, convincing most of those present that they touched the ground at the same time.[24]

The Scottish mathematician James Gregory, professor of mathematics first at Aberdeen and then Edinburgh, was among the first to attempt a mathematical treatment of motion in a resisting medium, namely air. His anonymous 1672 nine-page essay was published as an appendix to a miniature-size satirical work and dealt with pendular motion and the motion of projectiles. Gregory argued that the trajectory resulted from a uniformly accelerated motion due to gravity and a uniformly retarded one in the direction of projection. By "uniformly retarded" he meant a motion decreasing in such a way that if in the first second the body travels seven units, in the second it travels five, three in the third, and one in the fourth. In the meantime, the vertical component is uniformly accelerated. The resultant curve, according to Gregory, is still a parabola with the axis inclined to the horizontal. Flamsteed defended similar views in 1681.[25] Gregory's account was problematic in some respects, such that resistance occurs only in the direction of projection and motion decreases rather rapidly.

Two years later the dilettante mathematician Robert Anderson published *The Genuine Use and Effects of the Gunne*, a short treatise accompanied by extensive tables by the astronomer Thomas Streete. A key feature of Anderson's treatise was its claim that wind has no effect on shots. On the basis of experiments with steel bows he argued that shots above and below 45 degrees by the same angle had the same range. In a curious twist on Torricelli, Anderson defended his views about projectiles by means of an experiment with water spurting out of two holes in a tube. The holes were equidistant from the middle of the tube, and the experiment was analogous to one involving shots differing by the same angle from 45 degrees. The distance traversed by water jets is proportional to their horizontal speed and time of fall. Speed is proportional to the square root of the height of the water above the hole, whereas the time of fall is proportional to the square root of the height of the water below the hole, therefore both jets would hit the ground at the same place. It is curious that Torricelli used falling bodies and projectiles to account for water jets, whereas Anderson tried to do the opposite.[26]

Anderson was engaged in a pamphlet war against the gunners at the Tower of London as well as against Gregory. The latter was a friend of John Collins, an active promoter of mathematics at the center of an extensive correspondence network and a declared enemy of Anderson. As a result of Anderson's challenge, in September 1674 the Master of the Ordnance Sir Thomas Chicheley, the Surveyor of the Ordnance Sir Jonas Moore, Lord Brouncker, Hooke, and others including gunners went repeatedly to Blackheath to test parabolic trajectories, but it is not clear what they found. In 1675 Hooke performed another experiment trying to show the effect of air on resistance to motion.[27] In 1677 the Astronomer Royal John Flamsteed communicated in a letter to Collins the results of tests conducted with a steel crossbow. He found that ranges at elevations below 45 degrees were longer than those surpassing 45 degrees by the same amount, and the longest range was at an angle of about 42 degrees. In 1678 his patron Jonas Moore confirmed to the Royal Society that in his experience with mortars the longest range was at an elevation somewhat below 45 degrees.[28]

In the meantime Collins orchestrated a campaign in support of Gregory against Anderson, trying to elicit opinions from Newton at Cambridge and Wallis at Oxford. In 1674 he sent Wallis copies of the works by Gregory and Anderson, but Wallis doubted that trajectories were Galilean parabolas, because of air resistance, and also that they were inclined parabolas *à la* Gregory. In the same year Collins sent Newton a copy of Anderson's work. His reply constitutes a rare early statement on motion outside the domain of his private manuscripts. Newton argued that the horizontal component of motion for point-blank shots is in some decreasing geometric progression, but he also asked for his name to remain concealed because he did not wish to become involved in those matters.[29]

We saw in section 6.4 that the Paris Académie was heavily engaged in experiments trying to determine the power of moving fluids and therefore the resistance experienced by bodies moving through them. In October 1668 Huygens was able to determine the curve described by a body moving in a medium offering a resistance proportional to its speed. Beautiful as his solution was, it must have been disappointing for him to find out just a few months later that resistance was more likely to be proportional to the square of the speed. Huygens's later attempts to solve the problem with the quadratic resistance did not meet with success. Other academicians may have been unaware of Huygens's theoretical work.[30] In *L'art de jetter les bombes* Blondel claimed the effects of air resistance to be negligible and did not attempt to calculate them. His work provided a survey of ballistics prior to Galileo, tables of shooting, accounts of the works by Galileo, Torricelli, and Mersenne, and replies to the objections formulated against the theory of parabolic projection. His work seems to have had little impact on gunners because of its complexity, but it was also quite inadequate on theoretical grounds. For example, he still believed horizontal lines of projection to be circle arcs that can be taken as straight only as an approximation. By comparison, in a short treatise of 1677 the young Leibniz criticized Galileo and Kepler for accepting that curvilinear motion is natural. Blondel's work showed a considerable degree of cooperation with fellow academicians, and he proudly claimed that some of the experiments had been performed in the presence of the Dauphin, Blondel being his mathematics tutor. In another twist on Torricelli, Blondel too tried to use mercury jets to study the trajectory of projectiles.[31]

Leibniz

7.5 Huygens's Pendulum

By the time Fabri and Borelli were publishing their works on orbiting bodies and Borelli was disputing with degli Angeli and Riccioli about falling bodies, Huygens had already produced results far outstripping those of his contemporaries. Huygens had been first led to his research by a question posed by Mersenne, namely what distance a body falls in its first second. Huygens performed careful quantitative experiments, but as in other areas he also reframed the problem in an original and fruitful fashion. He combined Mersenne's interest in numerical values with Galileo's concern for a more theoretical approach to precision measurement. Huygens probably employed two methods, a clock regulated by a conical pendulum and a direct measurement. The former relied on the relation between period and the height of the cone described by the rotating cord. By keeping the cord at a constant angle of 45 degrees, the cone's height was in a fixed relation to the length of the cord. Since that length is as the square of the period, or T^2, the length of the seconds pendulum is to the length of an arbitrary pendulum as $1/T^2$. But $1/T^2$ is as $n^2/60^2$, or the square of the number of oscillations in 3,600 seconds or one hour; therefore, by counting the number of oscillations and the length of his pendulum, Huygens could determine the length of the seconds pendulum. As to the latter, Huygens connected a ball and a pendulum bob by a string making sure that when the string was cut they would be released at the same time. He then adjusted the height of fall so that the two objects would be heard hitting a sonorous target together. He found a value of just over fifteen feet, in close agreement with that determined by Riccioli in *Almagestum novum* and considerably greater than Mersenne's. In his work on the conical pendulum Huygens had to determine the cord's tension by taking into account gravity and centrifugal force, acting at different angles. He did so by imagining a contrivance of inclined planes and pulleys coming straight from the mechanical tradition and especially Stevin.[32]

Starting with Galileo, the pendulum was taking a prominent place in the study of motion and mechanics, both as a time-measuring device and as a tool for studying motion, force, gravity, and collision. Mersenne and others firmly denied on an experimental basis exact isochronism, as claimed by Galileo. In 1658 Huygens published a small treatise, *Horologium*, in which he used the pendulum to regulate the motion of a clock, which in turn sustained the oscillations and counted them. At one point Huygens thought of constraining oscillations between two metal strips or cheeks whose shape was found empirically by trial and error in order to make oscillations independent on their amplitude. Thus Huygens had invented a practical mechanical device with no theoretical dimension.[33]

At the end of 1659 Huygens's research began assuming greater theoretical significance. He was able to determine the curve along which oscillations are rigorously isochronous, namely the cycloid or the curve described by a point on a circle rolling without sliding on a plane, and also the shape of the constraining cheeks, which was also cycloidal. These were unpredictable results found purely via geometry. The cycloid was a new curve unknown to the Greeks, and its properties were being investigated by Huygens's contemporaries, such as Torricelli, Wren, Wallis, Pascal, and Roberval, purely as a mathematical exercise and without any notion that the results might have practical applications or a role in mechanics. From the manuscript record, it appears that Huygens did not deliberately set out to find the isochronous curve but rather was trying to solve problems to do with gravity and falling bodies.[34]

Huygens published these results in a classic of seventeenth-century mechanics, *Horologium oscillatorium* (Paris, 1673). This elegant volume in folio was a masterful combination of sophisticated mathematics and mechanics mixed with a range of practical applications culminating with a new clock aimed at resolving the vexing problem of longitude. Huygens had attained his chief results between 1659 and 1664 and had communicated many of them to his correspondents, who had tried to prove them in their own fashion. For example, Pardies published relevant results as an appendix to his *Statique*, and Lord Brouncker published his own in the *Philosophical Transactions*. Moreover, several clocks were built and tried at sea, with mixed results. Therefore his findings were circulating among the cognoscenti and did not have the impact of an entirely new work. Partly for this reason, partly for its mathematical difficulty, and partly also for a certain lack of direct cosmological implications, his masterpiece did not have a broad impact comparable to that of Galileo's *Dialogo* or Newton's *Principia*.[35]

Huygens's work is a highly polished product and cannot be taken at face value by the historian seeking to reconstruct his private road to discovery. His manuscripts reveal a remarkable skill and creativity in joining higher mathematics and mechanics. At times Huygens superimposed on the same diagram the physical space in which bodies move and the abstract spaces of mechanical and mathematical relations.[36] *Horologium oscillatorium* presents a complex stratification reflecting the various concerns that occupied Huygens over several years. Mersenne's original question had led him to the metrological problem of establishing a universal unit of length. The book had both an experimental and a theoretical dimension, involving the theory of cycloidal motion and the determination of centers of oscillations. These problems produced mathematical spinoffs far beyond what was needed for mechanics. In addition, there were the issues of having an actual clock constructed by reliable technicians and tested at sea to determine longitude.[37]

Horologium oscillatorium with its dedication to Louis XIV appeared at a time when French armies were trying to invade the Low Countries. It is not clear why the work appeared when it did, since Huygens had completed the theoretical research almost a decade earlier. (One could speculate that Huygens was prompted to publish his work when he did by Cassini's recent discoveries, in 1671 and 1672, of two more satellites of Saturn. Huygens, who had identified the ring and discovered the first satellite in the late 1650s, may have felt a need to reestablish his position at the Académie.) Huygens, however, wanted to have viable clocks and to solve the problem of longitude, not just to produce elegant theories, and by the time his work appeared, his attempts to provide evidence of successful sea-trials were far from complete or wholly satisfactory. In order to have his clocks constructed and then tested at sea, Huygens had to rely on a number of technicians and collaborators, such as Severyn Oosterwijk at The Hague and Isaac Thuret at Paris, though his relationships with them were not always easy. He tended to consider his own theoretical breakthrough as the crucial achievement while downplaying the ultimately important contributions made by instrument makers. Glory and fame were not the only issues at stake; Huygens was actively seeking patents in France, Britain, and the Low Countries. Those involved in the trials included Captain Robert Holmes, who tested Huygens's clocks on a trip to Lisbon, and Jean Richer, who took them to Cayenne in South America. Richer noted that the pendular oscillations were slower than expected, an effect later attributed to the rotation and shape of the Earth.[38]

Part 1 provides a description of and apologia for the instrument involving pinions and toothed wheels as well as reports of successful sea trials. As for those trials that had been unsuccessful, Huygens attributed the blame not to the clocks but to those in charge of them. He emphasized that oscillations are not isochronous unless they are constrained by cycloidal cheeks and provided easy rules for drawing cycloids (fig. 7.6).[39]

In part II Huygens set his proof of the isochronism of the cycloid within the tradition of bodies falling along different paths, rather than within that of periodic motion such as oscillations or vibrations of a string or spring. Huygens studied the second edition of *Discorsi* of 1656 and was dissatisfied with Galileo's proof that the speeds acquired by bodies falling along inclined planes of equal heights and different inclinations are equal. His attempt to improve upon Galileo's proofs

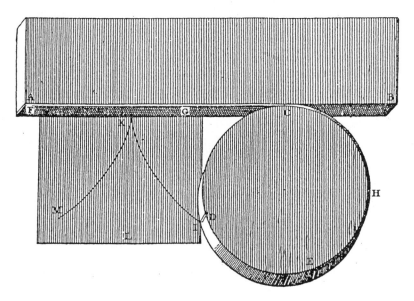

Figure 7.6. Huygens's rule for drawing a cycloid *(Horologium oscillatorium).* This largely self-explanatory illustration shows how to draw a cycloid. The diameter of the cylinder *CDEH* is equal to one-half the length of the pendulum. Huygens later provided yet another method for drawing the curve based on joining a series of points found by simple geometric means. His practical rules recall Galileo's recipes for drawing parabolas at the end of the second day of the *Discorsi,* where he introduced parabolic beams (see sec. 3.5 and fig. 3.15 above) and the hanging chain. In both cases the text describes easy ways of drawing relatively complex curves.

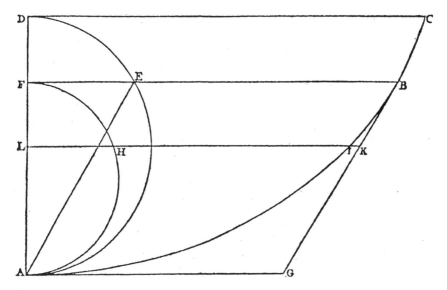

Figure 7.7. Huygens's cycloidal motion *(Horologium oscillatorium)*. This diagram accompanies the crucial proposition II, 25. *ABC* is the arc of a cycloid, *AD* its axis, and *AED* its generating circle. Huygens proved that the time required by a body released from an arbitrary point on the cycloid, such as *B*, to reach *A* is constant. In the previous proposition he had proved that the time of fall through the cycloidal arc *BA* is to the time of naturally accelerated fall through the tangent *BG* as the semi-circumference *FHA* is to *FA*, or as $\pi/2$. The line *BF* parallel to *AG* meets the semi-circumference *DEA* in *E*. *EA* is equal and parallel to the tangent to the cycloid in *B*, namely *BG*. It was this known property of the tangent of a cycloid that initially led Huygens to its identification as the isochronous curve. At this point Huygens referred to a theorem in the *Discorsi* in which Galileo had proved that "if, from the highest or lowest point of a vertical circle, any inclined planes whatever are drawn to its circumference, the times of descent through these will be equal." Thus Huygens could conclude that the times of descent through any cycloidal arc *BA* are the same and are to the time of fall through the diameter *DA* as the semi-circumference of a circle is to its diameter, or as $\pi/2$.

about falling bodies hinges on proposition 4: "If a heavy body begins to move upwards with the same velocity acquired at the end of a descent, then in equal parts of time it will cross the same distances upwards as it did downwards, and it will rise to the same height from which it descended. Also in equal parts of time it will lose equal amounts of velocity." From here Huygens sought to prove Galileo's axiom. After several preliminaries, in proposition 25 Huygens proved the isochronism of the cycloid (fig. 7.7). He was able to handle this curve with more sophisticated tools than cardboard, scissors, and a balance, as Galileo had done earlier in the century.[40]

Part III is devoted to the mathematical theory of the curves described by un-

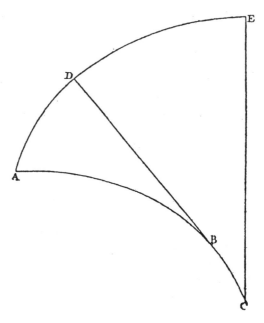

Figure 7.8. Huygens's evolution of curves *(Horologium oscillatorium)*. In this figure from part III, *ABC* is the starting curve, or evolute. If we imagine unrolling a taut string initially coinciding with *ABC*, we will describe another curve *ADE* called the involute, or evolvent. The tangent to the evolute meets the evolvent at right angles.

rolling a taut string around another curve called the evolute (fig. 7.8). The curve described by unrolling the string was called by Huygens "described by evolution" and has since been called involute or evolvent; it is important for the problem of rectification of curves. Huygens investigated this curve far beyond the needs of horology. He attained the chief results relevant to pendulum clocks in two propositions. In proposition 6 he proved that the evolution of a semicycloid, starting at the apex, generates another semicycloid equal to its evolute. It was this proposition that allowed him to establish the shape of the clock's cheeks. Together with proposition II, 25, it gave Huygens some crucial building blocks for his clock. In proposition III, 8, he found that the line whose evolution describes a parabola is a paraboloid known as a semicubic parabola. This result came to fruition in his study of the conical pendulum in part V, since the parabola was the isochronous curve for the conical pendulum.[41]

In part IV Huygens tackled the problem of centers of oscillation for the compound pendulum, which can be seen as a chapter in the study of rigid bodies (figs. 7.9 and 7.10). Just a few years before *Horologium oscillatorium*, the third volume of Descartes' letters (Paris, 1667) made public the contributions by Descartes and

Roberval to the problem of the center of oscillation, the problem Huygens had tackled unsuccessfully in the 1640s.[42] Huygens's new approach was based on two hypotheses or axioms: that the center of gravity of a system of weights moved by their own gravity cannot rise higher than its initial position, and that the center of gravity of a compound pendulum crosses equal arcs in descending and ascending. Huygens argued that his first hypothesis is valid also for liquid bodies, adding that by its means all Archimedes' results as well as other theorems of mechanics can be demonstrated.[43] As we shall see in the next chapter, while investigating collision Huygens relied on the pendulum, which enabled him to transform horizontal speeds into measurable heights both as an experimental device and as a conceptual tool. In a twist that is so typical of seventeenth-century mechanics, Huygens relied on collision in the investigation of the compound pendulum, imagining a rigid rod consisting of infinitely many connected identical bobs hitting a system of infinitely many unconnected identical bobs. While the rod comes to rest, the

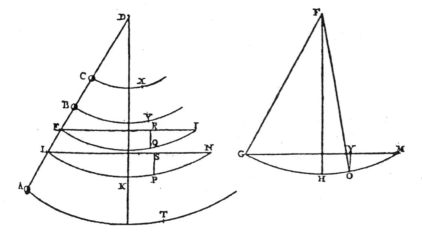

Figure 7.9. Huygens's compound pendulum: the formal presentation *(Horologium oscillatorium).* In proposition IV, 5, Huygens found a simple pendulum oscillating with the same period as a compound one composed of any number of weights w_i at the distance r_i from the axis of oscillation. He proved that the length of the simple pendulum is equal to the sum of all the $w_i r_i^2$ divided by WD, where W is the sum of all the w_i and D is the distance of the center of gravity of the compound pendulum from the same axis. The sum of all the wr^2, or in modern notation $\Sigma_i w_i r_i^2$, was later called moment of inertia. Huygens proved this proposition *per absurdum* with a compound pendulum with bodies A, B, and C, oscillating around D. E is the center of gravity and L the center of oscillation, marking the length of the simple pendulum oscillating with the same period, namely G suspended in F. Huygens first showed that if the speed of L at P is greater than that of G at O, then the compound pendulum would reach a height greater than that from where it was released. The opposite would be true if the speed of L at P is less than that of G at O.

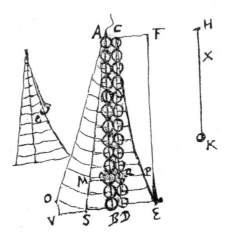

Figure 7.10. Huygens's compound pendulum: the manuscript evidence (*De centro oscillationis,* *HOC,* 16:421). The formal presentation in print differs from Huygens's private path. In one in-stance, for example, he considered a rod or compound pendulum AB consisting of an infinite num-ber of identical simple pendulums joined together. Starting from AO, on reaching the lowest posi-tion AB the compound pendulum hits a series of infinitely many identical simple pendulums CD. Following the impact rules, the rod comes to rest, and all the simple pendulums move upward with the speed they have acquired. Each unconnected bob CD can reach a height proportional to the square of the speed acquired, hence the parabola CPE on the right, where each horizontal segment is proportional to the square of the speed. Triangle ASB, where $SB = OV$, represents the initial heights of the infinitely many connected bobs, which are proportional to the distance from the point of suspension A. Huygens imposed the condition that the center of gravity of the system of the simple pendulums will reach the same height as the center of gravity of the rod in its initial position AO. Since the masses are identical throughout, they can be represented by a line equal to the diameter of one of the bobs. Initially Huygens based his calculations on the case in which the rod consists of only two bobs and then generalized the result. Since the height of each component of the rod is proportional to its distance from A, the height of the rod's center of gravity is as the area of triangle ASB. On the right, the height of the center of gravity of the system of unconnected rods can be found from the height each of them can reach, and is as the area of the parabola CPE. From the area of the parabola, Huygens found the length x of the simple pendulum with a period equal to AB, $x = \frac{2}{3}AB$.

unconnected bobs move upward, and the condition that their center of gravity reaches the same height from which the rod was released determines the center of oscillation of the compound pendulum. The determination of the center of os-cillation was a major achievement of seventeenth-century mechanics, but later in the century Huygens's result was still a matter of debate and controversy.[44]

In part IV, as in part III, Huygens again went far beyond the requirements for the construction of clocks and determined centers of oscillation for a wide range of solids with no horological applications. In addition to the theoretical interest of this subject, Huygens was motivated to work on it by the need to study the effects of adding a weight to a pendulum in order to regulate its period and to determine the center of oscillation of a sphere suspended from a weightless thread. Impressive as his results were when taken individually, Huygens ran into some difficulties when he had to put them together, much like a clockmaker who had crafted superb gears that did not quite fit together. This unease can be seen in proposition 24, claiming that "it is not possible to determine the center of oscillation for pendulums suspended between cycloids." Since the pendulum continually changes its length, its center of oscillation changes too and does not move along a cycloid, hence the oscillations of a compound pendulum constrained by cycloidal cheeks are no longer isochronous. Of course, air resistance too affects motion. Overall, however, Huygens had much improved the precision of clocks by providing a beautiful mathematical theory and a solution to the problem of centers of oscillation.[45]

In the last two proposition of part IV Huygens attempted to establish a universal measure of length based on the length of the seconds pendulum and the distance fallen by a heavy body in one second. The first problem had already been raised by the Royal Society in 1661 following Huygens's discovery of cycloidal cheeks. Huygens defined a new foot that is exactly one-third of the length of a seconds pendulum. His result relied on the determination of the center of oscillation for a sphere suspended from a weightless thread. About the same time *Horologium oscillatorium* appeared, data became available about the length of the seconds pendulum being shorter closer to the equator. In fact, it soon became generally accepted—by Huygens among the others—that the length of the seconds pendulum depends on the latitude and is by no means a universal measure.[46] The second problem presented great difficulties if tackled directly, because of the speed with which bodies fall. Huygens relied on proposition II, 25, whereby the period of an oscillation is related to the time of fall through the axis of the cycloid via π. The length of the seconds pendulum was determined partly using theory, partly through experiments. In this way Huygens found that the distance fallen by a body in one second was very nearly fifteen Parisian feet and one inch. His approach presents differences from modern methods because Huygens did not determine the parameter we are most familiar with, *g* or the gravity constant, which does not figure in a mathematics based on proportions. As in his preliminary attempts, discussed earlier in this section, he confirmed this result with a direct measurement involving an ingenious new device to synchronize the descent of a pendulum bob and the fall of a lead weight and to mechanize the measure of the distance fallen:

A half oscillation of a pendulum suspended from a wall or an erect board indicates the lapse of time used by a falling body. In order that the sphere of this pendulum be released at the same moment as the lead weight which is going to fall, each is tied to, and held by, a thin string, which is cut by a flame. But before this, another thin cord has been attached to the lead weight which is to fall. Its length is such that when all of it is stretched out by the falling weight, the pendulum has not yet hit the wall. The other end of this thin cord is attached to a prepared scale made on paper or on a thin membrane. This scale is attached to the wall or board in such a way that it is easy for it to follow the cord which pulls it, and to descend in a straight line along the length of the cord, passing the place where the sphere of the pendulum will hit the wall. When the whole length of the cord has been used up, the upper part of the scale will be pulled downwards by the falling lead weight before the pendulum hits the board. The size of this part is indicated when the sphere, which has been lightly covered in soot, makes its mark on the scale which is sliding by. If the length of the cord is added to this, then one has an exact definition of the space of the fall.

Although the details may not all be easy to visualize, it is clear that Huygens describes here an improved method over his previous attempts for obtaining a direct measurement without having recourse to "the eye or the ear." His sooty bob goes together with Galileo's sweaty bronze spheres and Newton's firmly compressed balls of wool that were mentioned in the introduction as indicators of their engagement with the material world of objects. Huygens's interest in precision and acumen are revealed once again by his effort to rely on two independent methods in remarkable agreement with each other. Whereas the former can be called, following Koyré, "theory incarnate," the other is squarely experimental. In privileging the former, Koyré offered a partial account, whereas the notion of "thinking with objects" presents a richer picture by emphasizing the interplay between theory and experiment, as well as between pendulums and falling bodies.[47]

In the concluding part V, Huygens presented a new type of clock based on the conical pendulum and stated without proof a series of theorems on centrifugal force, a term he coined. Clocks based on the conical pendulum present interesting features, for example the seconds pointer moves continuously and the clock makes no noise. Like the vertical oscillations of a pendulum, the rotations of a conical pendulum are not isochronous unless the bob moves on an appropriate curve, which in this case is a parabola (figs. 7.11 and 7.12). This can be achieved by adjusting the string of the conical pendulum on an appropriate support serving the same purpose as cycloidal cheeks. Huygens found by means of his theory of evolutes that the shape of the support must be a paraboloid, as mentioned above. The

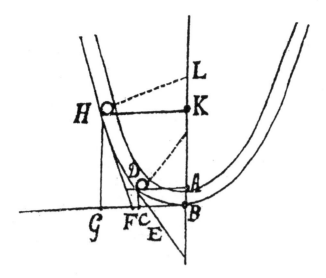

Figure 7.11. The parabola is the isochronous curve *(Horologium oscillatorium)*. If a body rotates parallel to the horizon in a container with a parabolic cross-section (the container is obtained by rotating a parabola around its axis), the rotation period is the same regardless of the body's height in the parabola. In a parabola *HB* the horizontal force necessary to keep a body in *H* from falling is to the body's weight as *HG* is to *GF*, where *HF* is the tangent at *H* and *HG* is perpendicular to *BG*. This result is obtained from the elements of mechanics applied to the inclined plane *HFG*. But *HG/GF* is the tangent to a parabola and is therefore as the abscissa, namely as *HK*. From the first theorem on centrifugal force, if a body moves along unequal circumferences in equal times, the centrifugal forces are as the respective radii, which is the condition for isochronism.

circular motion of the pendulum enabled Huygens to study centrifugal force in connection with the action of gravity.[48]

Huygens's theorems on centrifugal force must have appeared somewhat cryptic to most contemporary readers, since the full treatise *De vi centrifuga* was published only in 1703, though it had been composed in 1659.[49] Initially Huygens investigated centrifugal force in different ways from those he later used in his cosmology. There centrifugal force appeared as a key notion and as the cause of gravity, since gravity was simply the difference between the centrifugal force of a body and that of an equal volume of fluid. Thus centrifugal force was associated with motion, because every time a body fell, for example, an equal volume of fluid was extruded. In *De vi centrifuga* Huygens conceived centrifugal force in terms of the tension experienced by an observer on the rim of a rotating wheel holding a string with a body attached to it; therefore centrifugal force produced no motion, but only tension in the string (fig. 7.13). In all cases, however, centrifugal force was a

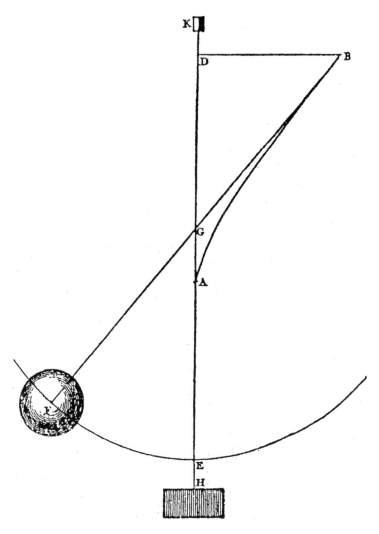

Figure 7.12. Huygens's isochronous conical pendulum *(Horologium oscillatorium).* The pendulum consists of a thread *BGF,* with a weight in *F,* rotating parallel to the horizon around the axis *DH.* The thread rests on the curved plate *BA,* and the faster it turns, the higher it rises. The plate *BA* is shaped like a paraboloid whose evolution describes the parabola *EF.* This condition makes the pendulum's rotation isochronous, much like cycloidal motion for the vertical pendulum.

real force due to the rotation of the body. Huygens was able to determine its measure by means of infinitesimal geometry. His conclusion was that centrifugal forces vary directly as the squares of the speeds of revolution in circumferences and inversely as their radii. By the time his theorems appeared in print in 1673, only one other mathematician had been able to find equivalent results. A series of

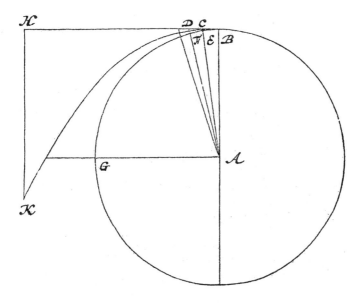

Figure 7.13. Huygens's quantification of centrifugal force *(De vi centrifuga)*. In the unpublished *De vi centrifuga* Huygens attempted a quantification of centrifugal force by superimposing a parabola on a circle. If a body moving along the circle *BEF* is released in *B*, it would move along the tangent *BCDH*, unless its course is deflected by gravity or other bodies. The reference to gravity provides a justification for using a parabola. Over very short distances around *B*, the parabola and the circle do not differ appreciably. Huygens's diagram looks like a modification of Galileo's study of extrusion as presented in day 2 of the *Dialogo* (fig. 3.11), where Galileo was also trying to compare gravity and the outward tendency due to the Earth's rotation. The geometric measure of gravity and centrifugal force is given by the tiny segments *EC* and *FD*, namely the distances between the tangent and the circle or the parabola. Notice also the conceptual and visual similarity with Descartes' analysis of circular motion and especially his image of the rotating sling (fig. 5.3). *EC* and *FD* ought to be arcs of the curve described by the evolution of the circumference, but over very short distances they can be taken to be straight. Much like gravity, centrifugal force produces acceleration. On the basis of the theory of proportions and elementary geometry, Huygens was able to prove that centrifugal force is as BE^2 or the square of velocity and inversely as the radius *AB*. If *C* is very close to *B*, CB^2 is [*Why?*] the mean proportional between *CE* and twice *AB*.

manuscripts by Newton dating from the 1660s show that he had reached similar conclusions by different means. How centrifugal forces should be calculated for motion along curves different from a circle, however, was not immediately clear.[50]

    Huygens's and Borelli's analyses shared common traits, but they also differed in significant respects. One can only wonder whether Borelli truly appreciated Huygens's achievements when he read his dedication copy of *Horologium oscillatorium*. Both conceived orbital motion as the result of opposite tendencies along the radius. Borelli believed that an orbiting body had an appetite or tendency to move toward the center and an outward tendency due to its circular motion. By contrast, Huygens rejected the tendency toward the center and tried to explain orbital motion in terms of the motion of a vortex, much as Descartes had done, as a result of the centrifugal force of a rotating fluid. Thus gravity was the difference between the centrifugal force of the vortex and that of the body, because if the vortex had a greater tendency to escape, it would push the body toward the center. The building blocks of their analysis of orbital motion differed.[51]

A notable feature of Huygens's work is that despite his strong commitment to a mechanical cause for gravity, his treatment was purely mathematical and did not include physical causes. Yet Huygens's views on the matter were very strong indeed, and in 1669 he participated in a debate at the Paris Académie des Sciences on the cause of gravity, together with Mariotte, Roberval, and other academicians. He also provided experimental evidence in support of his views by showing that solid bodies in a vessel of rotating fluid are pushed toward the middle. One problem with this interpretation already encountered by Descartes, however, was that gravity is directed toward the center of the Earth, not its axis. Thus Huygens had to hypothesize an implausible vortex whose particles rotate in all directions around the Earth. He published his views on the issue in 1690 as a response to Newton.[52]

7.6 English Approaches to Orbital Motion

In the meantime, from about the mid-1660s the English mathematicians Wren, Wallis, and Hooke had been examining curvilinear motion from a different perspective. They attempted to establish an analogy between orbital motion and the motion of the bob of a conical pendulum. In this they probably found inspiration in the papers by the astronomer Jeremiah Horrocks, which were being edited at that time and had been perused by Wren and Wallis. At a meeting of the Royal Society in May 1666 Hooke argued that for a planet to move along a circular or elliptical orbit, something retaining it from flying along the tangent in a straight line was required. Probably relying on a previous claim by Wren, he drew an analogy with the pendulum, cautioning, however, that the analogy was imperfect because the attraction of the sun was likely to decrease with distance, whereas in the pendulum the retaining force increases. Indeed, in the pendulum the endeavor to descend is proportional to its distance from the axis. He proceeded to prove his statement with the help of the inclined plane. Hooke, however, stated that circular or orbital motion resulted from an "endeavour by a direct motion by the tangent, and of another endeavour tending to the center," due to an "attractive principle." At times Hooke had talked also of a continuous bending of a body toward a center, or an "inflection," a term derived from optics. Hooke's paper followed one by Wallis, in which he had tried to develop Galileo's theory of tides presented in the last day of the *Dialogo*. Wallis accepted the basic structure of Galileo's reasoning but argued that rather than considering the Earth alone, it was necessary to take into account the center of gravity of the Earth-moon system. This would add an additional motion to the Earth besides the daily and annual rotations, namely a further rotation about the common center of gravity, thus refining Galileo's explanation. In response to the objection that the Earth and moon are not physically connected, Wallis argued that it may not be clear how they were connected, but they certainly were so. To illustrate the analogy further, Hooke attached a smaller pendulum to the bob in an attempt to illustrate the motion of the moon. This example follows a pattern we have already encountered in moving from a system with a material link such as the pendulum or a simple machine, as in Torricelli's principle, to a system without material constraints.[53]

English mathematicians further discussed the motion of comets, which was conceived as a perturbed rectilinear motion, the perturbations being due to a fluid or vortex in the interplanetary spaces, and the attractive power of celestial bodies, such as the sun and the Earth. Their discussion was stimulated by the appearance of two comets in 1664–65. A prominent view about their trajectories argued that in a first approximation they were straight lines, which were subsequently bent by the attraction of other celestial bodies. In 1677 Hooke published a Cutlerian lecture, *Cometa*, part of a series endowed at the Royal Society by Sir John Cutler for Hooke. In this lecture he put together a not entirely consistent series of his views on comets, many dating from the mid-1660s.[54]

In another of his Cutlerian lectures, *An Attempt to Prove the Motion of the Earth* (London, 1674), Hooke put forward three powerful and suggestive suppositions about celestial motions. According to the first, "All Coelestial Bodies whatsoever, have an attraction or gravitating power toward their own Centers, whereby they attract not only their own parts, and keep them from flying from them, as we may observe the Earth to do, but that they also attract all the other Coelestial bodies that are within the sphere of their activity." Clearly Hooke did not consider attraction to be universal. The second supposition states that once a body is set in motion, it will continue its motion in a straight line until deflected by some external power so as to describe a circle, an ellipse, or a more complex curve. The third argues:

> These attractive powers are so much more powerful in operating, by how much the nearer the body wrought upon is to their own Centers. Now what these several degrees are I have not yet experimentally verified; but it is a notion, which if fully

prosecuted as it ought to be, will mightily assist the Astronomer to reduce all the Coelestial Motions to a certain rule, which I doubt will ever be done without it. He that understands the nature of the Circular Pendulum and Circular Motion, will easily understand the whole ground of this Principle.

A few years later Hooke was able to inform Newton that gravity or the attractive power of celestial bodies is inversely proportional to the square of the distance from the center, a view apparently held also by Wren.[55] Although these three suppositions sound suggestive to the modern reader, their status in the 1670s was that of reasonable and plausible guesses, not proofs. Let us examine them in turn.

The first supposition is akin to a principle of universal "gravities." Although Hooke talked about all celestial bodies, there is no indication that the intensity and dependence on distance of all the "gravities" would follow the same rule. The second supposition presents curvilinear motion as the result of the combination of a rectilinear uniform motion and a deflecting power. The conical pendulum and comets would have been notable sources of inspiration for his claims. Thus Hooke's approach to curvilinear motion and English approaches in general differed from Continental ones with respect to the outward tendency or centrifugal force, which was considered by Descartes, Borelli, and Huygens as part of their study of curvilinear motion. The third supposition deals with the attractive powers with regard not to their physical causes but to the mathematical law of their decrease from the center. Hooke's reference to an experimental verification suggests that he was hoping to find on the Earth how gravity decreases with height and then apply this result to other cases. Indeed, we know that he performed experiments in Westminster Abbey and the old cathedral of St. Paul's, as well as "some considerably deep wells," all without appreciable results.[56] Thus regardless of the actual beliefs about the cause of gravity or attraction, at this stage the focus was on its decrease with distance.

Between 1679 and 1681 Newton engaged in a correspondence with Hooke, who had become secretary of the Royal Society, and John Flamsteed, the Astronomer Royal at Greenwich. Since 1669 Newton had been the Lucasian professor of mathematics at Cambridge University. Around 1680 he was known to experimental philosophers and mathematicians mainly through his optical publications of the 1670s on light and the reflecting telescope. Although he had been speculating on motion and mechanics since the 1660s, his work had remained in manuscript form. His extraordinary skill in mathematics, including his discovery of calculus in 1666, was known to a small number of mathematicians in England and on the Continent. Like most of his contemporaries, Newton believed that the heavens were fluid, and he tried to account for celestial phenomena in that Cartesian fashion well into the early 1680s. In a paper read at the Royal Society in December 1675, for example, he argued that the fact that a pendulum in a glass container nearly exhausted of air oscillates "almost as quickly as in the open air, is no inconsiderable argument" in support of the existence of subtle fluids.[57]

With Hooke, Newton discussed issues relating to curvilinear and orbital motion. In the 1660s Newton, like Continental mathematicians, had conceived curvilinear motion to generate outward tendencies and had been able to provide a quantitative measure of the outward pressure of a ball rolling inside a hollow circular container as well as in other circumstances. On 24 November 1679 Hooke asked Newton to consider his opinions about planetary motions as resulting from a straight motion in a tangent and an attractive one toward the central body. As it appears from a later letter, Hooke expected his opinions to be of potentially great significance for the determination of lunar motion and the solution of the problem of longitude.[58]

In his reply Newton astonishingly claimed that he had nothing to say about Hooke's opinions but proceeded to outline a fancy of his about determining the motion of the Earth by dropping bodies from suitable places, such as a high church tower. He claimed that their fall would describe a path that deviated to the east, but he provided no estimate of the deviation from the perpendicular in relation to the height of fall. Newton went on to venture a spiraling trajectory for an object falling inside the body of the Earth toward its center.[59] Hooke's response agreed with the eastward deviation but challenged Newton's proposed trajectory. He claimed that, neglecting all impediments, the body would not reach the Earth's center but would rather move in a kind of "Ellepteuid." Only the resistance of the medium would make the body progressively approach and eventually reach the center.[60] It is not surprising that in his reply Newton thanked Hooke for having stimulated him to consider the features of the curve, since his initial claim about the trajectory, and especially the body reaching the center of the Earth, was quite naive. Newton now provided a much more sophisticated analysis of the trajectory, arguing further that if "gravity be supposed uniform [the body] will not descend in a spiral to the very center but circulate with an alternate ascent & descent made by it's *vis centrifuga* and gravity alternately overballancing one another."[61]

In the last significant letter in this crucial exchange, Hooke told Newton that "the Attraction always is in a duplicate proportion to the Distance from the Center Reciprocall," deriving thence via dubious steps that since the speeds are as the square root of the attractions, they are inversely as the distance, "as Kepler supposes." In the same letter Hooke also reported that preliminary trials of the experiment suggested by Newton with three balls showed them to fall toward the southeast by differing amounts. Traditionally Hooke is considered to have held the "right" approach to the study of curvilinear motion, whereas Newton held "wrong" views because of his recourse to centrifugal force. This view presupposes that there is only one way to adequately study curvilinear motion, whereas already in the late 1680s Leibniz, for example, argued that the matter could be investigated in different ways. Ironically, the two interlocutors induced each other to change approach. Hooke was stimulated by Newton to incorporate centrifugal force into his picture of orbital motion and in his 6 January letter talked of an "Indeavour of Recesse." In a deleted passage in the same letter, Hooke had originally written, "but I had not then thought of this Increase of the centrifuga virtue from the vertiginous motion of the Earth."[62] Conversely, under the stimulus of Hooke's suggestions, Newton started to consider orbital trajectories as resulting from rectilinear uniform motion and an attraction toward a center inversely proportional to the distance from the center. Although Newton later adopted this approach, in his published works he did not explain why the outward or centrifugal force could be ignored in the calculations. The clear and explicit notion that centrifugal force depends not on the circular motion of a body, but on our way of describing it with a rotating frame of reference, was alien to seventeenth-century studies on motion. Around 1680 Newton was apparently able to establish a connection between motion along an ellipse and attraction toward a center following the inverse-square law, but historians are in disagreement as to whether copies of his early proof are extant.[63]

In 1680–81 two huge comets appeared in the heavens and stimulated astronomers and mathematicians across Europe to reflect on their nature and trajectory. Flamsteed and Newton had an important exchange via James Crompton, a fellow at Jesus College, Cambridge. The analysis of the comet's path involved a great deal of sifting through data and observations, but it also stimulated Newton to reflect on orbital trajectories and solar central attraction. I will focus on some key points relevant to the analysis of orbital motion.

Flamsteed argued that the two comets were in fact one, which had turned in the proximity of the sun because of its magnetic properties and was then repelled by the sun. At that stage Newton refused to accept this suggestion and insisted that there were indeed two comets. In a series of objections raised *ex suppositione,* or without necessarily being committed to them, Newton doubted the sun's magnetic powers because of its heat. But even if the sun were a magnet, he claimed that it "would act on the Comet as a great magnet does on a little one floating in a walnut shell or other little boat on water," namely turning it around and attracting it. Thus he questioned that the comet was attracted in its approach and repelled in its recess. Much like Borelli and other scholars, Newton was reasoning about celestial motions by analogy with objects of common experience. Elsewhere he compared the comet to a bullet shot "from west to east" on the Earth's surface, or a "hand ball tossed up." These comparisons between the motions of projectiles and celestial bodies are of considerable interest. He also challenged Flamsteed's notion that the comet had turned before reaching the sun. Rather, Newton argued that the comet was always attracted by the sun and that it had turned moving around it, "the *vis centrifuga* at C [perihelion] overpow'ring the attraction & forcing the Comet there, notwithstanding the attraction, to begin to recede from the Sun."[64]

At some point after that exchange, Newton drafted a short set of propositions on comets in which he argued that the heavens are fluid and the celestial aether carries the planets in their course. He also stated that there is a gravity toward the sun and the planets inversely proportional to the square of the distance; that if the cometary path were closed, it would be an oval; and that in 1680–81 there was indeed only one comet, which had turned around the sun. This manuscript represents a crucial document on the transition of Newton's thinking on these matters. Notice that at that late date he still accepted the fluidity of the heavens and was somewhat uncertain about the link between inverse-square forces and conic sections for cometary paths.[65]

In this chapter we have repeatedly seen the conical pendulum being invoked in the debates between Borelli and degli Angeli, in Huygens's work on falling bodies, the clock, and centrifugal force, in Hooke's work on the clock, and in the passages quoted in this section. It appears that Borelli's systematic use of broad-ranging analogies did not serve him well in those investigations. His analogies were often imprecise and lacked rigor, a useful reminder that objects provided no automatic solutions. Adopting the conservation of speed in circular motion did not help either. By contrast, Huygens was able to support his intuitions—of the analogy between gravity and centrifugal force, for example—with rigorous proofs, displaying a mathematical ability Borelli could not match. But in many investigations mathematical prowess was as crucial as careful experimentation. Moreover, Huygens combined experiments with an adroit choice of principles, including modifications and generalizations of those put forward by Galileo and Torricelli that would have been very familiar to Borelli. On the English scene, Hooke attempted a mathematical treatment of the conical pendulum when dealing with the isochronous clock. He, Wren, and Wallis all invoked the conical pendulum as a useful, if imperfect, analogue to help conceptualize certain features of orbital motion. That so many had recourse to the same tool, whether using it conceptually, mathematically, or experimentally, highlights both its versatility and its centrality in the study of motion and forces.

CHAPTER EIGHT

Colliding Bodies, Springs, and Beams

The Particles therefore that compose all bodies I do suppose to
owe the greatest part of their sensible or potential Extension to a
Vibrative motion.

Robert Hooke

8.1 The Emergence of Elasticity

Beginning about the last third of the seventeenth century, we observe a cluster
of publications on themes that had been investigated in the 1630s and 1640s, such
as collision, the motion of strings and springs, and the resistance of materials.
These were themes central to the history of mechanics, leading to new rules of
impact, a mathematical formulation of the oscillations of springs, and a deeper
understanding of the behavior of loaded beams. The three areas covered in this
chapter were studied mathematically, experimentally, and philosophically as important topics in their own right. Although these areas are conceptually quite
different from each other, in all three the study of the properties of solid bodies,
especially elasticity, entered the mathematical description of phenomena. Thus by
investigating the research on colliding bodies, springs, and beams, this chapter
also charts the changing understanding of the properties of solid bodies.

Interest in the emergence of elasticity stems largely from the barometric experiments and published works of Otto von Guericke, Boyle, and their contemporaries. Boyle was very conscious of his disciplinary affiliations and objected to his
work being understood as part of the mathematical disciplines; therefore this area
pertains more to the history of experimental philosophy than of mechanics.
There are a few aspects of his contributions deserving attention here, however, especially the prominence gained by elasticity through the debates and controversies it attracted, and the use of tables of experimental data by Boyle and others.
Section 8.2 contains a brief analysis of the works by Boyle and his contemporaries
on the elasticity of the air. Section 8.3 deals with colliding bodies in the works by

Borelli, Huygens, Wren, Wallis, and some of their contemporaries. Those researches stimulated reflections on the properties of solid bodies, helping to redefine some of them, such as bulk or *moles*, as well as the meaning of terms such as "hard," "soft," and "elastic." Section 8.4 focuses on the work by Fabri, Huygens, and Hooke on springs and their properties, with special emphasis on their attempts to prove the isochronism of oscillations. Finally, section 8.5 returns to the problem of the loaded beam pioneered by Galileo (see sec. 3.5 above). I explore the works by Blondel, Marchetti, Fabri, and Parent, ending with the attempts by Mariotte and Leibniz to include elasticity.

The pendulum served both as a conceptual tool to explore oscillations and vibrations, or more generally periodic motions, and as an experimental tool to investigate collision. Additionally, colliding bodies and springs had broader implications about the universe and its workings, whereas springs and beams had technological implications for horology and the resistance of materials. We saw in the previous chapter that in Huygens's studies of the pendulum higher mathematics entered the scene of mechanics with the cycloid. The theory of proportions figures among the mathematical tools in this chapter, but in addition algebra was used in some studies of collision, the method of indivisibles figured in Hooke's work on springs, and Leibniz relied on simple quadratures in his study of elastic beams.

8.2 Boyle and Elasticity

We have already briefly encountered some examples of elastic properties, such as in the case of the wind gun. Moreover, Torricelli's barometric experiments prompted much debate and were followed by further trials, most of them initially undertaken in France by such investigators as Pierre Petit (the engineer in charge of the Rouen fortifications), Roberval, Mersenne, Pascal, and Perier. In addition, Mersenne and the physician Jean Pecquet publicized many debates and experiments. Pecquet was explicit about the elasticity of air and introduced the terms *elater* and *elasticus* to describe it in his 1651 *Experimenta nova anatomica*. The terms were rendered into English as "elatery" and "elastick," but, quite appropriately from my perspective, the term "elatery" was also rendered as "spring," as in "the spring of the air." It was largely through this work, which was translated into English in 1653, that experiments and debates on air became known in England. Thus besides weight and pressure, elasticity started to emerge as an important property of air and was about to become the subject of mathematical and physical inquiry. Pecquet is not the only medical man to appear in this story; Walter Charleton and Henry Power in England were physicians as well. The former

published an influential account of debates on the vacuum, whereas the latter performed experiments at Cambridge and his native Halifax.[1]

Three years after the Jesuit Gaspar Schott published an account of Otto von
Guericke's experiments with the air pump in *Mechanica hydraulico-pneumatica*
(Frankfurt/M, 1657), Boyle opened his series of publications on the subject with
New Experiments Physico-Mechanical, Touching the Spring of the Air (Oxford,
1660). Boyle's work attracted a huge amount of attention, both critical and lauda-
tory, from the Jesuit Francis Line, Hobbes, Power, and the Cartesian natural
philosopher Richard Towneley. The work by the last two, authored by Power, was
published as *Experimental Philosophy* (London, 1664). Before that date, however,
they had been in contact with Boyle and sent manuscripts of their experimental
work to him. When Boyle published a new edition of his work with additions, one
of them contained his reply to Line, *A Defence of the Doctrine Touching the Spring
and Weight of the Air* (London, 1662), with a quantitative formulation of the proportionality between the spring of the air and its density. This formulation was in
the form of a table wherein Boyle compared his experimental data with theoretical calculations (fig. 8.1). Boyle used a J-tube open at the tall end and sealed at the
short one. The experiment consisted in pouring mercury into the long open leg
and measuring both its height and the diminishing volume occupied by the air in
the closed short leg.[2]

Boyle also performed a symmetrical experiment trying to show the effects of a
decreasing spring of the air (fig. 8.2). He immersed a glass pipe with the diameter of "a swan's quill" into a container about seventy inches deep filled with mercury, until the pipe emerged by only one inch. He then sealed with wax the open
end of the pipe above the mercury and started raising the pipe, measuring on a
graduated scale attached to it the space occupied by the air and how much the
mercury had risen. Boyle gives precise instructions on how to perform the experiment and especially on how to seal the pipe. The results were again tabulated and
compared with theoretical predictions, showing that the density is as the spring of
the air.[3]

Boyle's experiments, as well as those of his contemporaries, made elasticity a
key notion in the experimental and quantitative investigation of nature. His works
went through many reprints and translations. In 1665 Hooke included in his *Micrographia* tables of the "elastick power of the air," showing a remarkable agreement between experiments and the "hypothesis." In the following years elasticity
and its mathematical laws entered the analysis of different phenomena.[4]

A Table of the Condenſation of the Air.

A	A	B	C	D	E
48	12	00		29 2/16	29 2/16
46	11 1/2	01 7/16		30 9/16	30 6/16
44	11	02 11/16		31 11/16	31 12/16
42	10 1/2	04 6/16		33 6/16	33 7/8
40	10	06 3/16		35 5/16	35 --
38	9 1/4	07 14/16		37 --	36 19/16
36	9	10 2/16		39 2/8	38 8/8
34	8 1/2	12 8/16		41 2/16	41 2/17
32	8	15 1/16		44 2/16	43 11/16
30	7 1/2	17 15/16		47 2/16	46 3/5
28	7	21 1/16		50 2/8	50 --
26	6 1/2	25 2/16		54 2/16	53 10/13
24	6	29 11/16		58 11/16	58 2/8
23	5 1/2	32 2/16		61 2/16	60 23/23
22	5 1/2	34 4/16		64 2/16	63 11/17
21	5 1/4	37 12/16		67 2/16	66 6/7
20	5	41 4/16		70 2/16	70 --
19	4 1/2	45 1/2		74 2/16	73 11/19
18	4 1/2	48 12/16		77 11/16	77 3/8
17	4 1/4	53 11/16		82 11/16	82 2/17
16	4	58 2/16		87 12/16	87 7/8
15	3 1/2	63 15/16		93 2/16	93 3/8
14	3 1/2	71 2/16		100 2/16	99 6/7
13	3 1/2	78 4/16		107 12/16	107 1/13
12	3	88 7/16		117 7/16	116 8

(C column, rotated: "Added to 29 1/8 makes")

A A. The number of equal ſpaces in the ſhorter leg, that contained the ſame parcel of Air diverſly extended.

B. The height of the Mercurial Cylin-der in the longer leg, that compreſs'd the Air into thoſe dimenſions.

C. The height of a Mercurial Cylinder that counterbalanc'd the preſſure of the Atmoſphere.

D. The Aggregate of the two laſt Co-lumns *B* and *C*, exhibiting the preſſure ſuſtained by the included Air.

E. What that preſſure ſhould be accor-ding to the *Hypotheſis*, that ſuppoſes the preſſures and expanſions to be in reciprocal proportion.

Figure 8.1. Boyle's first table *(A Defence).* This table compares experimental results against theoreti-cal predictions. The first two columns *AA* give the spaces occupied by the increasingly compressed air. The values in the two columns are in the ratio of 1 to 4. Column *B* gives the height of the mer-cury compressing the air into the spaces of the previous column. Column *C* gives the height of the mercury counterbalancing atmospheric pressure, namely 29⅛ inches. Column *D* gives the sum of the values in *B* and *C*, namely the pressure of the enclosed air. Supposing the pressure to be in-versely as the volume, column *E* gives the pressure according to the hypothesis. Thus if the space is halved from 48 to 24, for example, pressure is doubled from 29²⁄₁₆ to 58⅝.

8.3 The Transformation of the Impact Rules

Whereas Marci's work remained little known, Descartes' impact rules helped shape a research program pursued by several scholars for several decades. Des-cartes' rules were widely rejected, but others too believed that to some extent im-pact could be subjected to mathematical investigation. Although not all could be grouped unproblematically, or in some cases at all, as defenders of the mechanical philosophy, the topic they worked on did stem from its fold. They included the mathematicians Borelli, Huygens, Wren, and Wallis, the experimentalist Mari-otte, and the French Jesuits Fabri, Dechales, and Pardies, among others. Besides coming from a wide range of philosophical positions, those scholars studied and

A Table of the Rarefaction of the Air.

A	B	C	D	E
1	00 $\frac{0}{0}$		29 $\frac{1}{4}$	29 $\frac{1}{4}$
1$\frac{1}{2}$	1C $\frac{1}{8}$		19 $\frac{1}{8}$	19 $\frac{3}{8}$
2	15 $\frac{1}{8}$		14 $\frac{1}{8}$	14 $\frac{7}{8}$
3	2C $\frac{2}{8}$		9 $\frac{7}{8}$	9 $\frac{11}{12}$
4	22 $\frac{1}{8}$		7 $\frac{1}{8}$	7 $\frac{1}{16}$
5	24 $\frac{1}{8}$		5 $\frac{7}{8}$	5 $\frac{17}{20}$
6	24 $\frac{7}{8}$		4 $\frac{7}{8}$	4 $\frac{11}{24}$
7	25 $\frac{5}{8}$		4 $\frac{1}{8}$	4 $\frac{1}{4}$
8	26 $\frac{0}{0}$		3 $\frac{7}{8}$	3 $\frac{11}{32}$
9	26 $\frac{1}{8}$		3 $\frac{1}{8}$	3 $\frac{3}{16}$
10	26 $\frac{5}{8}$		3 $\frac{2}{0}$	2 $\frac{12}{40}$
12	27 $\frac{1}{8}$		2 $\frac{1}{8}$	2 $\frac{11}{48}$
14	27 $\frac{7}{8}$		2 $\frac{1}{8}$	2 $\frac{1}{8}$
16	27 $\frac{7}{8}$		2 $\frac{0}{8}$	1 $\frac{11}{64}$
18	27 $\frac{7}{8}$		1 $\frac{3}{8}$	1 $\frac{4}{7}$
20	28 $\frac{4}{}$		1 $\frac{5}{8}$	1 $\frac{8}{10}$
24	28 $\frac{8}{8}$		1 $\frac{5}{8}$	1 $\frac{31}{90}$
28	28 $\frac{4}{8}$		1 $\frac{1}{8}$	1 $\frac{3}{16}$
32	28 $\frac{2}{8}$		1 $\frac{3}{8}$	0 $\frac{11}{12}$

(Column C label, set vertically: Subtracted from 29 $\frac{1}{4}$ leaves)

A. The number of equal spaces at the top of the Tube, that contained the same parcel of Air.

B. The height of the Mercurial Cylinder, that together with the Spring of the included Air counterbalanced the pressure of the Atmosphere.

C. The pressure of the Atmosphere.

D. The Complement of *B* to *C*, exhibiting the pressure sustained by the included Air.

E. What that pressure should be according to the *Hypothesis*.

Figure 8.2. Boyle's second table *(A Defence)*. In this table column *A* gives the spaces occupied by air in the sealed pipe. Column *B* gives the height of mercury in the pipe. This height, together with the spring of the air, counterbalances atmospheric pressure, whose value is given in column *C*, namely 29¾ inches. The numbers in column *D* are obtained by subtracting the values in *B* from 29¾. Column *E* gives the result obtained by dividing 29¾ by the values in column *A*. Column *E* contains some small inaccuracies of computation.

presented their works on collision in radically different ways. Their investigations provide us with rich material for understanding how motion and mechanics were investigated in the third quarter of the seventeenth century. While Descartes presented his rules of collision as self-evident, others challenged those rules on conceptual and empirical grounds and presented their results in different ways ranging from axiomatic formulations to experimental reports.[5]

These investigations involved two intertwined themes: rules about collision, as well as an understanding and classification of the nature and properties of colliding bodies. According to many Peripatetics, a body's *pondus* (or weight) and *moles* (or bulk) had a key role in the investigation of falling bodies. Their role became more problematic at the time of Galileo, but they regained a significant place in the investigation of impact, requiring a fresh understanding and redefinition of many terms and notions. Historians have often rendered earlier notions used to classify bodies, such as hard or soft, in different terms as if to uncover what the scholars must have truly meant, taking for granted the notion of elasticity. In earlier investigations by Marci and Descartes, however, elasticity did not figure at all, and it emerged only slowly in the 1660s and 1670s as a chief feature of colliding bodies. Thus those investigations contributed to reshaping the understanding not just of conservation laws but also the properties of solid bodies and the notion of mass.[6]

Borelli dealt with the problem of impact in *De vi percussionis* (Bologna, 1667), starting like other scholars at the time from several propositions about motion and its properties. In one of those propositions he provided a colorful refutation of the Aristotelian notion that a projectile, once released, continues its motion because of the fluid medium in which it moves. Borelli threw a ball to which he had attached long hairs and showed in a delightful drawing that the hairs do not protrude forward but point in the direction opposite to motion, a clear sign that the medium opposes motion (fig. 8.3).[7]

Borelli considered *vis motiva* as the key notion in the study of impact and took it to be in the compound proportion of *velocitas* and *corpus*, by which he understood a body's bulk.[8] He stated that in order for his rules to apply, impact had to occur in the middle of the bodies and had to be straight, presumably in order to avoid spinning. The cornerstone of his investigation was the notion of absolutely hard bodies, whereby he meant that they were inflexible or nonelastic. Only *prima corpora*—I believe Borelli had atoms in mind here—are absolutely hard, whereas the bodies of our experience are always compressible and flexible to some degree. This feature can be evinced from the sound of percussion, resulting from compression and elasticity. Given his profound dislike for Descartes, it is not surpris-

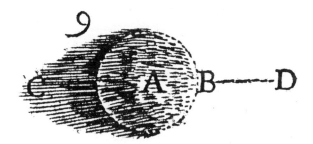

Figure 8.3. Borelli's refutation of Aristotle *(De vi percussionis)*. Hairs attached to sphere *A* moving in the direction *BD* tend backward toward *C* rather than forward, showing that the medium does not propel the sphere forward but rather impedes its motion.

ing that Borelli began his treatment of impact by refuting Descartes' fourth rule, whereby a body in motion hitting a larger body at rest rebounds with the same speed, while the body at rest does not move. By contrast, Borelli argued that what matters in impact is *vis motiva* rather than *moles;* therefore a body at rest can be moved by any *vis motiva,* even an arbitrarily small one.[9] Borelli believed *vis motiva* to be conserved in impact. Curiously, he did not state this important point in an axiom but rather mentioned it only in passing, adopting a terminology suggesting that he conceived impact as being like rarefaction or gas diffusion. In the case of an impact wherein a moving body hits one at rest, for example, he argued that the *vis motiva* initially compressed in the impinging body is diffused and partly transferred to the body at rest (fig. 8.4). Thus according to him a body in motion hitting one at rest does not lose or communicate all its speed. Borelli formulated his key impact rule as a proportion and generally relied on this mathematical tool, which is actually unsuited to the study of impact and makes his work look awkward to modern eyes.[10]

Although Borelli admitted that all bodies of our experience are to some degree elastic, in the formulation of his impact rules he first considered them deprived of elasticity; thus they do not rebound. Subsequently, he was able to generalize his results to bodies with different initial velocities by using the principle of relativity of motion. Borelli was trying to address the problem that experience seems to con-

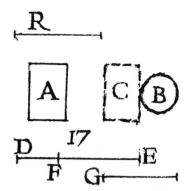

Figure 8.4. Borelli on impact *(De vi percussionis).* Body A with *vis motiva* R arrives in C, where it hits body B at rest. DE is the speed of the impinging body A before impact and DF its *retardatio,* namely the amount by which the speed of A decreases after the impact. Borelli's proportion is $DE{:}DF{::}(A + B){:}B$. For Borelli, all magnitudes are positive. Abandoning Borelli's proportions, from the conservation of *vis motiva,* $A \times DF$ is equal to the *vis motiva* acquired by B after the impact. After the impact both bodies have the same speed $G = A \times DE/(A + B)$, and the conservation of *vis motiva* gives:

$$A \times A \times DE/(A + B) + B \times A \times DE/(A + B) = A \times DE.$$

tradict his proposition, as with balls rolling and colliding on a plane, where reflection may occur or the impelling ball may stop. The reasons for this are that the balls are not absolutely hard but rather flexible and compressible, they may not be perfect spheres, the plane may not be perfectly smooth, and therefore the balls rotate on it like cogwheels. Whereas impacts between perfectly hard bodies are instantaneous, impacts between elastic bodies occur in a finite time. In his treatment of elasticity Borelli was rather verbose and did not include impact rules for elastic bodies, probably because he thought them not amenable to a general mathematical treatment. Rather, he seems to have been interested in comparing elasticity with the force of percussion.[11]

In chapter 15 Borelli tackled a rather awkward problem in his framework, namely reflection after percussion. Since his rules apply to hard and inflexible bodies, Borelli was at pains to explain whence reflection arises. He tried to argue that its cause was the impetus of the incident body and established an analogy with the motion of a pendulum that changes direction if one interposes a nail about halfway down the thread. No reflection occurs between hard and inflexible bodies indifferent to motion. If the body at rest offers resistance to motion, however, reflection can occur.[12] It is in this context of hard and inflexible bodies that Borelli described what we call elastic collision, but it seems peculiar to have general mathematical rules depending on such an unpredictable variable as a body's resistance to motion, which in this context is something quite different from bulk. If equal bodies collide with opposite speeds, they rebound with their speeds reversed. If hard and inflexible bodies with opposite speeds and equal *virtutes motivae* collide, they too rebound with their speeds reversed.[13] When dealing with more elaborate cases, however, Borelli ran into problems stemming from his lack of algebra and sound principles. For example, if two hard and inflexible bodies with opposite speeds and different *virtutes motivae* collide, the outcome can be quite complex depending on the relative speeds and *moles*. Borelli did not provide a way of determining the respective speeds after the impact but simply tried to predict whether the bodies would rebound or not. He argued correctly that the body with the smaller *vis motiva* always rebounds with a greater speed. As to the body with the larger *vis motiva*, he distinguished two cases depending on the difference in *moles*. If it is larger it is never reflected, whereas if it is smaller it is not always reflected. His conclusion did not specify the conditions for reflection and did not determine the speeds after the impact.[14]

Although Borelli was the first of a new generation of scholars to publish on the rules of impact in a widely circulating work, Huygens was the pivotal figure in this area. He began his investigations in the early 1650s and communicated some of his results to certain correspondents.[15] Before long he had produced a master-

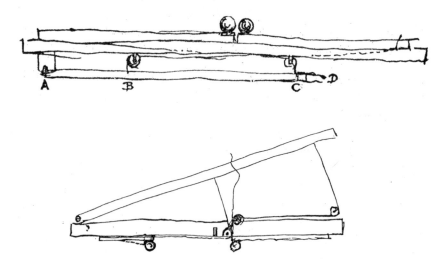

Figures 8.5. Huygens's devices for investigating impact (Huygens's manuscripts as reproduced in *HOC*, 16:158 and 185). These are some of the devices Huygens invented to assist in his investigations of impact.

ful axiomatic theory of impact for what he called hard bodies, *De motu corporum ex percussione*, which remained unpublished in his lifetime, possibly because Huygens wanted to extend it to the force of percussion. The axiomatic formulation gives his treatise a classical look hiding the heuristic path, much as in Galileo's case. Huygens conducted careful quantitative experiments with pendulums as well as with special devices he had invented (fig. 8.5). The rules of pendular motion were known, and in addition the pendulum provided a convenient way of studying central impacts that avoided the complications arising on Marci's billiards table. Despite the extensive manuscript record, however, it is unclear whether Huygens relied more on experiments or on theory. We do know, however, that he used the new algebraic methods of solving equations whereby the unknowns are indicated with the letters x and y.[16]

Huygens's formulation is based on three general principles: relativity of motion, the reversibility of impact, and a variant of Torricelli's principle. The first states that the outcome of impacts should be invariant when it is witnessed by observers in uniform rectilinear motion, such as those moving on a barge along a canal. This principle refuted several of Descartes' rules. The second states that if two bodies collide again with the speeds they have acquired after the first impact, they will acquire the same speeds they had before the first impact.[17] The third states that, converting the speeds of colliding bodies from horizontal to vertical, their center of gravity cannot rise after the impact. This principle looks quite natural in the case of colliding pendulums, where the horizontal speeds are indeed converted along the vertical. Torricelli's principle concerned joined bodies, whereas Huygens's colliding bodies were not joined. The editors of the *Oeuvres complètes* suggest that this was indeed Huygens's heuristic path.[18] Their interpretation fits well with Huygens's earlier attempts to defend Galileo's law of falling bodies, where he did not have recourse to experience but rather sought an abstract general principle whereby the law of fall ought to be independent of the time unit adopted. It is tempting here to identify a Huygensian style, inspired by Archimedes, in which the search for general principles and axioms takes center stage and experience plays a secondary role as part of the scaffolding rather than part of the formal presentation.[19]

We can get interesting glimpses of Huygens's way of proceeding from his correspondence with the mathematician René François de Sluse, Canon at Liège. In November 1657 Huygens stated that he had come to doubt Descartes' rules because they contradict experience. In addition, he argued that Descartes' rules 2 and 5 were problematic. In rule 2, B is larger than C, and the two bodies collide with equal and opposite speeds. After the impact, C is reflected with its initial speed reversed, whereas B continues with the same speed; therefore after the impact they move with the same speed in the same direction, in accordance with the conservation of Cartesian quantity of motion. In rule 5, B is larger than C, but C is now at rest. After the impact, the two bodies move together with a speed smaller than B's speed before the impact, in accordance with the same conservation law. Therefore, B's motion would be hindered more by a body at rest than by the same body moving with an opposite speed, a result Huygens deemed utterly absurd. This may well have been the example that alerted Huygens to the defects in Descartes' conservation law and prompted him to seek a better one. Huygens claimed that he had proofs of his own rules and that they agreed perfectly with experience ("experientiae exquisite consentiunt"). Later, however, Huygens warned Sluse not to think that he was following experiences, which are uncertain, or in his word "slippery."[20]

Huygens disclosed his ideas in Paris in 1660, and while in London in 1661 he discussed problems of impact at the Royal Society with Wren, Lawrence Rooke, Wallis, and others. They observed impacts between the bobs of two pendulums of equal length and set themselves various problems to solve. One problem was to predict what would happen when a bob of one pound was displaced from the equilibrium position by 48 degrees and hit a bob of half a pound at rest. Huygens was able to solve impromptu several such problems proposed by William Brouncker, president of the Royal Society. Huygens communicated only his results, however; he kept his rules to himself and even blotted out his calculations. It was with some irritation that he pointed out in 1669 that on that occasion Wren and Rooke had not yet found satisfactory rules.[21]

In 1666 the Royal Society addressed the issue of impact as part of its concern with motion, and several experiments and discussions took place at its meetings.[22] On 4, 11, and 18 January 1668 the French Académie held a discussion on impact where Huygens presented his rules and discussed those of Descartes and Fabri.[23] Later in the same year the Royal Society invited contributions to the problem of impact, to which Wallis, Wren, and Huygens offered solutions. Hooke too took part in those discussions, showing his concern with the notion of elasticity by arguing that if there were an absolutely hard body, it would not rebound.[24] The essays by Wallis and Wren appeared in the *Philosophical Transactions* at the end of 1668. Huygens's essay appeared in the *Transactions* early in 1669, after he had already published it in the *Journal des sçavans*. Wallis began with some generalities about motion before tackling impact. He treated collisions between what he called "absolutely hard" bodies, by which he meant nonelastic ones. Wallis stated, for example, that if a body with weight *(pondus)* P and speed C collides with another body at rest with weight mP, after the impact they would both move with a common speed $\frac{1}{1+m}C$. He explained that if the colliding bodies were elastic they would rebound by an amount dependent on their elasticity, but he provided no rules for elastic impact.[25] Wallis's insistence on bodies being hard or elastic means that he did not provide a general treatment of impact for all bodies, possibly because he did not believe that soft bodies were subjected to mathematical regularities.

In that same issue of the *Philosophical Transactions* Wren put forward his "law of nature in the collision of bodies," though he failed either to specify the conditions under which his law applied or to provide any justification for it. He stated that colliding bodies have "proper" and "natural" speeds when they are inversely proportional to the bodies themselves (or, as we would say, to their masses). When this is the case, after the impact the bodies rebound with the same speed, a case Wren considered analogous to a balance oscillating around its center of gravity. For cases in which the colliding bodies do not have their natural speeds, Wren and Huygens both provided a graphical method for solving the problem. They provided similar diagrams for examining impacts between what they called hard bodies. Wren, however, did not state a conservation principle, either of quantity of motion in one direction or of the product between body and the square of the speed, and tried instead to justify his account by analogy with the lever (fig. 8.6). Once again, the lever emerges as a creative tool of investigation. Wren concluded his essay by noting that nature follows the algebraic rules of addition and sub-

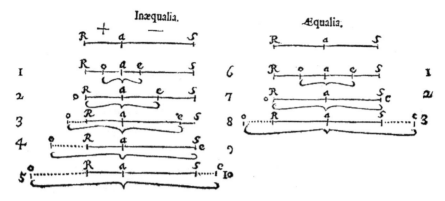

Figure 8.6. Wren on impact ("Lex naturae de collisione corporum"). *R* and *S* are the colliding bodies and *a* their center of gravity. If the bodies have what Wren called "natural" speeds, those speeds
will be proportional to the arms of the balance and will stay so after the impact. In all other cases,
let *e* be the point whereby the speeds *Re* and *Se* are as the respective segments. After the impact, the
speeds will be as *Ro* and *So*, where *oa* = *ae*.

traction taking into account the sign of the speed, but this important point was not
fully developed or explained.[26]

Not only was Huygens presumably a major source for other investigations, but
his work also provided the clearest conditions of applicability and attained the
most general conclusions. Huygens distinguished two types of bodies, soft and
hard. In a manuscript draft published in *Oeuvres complètes* Huygens claimed that
hard nonelastic bodies rebound like elastic ones. The difference between them is
that elastic impact occurs in a finite time, whereas the impact between perfectly
hard bodies is instantaneous. Examples of perfectly hard bodies are atoms, whose
existence Huygens was inclined to support, and Descartes' subtle matter.[27] These
observations highlight the profound connections between matter theory and the
study of collision. One of the key issues was the conservation of motion, an issue
clearly inspired by Cartesian concerns. While asserting that the state of the center
of gravity of colliding bodies is not affected by their impact and that quantity of
motion in one direction is conserved, Huygens wished to spell out that quantity of
motion in the Cartesian sense could be increased or diminished by impact.[28] No
restrictions are given for his own rule, thus suggesting its universality. He formulated a more general rule for direct and oblique impacts, stating that the center of
gravity of the colliding bodies will move with the same speed before and after the
impact.[29] Huygens then stated that the sum of the products of each body times the
square of its speed is conserved only for hard bodies. In his presentation Huygens

gave similar emphasis to that proposition and to one stating that in the impact between hard bodies one of which is at rest, the transfer of motion is maximized by
the interposition of a third body mean proportional between them. The former
claim was to assume a major philosophical significance as an expression of the
conservation of living force, whereas the latter remained a simple curiosity.[30]

Subsequent works did not surpass Huygens's. Several debates and experiments
followed at the Royal Society, involving the anatomist and mathematician William Croone, the naturalist Francis Willughby, and especially the mathematician
William Neile, whose dispute with Wallis went on through innumerable exchanges. Neile believed that since bodies are indifferent to motion or rest, bodies
at rest offer no resistance to the motion of others; therefore if a body in motion hit
an equal one at rest, both would continue with the speed of the impinging body.
He also believed that if two equal bodies collide with opposite and different
speeds, they would come to stop after the impact. He was concerned with the physical cause of the communication of motion, a topic Huygens found less pressing:
"The question of whether motion is communicated to bodies by the elasticity of
the parts, or by that of the air or of some other substance squeezed between them,
or by hardness alone has not yet been resolved. But in the demonstration of my
rules it does not matter at all which of the three one assumes." A few months later
he reiterated the point: "As for Mr. Neile's question, what is the reason why a body
puts another which it meets into motion, I do not think that this can be found out
by any better known principles." His answers sound very much like those he was
going to get from Newton in about two decades.[31]

In 1671 Wallis included an extensive analysis of impact in the third part of his
Mechanica, in chapters 11 ("De percussione") and 13 ("De elatere, & resilitione seu
reflexione"). He still provided a three-way classification of bodies: hard, soft, and
elastic. Soft bodies are made of clay, wax, lead, or even fluids, for example, whereas
elastic ones are made of steel or wood. Wallis provided no examples of hard bodies, but it is those that are the subject of the first part of his study of impact. Soft
bodies are explicitly excluded because they are deformed in impact and therefore
a portion of their *vis* or *momentum* is lost. For Wallis, *vis* and *momentum* have the
same dimensions, but the former is often qualified as *vis impellens*. Thus Wallis
did not formulate a universal conservation rule of momentum joined with direction. He claimed also that breakable bodies waste some portion of their force in
the process of being broken. Wallis admitted that absolutely hard or elastic bodies
are abstractions and was prepared to talk of deviations from the rules he has provided as *accidentia*. Thus the less than total accuracy of his laws did not prevent
Wallis from seeking a mathematical formulation whereby accidents could be

treated separately. This attitude and the use of the term *accidentia* appear to be inspired by Galileo.[32]

While treating impact between hard bodies Wallis introduced the notion of *ictus*, or blow, in an attempt to measure the strength of the impact. In the case of
two bodies, for example, whether equal or different, colliding head-on with the
same quantity of motion, the blow is equal to the sum of their quantities of motion, or twice that of one of them. In this particular case, if the bodies are hard or
nonelastic, they would both come to stop. The notion of *ictus* introduced in chapter 11 is employed in chapter 13 as a bridge to dealing with elastic impact, because
the *elateris vis restitutiva*, or restoring elastic force, is equal to the *ictus vis*, or force
of the impact. For example, in the case of an elastic impact between two bodies,
whether equal or different, colliding head-on with the same quantity of motion,
they are subjected to a *vis restitutiva* equal to the *ictus* equally divided between
them; therefore they will rebound each with its own speed. Generally Wallis provided different proofs for each proposition using a range of tools, including the
principle of relativity of motion.[33]

Thus between 1668 and 1671 Wren, Wallis, and Huygens provided only partly
equivalent treatments of impact. Laws or rules could be formulated in different
ways, and whereas Wren and Wallis emphasized the classification of different
cases and treated them individually, Huygens sought a more general formulation.
It was Huygens who stated the conservation of the product of the bulk or mass
times the square of the speed for hard bodies, and the conservation of quantity of
motion in one direction for all bodies.

Soon after the publications of 1669, Fabri and Pardies, a professor of mathematics at the College of Clermont in Paris, published works on collision. Fabri discussed the issue in *Dialogi physici* (Lyons, 1669) criticizing Borelli's *De vi percussionis*,[34] whereas Pardies in *Discours du mouvement local* (Paris, 1670) showed
himself to be aware of the recent works by Wallis, Wren, and Huygens.

In his preface Pardies characterized local motion and impact rules as part of
mechanics. He claimed that Descartes had provided false rules of impact, with the
exception of the first one, but he agreed with Descartes that the problem was best
tackled by trying to derive from metaphysics which laws nature must observe.
Thus he addressed the problem without having recourse to experience, drawing
two key distinctions: between impacts occurring in the void or in a space filled
with a fluid matter, and impacts between equal versus different bodies. Only impacts in a fluid are related to our experiences. Pardies began by giving rules for
equal bodies colliding in the void, but then in a dramatic twist he argued that the
same rules apply even to different bodies, because bodies are indifferent to mo-

tion.[35] In a fluid the situation is quite different, and Pardies refrained from providing rules of impact for different bodies, arguing that no general rule could be formulated. He made an exception for the impact between equal bodies, for which he gave rules similar to Huygens's, namely the same rules holding in the void. In what appears to have been a last-moment addition, however, Pardies referred with tentative approbation to Huygens's solution as published in the *Journal des sçavans* in 1669. Huygens expressed astonishment that how a scholar "non vulgaris ingenij" like Pardies could hold opinions so far removed from the received ones. At the end of his *Discours* Pardies added some defensive remarks that give us a sense of the pressure a Jesuit was under at the time: he tried to distance himself from Descartes and made an attempt to attribute to Aristotle the view that in a vacuum motion would be conserved.[36]

There are some similarities between Pardies' work and Leibniz's first analyses of impact, which were developed independently of the French Jesuit, however. Leibniz wrote a *Theoria motus abstracti* (Mainz and London, 1671), dedicated to the French Académie des Sciences, and a *Hypothesis physica nova* or *Theoria motus concreti* (Mainz and London, 1671), dedicated to the Royal Society. The former relies on Hobbes's notion of *conatus* and deals with impact independently of experience, arguing that the mass or size of bodies plays no role in collision. The latter relies on Huygens and Wren and considers collision in the real world, where the mass or size of bodies does play a role.[37]

In 1673 the topic was taken up again by Edme Mariotte, a colleague of Huygens's at the Académie, in *Traitté de la percussion*. The title page identifies Mariotte as a member of the Académie, which had served as the physical and intellectual venue for the experiments. The title of the first edition, though not the third, states that the work argues against the rules put forward by Descartes and a certain other modern author, probably meaning Pardies. Possibly in contrast to Pardies, Mariotte's treatise is very much based on experience. For example, he described an apparatus for studying the collision between pendulum bobs and often used the expression "principle of experience" for his propositions. Both the experimental apparatus and the rules of impact were clearly inspired by Huygens, who accused Mariotte of plagiarism for having failed to mention his name.[38]

At the outset of the *Traitté* Mariotte claimed that all bodies are flexible, but some (such as inflated balls, a ring of tempered steel, or a tense cord made of gut) rebound while others (such as wax or wet clay) do not. Thus Mariotte excised the notion of perfectly hard bodies that had been at the center of previous investigations and made the ability to rebound the distinguishing feature between only two types of bodies. The sound produced by colliding bodies provides evidence that all bodies are to some degree elastic. Mariotte did not start from the conservation of quantity of motion. Rather, after having claimed that a body set in motion continues its motion in the same direction with the same speed,[39] he proceeded to two suppositions on pendulums. He claimed that for falling and rising bodies the height is as the square of the speed, and that a bob pushed with the same force horizontally or vertically reaches the same height. Small pendular oscillations are nearly isochronous and could be taken to be so for his purposes. In his study of impact Mariotte used the pendulum foundationally, not simply to confirm rules previously established. While presenting the very same propositions, Huygens in many cases had tried to use a priori principles, from symmetry and relativity of motion to the impossibility of perpetual motion, whereas Mariotte chose experience as the foundation of his work. He provided rules for both elastic and inelastic impact, but here I wish to focus on the way he dealt with the problem of impact in general.

Mariotte compared elastic bodies (or bodies *à ressort*), such as steel, marble, glass, ivory, and jasper, to stretched strings and to oscillating pendulums, adding an illustration of a stretched string to make his point.[40] Much as dal Monte had seen a lever in the pulley and in the other simple machines, Mariotte saw a string or oscillating pendulum in colliding bodies exemplifying *ressort*. Much like Borelli, Mariotte dealt in the final part of his work with the force of percussion, the determination of the centers of vibration, agitation, and percussion. His work went through several editions. In the third edition (Paris, 1684) Mariotte expanded his treatment of impact (adding an experiment with a recoiling gun) and of fall in a resisting medium.[41]

Claude Dechales in the posthumous *Traitté du mouvement local, et du ressort* (Lyon, 1682) examined the problem in a way that reminds us of the complex issues involved and the many links with other themes, such as the causes of motion and the nature of *ressort*. Around 1690 Huygens planned a treatise on impact and centrifugal force whose preface contains a valuable contemporary survey of the field. His opinion of Dechales' treatise was that it was based either on Wallis's work or on a secondhand account of what Huygens himself had stated at the Académie in 1668. In his treatise Dechales considered both elastic bodies and inelastic ones, the latter being divided into soft and inflexible.[42] The work provides impact rules for bodies without *ressort* similar to Huygens's, but it also contains a philosophical analysis quite different from Huygens's. When dealing with elastic impact in book 5, Dechales made some interesting statements but also committed elementary mistakes. For example, he claimed that when elastic bodies collide with speeds inversely as their weights, the quantities of motion destroy each other, and the bodies rebound from rest as if one had placed a spring between them. He also mixed quantity of motion and speed, and argued that after the impact the speeds would be as the squares of the speeds before the impact.[43] He denied that respective speed, or the speed of a body with respect to the other, is preserved in the impact between elastic bodies, and he further denied that the motion or rest of the center of gravity of colliding bodies is always unaffected by the impact. Dechales considered the case of an elastic impact between bodies with speeds inversely as their weights, and claimed that after the impact their center of gravity would move, whereas before the impact it was at rest.[44]

In a yet later posthumous work, namely the second 1690 edition of his gigantic *Cursus* brought into print by his fellow Jesuit Aimé Varcin, Dechales repeated his earlier claims, this time explicitly identifying Mariotte as his source.[45] Dechales' treatment of collision under the heading *Mechanices* in an encyclopedic work on the mathematical disciplines attests to the extension of mechanics to new domains.

About three decades after Marci and Descartes posed the new problem of the mathematical formulation of impact rules, several scholars tackled it producing a broad range of solutions. Through correspondence, oral communication, and publication, Huygens's work became the main reference point and the focus for further debates.

8.4 Springs between Technology and Cosmology

Following the works by Boyle on the spring of the air, elasticity became a major area of research. In 1669, for example, the Jesuit Fabri began publication of his gigantic work in ten treatises, *Physica, id est, scientia rerum corporearum*, in which he covered, in thousands of propositions, everything from the metaphysical properties of bodies to plants, the generation of animals, and man. Several portions of Fabri's treatise on *physica* were devoted to physico-mathematical subjects, especially in the first volume in two treatises dedicated to Cardinal Leopoldo de' Medici. The first treatise includes a book on *De compresso et tenso* in which he discussed objects such as cords, beams, and the wind gun. Fabri introduced the notion of elasticity as an intermediate moving power in that it is neither like the *potentia motrix* of moving bodies (which he called natural) nor like that of animals (which he called animal or vital) but is intermediate between them. A portion of the discussion is based on common natural philosophical debates as to the physical cause of elasticity and the denial that dilatation or tension occurs because of tiny empty spaces.[46]

Fabri claimed that the same cord can be taut in different ways and that its tension is proportional to the force applied, a statement based on the metaphysical principle of the proportionality between cause and effect. Moreover, he argued

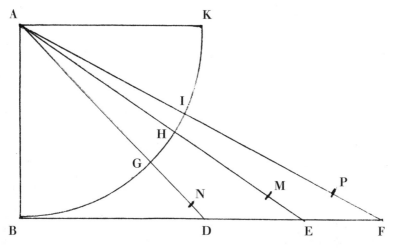

Figure 8.7. Fabri's attempt to prove isochronism *(Physica)*. In the quadrant *ABGHIK* the radius *AB* represents the length of the cord not subjected to tension. The cord can be taut from *G* to *D*, from *H* to *E*, or from *I* to *F*, for example, which are as 1, 2, 3. Fabri claimed that if in a given time *D* moves to *N*, in the same time *E* will move to *M* such that *EM* = 2*DN*. He justified this claim by arguing that motion is proportional to the displacement, and therefore the motions along *EM* and *DN* are as *EM* and *DN* themselves. Fabri divided *GD* into four equal parts, the first one being *DN*, arguing that once released, the string will move through *DN* in the first instant and through *NG* = 3*DN* in the second. Likewise, in the same time *E* will move first to *M* and then to *H*, where *HM* = 3*ME*. The same applies to *AIF*, for symmetry reasons. Fabri sought to prove his proposition with proportions and finite, rather than infinitesimal, time intervals. He was aware that he was applying Galileo's odd-number rule to a case in which acceleration is not constant, but he argued that regardless of how acceleration increases, motion is always proportional to the displacement, and therefore oscillations are isochronous.

that the tension of a cord is proportional to its displacement from rest. From these propositions Fabri stated the isochronism of a vibrating cord, by which he meant something more akin to the vibrations of a spring pulled longitudinally than to a cord plucked transversally. His proof (fig. 8.7) is based on a comparison between accelerations and displacements.[47]

It is not clear how widely Fabri's work circulated and, especially, whether it was known to Huygens, who had crossed swords with Fabri over the ring of Saturn about a decade earlier, though relatively soon after the publication of *Horologium oscillatorium* Huygens reexamined the matter from a different perspective. As we saw in the previous chapter, the original contexts for Huygens's work were the determination of the distance traveled by a falling body in one second, the determination of centrifugal force, and the cycloidal cheeks that made oscillations isochronous. After publication Huygens apparently started to see the matter in a different light and to ask which feature of the cycloid made oscillations isochronous. The answer was that if a body is placed in any point B of an inverted cycloid ABC (see fig. 7.7), the component BKG of its weight tangent to the curve is proportional to the length of the cycloidal arc BIA from the cycloid's vertex to the body. From here Huygens was able to generalize the result that if the force is as the displacement, oscillations are isochronous, and to apply it to a number of cases ranging from springs to strings. Working by analogy with the pendulum, he was able to retrieve the proportion whereby the period of a vibrating string is as the square root of the tension or the weight attached. Although Huygens had grasped some key principles of a general mathematical theory of elasticity, all he put into print was a brief essay in the *Journal des sçavans* for February 1675, in which he announced his discovery of a new type of portable watch regulated by a spiral spring. It seems that he was more interested in constructing a viable portable watch than in publishing a treatise. His work on the portable watch was done in collaboration with the clockmaker Isaac Thuret, although their collaboration rapidly turned sour because Huygens and Thuret disagreed as to the value of their respective contributions.

Huygens's manuscripts reveal a wealth of investigations on the matter, but here I wish to highlight the shift in his perception of the pendulum clock: from constrained motion of a body under the action of gravity to something analogous to an oscillating spring or string where the force (or, as Huygens called it, the incitation) is as the displacement. Besides elasticity, examples of incitation include gravity, wind, and magnetic attraction. In my interpretation centrifugal force would not be among them, because by "incitation" Huygens understood an external action, whereas centrifugal force was a tendency internal to a rotating body. Although gravity was caused by centrifugal force, it was an instance of incitation because it resulted from the action of the vortex on the body.[48]

Robert Hooke too was interested in the behavior and constitution of bodies with regard to their spring or elasticity, partly in connection with horology. In the 1670s he engaged in a major priority dispute with Huygens over the invention of the spring-regulated watch.[49] Much as Huygens did with the pendulum clock, Hooke devised an elaborate instrument from a simple object whose mathematical properties he could investigate. Despite his considerable manual and technical skills, Hooke relied on the help of technicians, especially the London instrument maker Thomas Tompion. Unlike Huygens, Hooke did not write a *magnum opus* but published his findings in articles and short treatises. His main reflections on elasticity can be found in *Lectures de potentia restitutiva* (London, 1678), part of the Cutlerian lectures. In his brief (just over two dozen pages) and ambitious work in English, he quickly moved from a careful experimental investigation of different types of springs to a mathematical analysis of their oscillations and to an explanation of certain profound features of nature, such as the "congruity" and "incongruity" of bodies.

Making his usual claim of priority and of having arrived at the "Theory of Springs" eighteen years before, Hooke recounted the experiments he had performed in before Charles II, including the one involving his spring watch, about three years earlier. Two years earlier Hooke had printed an anagram that was now solved as "ut tensio sic vis," meaning that the force or "the Power of any spring is in the same proportion with the Tension thereof: That is, if one power stretch or bend it one space, two will bend it two, and three will bend it three, and so forward." Likewise, "the force or power thereof to restore it self to its natural position is always proportionate to the Distance or space it is removed therefrom." Hooke used three types of spring to "try" his "rule or law of nature": a wire coiled around a cylinder, a wire coiled in a spiral as in a watch, and a simple straight wire hanging from a nail (fig. 8.8). In his discussions of springs Hooke referred to a wide range of materials, including horn, wood, many types of metal, and air. In his discussion of wood he mentioned a beam where one end is fixed and the other is loaded, recalling the example in Galileo's *Discorsi* (see fig. 3.13). Later on Hooke also showed that in a bent beam some parts are extended and some are compressed (fig. 8.9).[50]

Hooke added that the tension or displacement of a spring by a body is proportional to the body's weight, and he proposed a way to weigh the body by measuring the displacement of a spring on a graduated scale. Probably exploiting the analogy with the barometric experiments, he proposed as well to investigate whether weight changes with altitude by carrying out experiment on the "Pike of Teneriff." He also argued that from his principles one could easily explain why the vibrations of a spring or a string are isochronous, regardless of their amplitude, thus establishing an interesting link with the study of sound.[51]

One of the most intriguing aspects of Hooke's tract is the ease with which it moves from references to catapults and fire engines to far-reaching discussions of matter theory. In this work Hooke tried to explain in mechanical terms the notions of "congruity" and "incongruity" of bodies, a distinction he had introduced earlier in his explanation of capillary actions. The term "mechanical" here should be understood in two senses: as being in accordance with the mechanical philosophy, and as having to do with mechanics. The tool enabling Hooke to effect this union was the notion of elasticity. In Hooke's account, bodies in motion are the constituents of the sensible universe, but the notion of body itself is redefined as

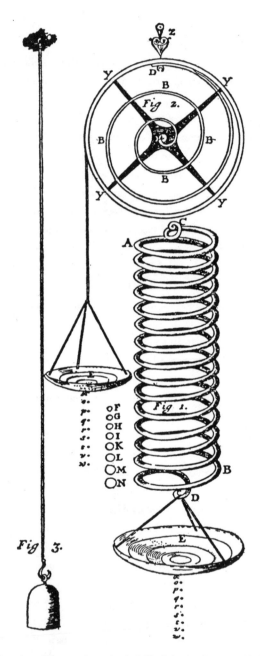

Figure 8.8. Hooke's springs *(De potentia restitutiva).* Hooke's experiments with different types of springs showed that the displacement is as the power, or force, applied.

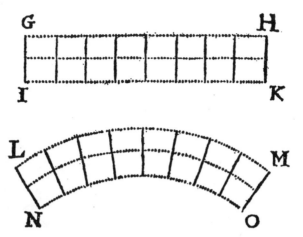

Figure 8.9. Hooke on compound bending *(De potentia restitutiva)*. The picture shows a straight solid body *GHKI* and then the same body bent *LMON*. The side *LM* is extended, while *NO* is compressed.

something "receptive or communicative of motion." Notions such as extension, quantity, hardness and softness, fluidity and solidity, and rarefaction and condensation are also redefined in terms of motion.

Hooke initially defined congruous bodies as those "whose particles have the same magnitude, and the same degree of velocity, or else an harmonical proportion of magnitude, and harmonical degree of velocity," the reverse being true for incongruous bodies. Hooke later qualified this definition, arguing that the greatest part of the extension of bodies is due to their vibrations. For example, a thin flat plate of iron of one square foot vibrating very swiftly over the length of one foot occupies one cubic foot. Indeed, for Hooke this is what a cubic foot of iron *is*. On this basis he compared all parts of matter to strings endowed with their own characteristic vibratory motions or frequencies, which are transmitted by a pervading fluid. Hence the emphasis on terms such as "harmony" and "unison." Congruity and incongruity are now explained in terms of the bodies' vibratory motions. The analogy with the vibrating string emphasizes that this pan-elastic cosmology is explicable in terms of the mathematical laws of mechanics. We see in Hooke how terms like "congruity" and "incongruity," which in the tradition of Renaissance philosophy might have suggested ideas of sympathy or antipathy, have been reinterpreted in strictly mechanical terms by means of analogies with simple objects such as springs and strings.[52]

In *De potentia restitutiva* Hooke provided a somewhat more detailed account of the spring of the air, gesturing toward a micro-explanation of Boyle's law, though without mentioning Boyle. Hooke claimed that the "vibrative spaces" of air are very large and that the motion is very swift. If the space in which a quantity of air is enclosed is compressed in half, "supposing the heat the same," the air will preserve its motion, and the number of its vibrations and impacts (or "occursions") against the container will double. If the space is doubled, the number of vibrations and impacts will be halved, "and consequently its indeavours outward will also be weaker by half."[53] Hooke's words suggest an inverse relation between pressure and volume.

When it came to investigating geometrically elastic vibrations, Hooke did not try to establish a link with physical actions. Rather, he drew conclusions from the nature of the continuum, as when he claimed, "Every point of the space of flexure hath a peculiar power, and consequently there being infinite points of the space, there must be infinite degrees of power."[54] This passage may be helpful in clarifying a terminological and conceptual problem having to do with dimensions. Hooke talked of a force or power as being responsible for, say, the compression of a spring. The sum or aggregate of all the powers acting over a given space is still called a power, much as in Galileo's triangle of speeds the aggregate of all the speeds is still considered to be a speed of sorts, whereas we tend to think that dimensions change in the process. This is why in *Lampas*, another Cutlerian lecture published in 1677, Hooke could state that "the proportion of the strength or power of moving any Body is always in a duplicate proportion of the Velocity it receives from it." The statement is true for pendulums, musical strings, springs, perpendicularly or obliquely falling bodies, projectiles shot from any weapon, and any other type of motion. Hooke invoked the example of the pierced cistern, where the height is as the square of the speed, namely if we want to double the speed, we must quadruple the height.[55]

Despite the dimensional problems, the results Hooke presented were genuinely important. His attempt at proving the isochronism of vibrations, however, was marred by his taking the body's maximum rather than average speed (fig. 8.10). A brief but elegant proof of the isochronism of elastic oscillations was to be found a few years later in a few propositions and corollaries in Newton's *Principia mathematica* (see section 9.5).[56] Whatever its problems, Hooke's mathematical analysis was a significant improvement upon Mersenne's views on oscillation and vibration. Hooke applied more sophisticated mathematical tools and grasped important relations between speed and acceleration that had eluded Mersenne. For example, whereas Mersenne believed that maximum speed occurred when a string was released, Hooke believed speed to increase from the point C where the string is released (fig. 8.10) to a maximum (in A) corresponding to the rest position. Hooke concluded his remarkable essay by emphasizing the uses of elasticity, especially for the construction of mechanical devices based on compressed air.[57]

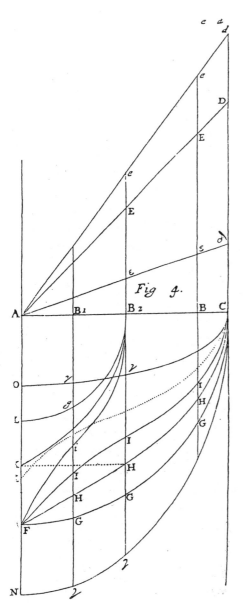

Figure 8.10. Hooke's attempt to prove isochronism *(De potentia restitutiva)*. AC represents the displacement of the spring from the initial rest position in *A*, where *C* is the maximum displacement. The power or *vis* at each point *B* is represented by *EB*, its maximum being *CD*. The aggregate of all the powers is represented by the triangle *ACD*, which is proportional to AC^2 or to the square of the displacement. If a body is released from *C* to *B*, it will receive the aggregate of all the powers in the trapezium *BCDE*. Hooke stated that the speeds of moved bodies are as the square roots of the aggregates or sums of the powers by which they are moved, a result he derived from pendular motion and the pierced cistern. Since, as we have seen, the displacements too are as the square root of the same aggregates, Hooke concluded that the speeds are as the displacements and the times are equal; therefore oscillations are isochronous. Although this conclusion is correct, Hooke's reasoning is based on the assumption that he could take the speed at *A*, rather than the average speed over the whole interval *AC*, in order to find the time. In the lower portion of the diagram, the circle arc *CGF* is the curve of speeds, and *CIF* represents time.

8.5 Bending and Breaking Beams

Unlike his new science of motion, Galileo's science of the resistance of materials attracted the attention of few scholars upon its publication, notable exceptions being Baliani and Mersenne. From the 1660s onward, however, several works appeared that focused on issues such as the principles of the latter new science, its

empirical adequacy, the shape of solids of equal resistance, and the related problems of Galileo's coefficient and the bending of the beam before rupture due to its
elasticity. In these works, beams were regarded as springs, Galileo's own views on
the causes of cohesion apparently having attracted little attention or favor.

In the 1650s and 1660s the Swedish baron Paul Wurtz and the mathematicians
Blondel and Jacques Buot debated the issue from experimental as well as theoretical perspectives. Wurtz gave the beam a parabolic shape and expected it to offer
equal resistance everywhere, but apparently his experimental results did not confirm his expectations. In an extremely rare pamphlet published in 1661 as a letter
to Wurtz, Blondel argued that Galileo had made a mistake in extending a theory
based on beams infixed in a wall to beams standing on two supports. Blondel found
that the profile of a beam resting on two supports, such that any point on its axis
is equally resistant, is elliptical. Soon thereafter the mathematician Jacques Buot
challenged Blondel's attack on Galileo, and Blondel replied with another letter
dated that same year but published only in 1673. Blondel's objection was only
partly justified, because "equal resistance" can mean different things in different
contexts. Although Galileo had claimed that he wanted to design beams for architecture and shipbuilding, he calculated the shape for a cantilever with a fixed
load. By contrast, Blondel and others after him were working with beams supported at both end and offering equal resistance to a load located at any point along
the axis of the beam. The two cases pose different legitimate questions and give
different legitimate results. Given Galileo's declared aims, however, Blondel was
not entirely unjustified.[58]

In 1669 the mathematician Alessandro Marchetti published the treatise *De resistentia solidorum*, in which he extended Galileo's doctrine along similar lines to
Blondel. Marchetti was a student of Borelli's, who is mentioned in the book's introduction. The book was dedicated to Cardinal Leopoldo de' Medici and is clearly
a product of the Galilean tradition. Marchetti was quite careful in dealing with
Galileo, arguing that his results for a beam without weight with one extremity
fixed perpendicularly in a wall were correct. On the basis of tenuous textual evidence, Marchetti interpreted Galileo as suggesting that the same result applied to
a heavy beam supported at both ends. But "facto itaque periculo" (having made
the test, by which Marchetti meant a mathematical calculation rather than an experiment), he realized that Galileo's alleged inference was incorrect.[59] Marchetti
was careful in establishing a theory based on clear definitions and propositions. He
started in book 1 with solids infixed in a wall, then in book 2 compared them with
solids supported on both sides, finally concluding that the resistances of elliptical
or circular solids supported on both sides were equal at any point on the axis (fig.
8.11). The same result had already been proved by Blondel.[60]

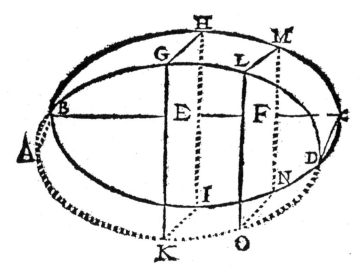

Figure 8.11. Marchetti on the resistance of materials *(De resistentia solidorum).* ABCD is a solid
body with an elliptical profile. Marchetti's proposition argues that the resistance in E is to the resistance in F as $FB \times FC \times HI^2$ is to $BE \times CE \times MN^2$. Since in an ellipse HI^2 is as $BE \times CE$ and
MN^2 as $FB \times FC$, it follows that the ratio between the two resistances is the unity. Therefore a
beam with an elliptical, or circular, profile offers at any point the same resistance to a given load.

As we saw in the previous section, in the same year as Marchetti, Fabri began
publishing the treatises of his *Physica,* including a book *De resistentia corporum*
in which he loosely translated relevant material from the first two days of Galileo's *Discorsi.* Fabri argued that since Galileo's pioneering work was in Italian, it
had not been read by many, and he also wished to add to it.[61] It is remarkable that
such a mathematical subject found a place in a *physica* treatise. In the first day of
the *Discorsi,* however, Galileo, had tentatively discussed the physical causes of the
cohesion and connection of the parts of solid bodies, arguing that they are due to
innumerable empty spaces interspersed within them. In *Physica* Fabri rejected
Galileo's conjecture, invoking instead the *unio continuativa* as a cause for resistance to fracture. Although Fabri had mentioned elastic beams in book 2 of his
first treatise, he did not attempt to formulate a mathematical theory of their resistance.[62] He did apply mathematics to the study of solids of equal resistance, expressing amazement that the lynx-eyed Galileo had failed to identify an obvious
one, namely a prism with a triangular base parallel to the horizon (fig. 8.12).
Moreover, this shape is easier to cut and is more convenient than a parabolic beam
because the weight is reduced to one-half rather than two-thirds.[63]

In 1673 Fabri's friend and fellow Jesuit Ignace Gaston Pardies published *De la
statique,* in which he investigated a number of themes including simple machines

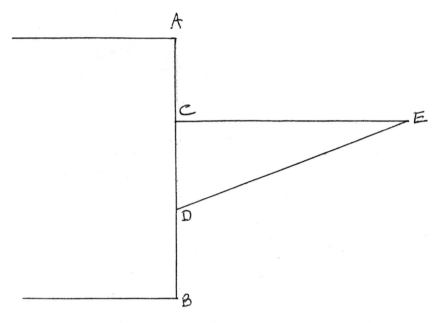

Figure 8.12. Fabri's prism (adapted from *Physica*). *CDE* is a prism infixed in a wall *AB* seen from the top with a triangular base. Fabri's prism is a solid of equal resistance and is both simpler and lighter than Galileo's parabolic beam. By keeping a constant height, Fabri had simple as opposed to quadratic proportions among variables, notably between width and length.

and related problems in statics, the equilibrium of cords, and the resistance of solids. Pardies had profound intuitions that were often followed up by later scholars, but many of his proofs were mere sketches. His main results have to do with the problem of the catenary, regarding which he put forward a remarkable proposition and disproved Galileo's claim that its shape is parabolic. If equal weights are suspended at equal horizontal intervals from a weightless string, the points where they are suspended describe a parabola, but hanging chains and cords do not describe parabolas (fig. 8.13). In some propositions Pardies timidly introduced the notion of elasticity by talking of strings that can be lengthened, but he did not attain significant results in this area. Similar remarks apply to the resistance of beams and solids of equal resistance.[64]

The problem of the resistance of materials was debated both at the Royal Society and at the Paris Académie. In London, experiments were carried out with wooden beams in order to determine the ratio of the weights necessary to break them transversally and longitudinally, very much in Galileo's tradition.[65] In Paris, Huygens considered the case of a beam infixed in a wall at an angle different from the straight, but he did not consider that the beam bends before breaking.[66] His *prop ?*

colleague Mariotte subjected Galileo's propositions to experimental tests witnessed by Huygens, Roberval, and the mathematician and Royal Librarian Pierre
de Carcavi. Meanwhile, Mariotte was in correspondence with Leibniz.

Mariotte's work appeared in his treatise on the motion of water, because he was
led to the problem of the resistance of materials by his interest in the resistance
of water pipes. He was concerned with Galileo's coefficient and argued that the resistance of a beam is proportional not quite to half its height but to more like a   *?*
quarter or a third. The reason is that the fibers of the beam stretch before breaking, and rupture occurs when the fibers reach their maximum elongation. The
elongation of a spring is proportional to its load, therefore also beams behave like
springs. Mariotte accepted Galileo's proportions, and the focus of his empirical
concerns, unlike Wurtz's, was the neglect of bending and the related proportionality factor. Mariotte started with a thought experiment whereby one side of a balance is tied with three equal strings at different distances to a fixed block (fig.

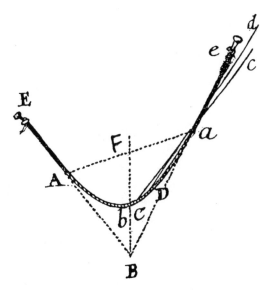

Figure 8.13. Pardies' catenary *(De la statique)*. Pardies claimed that given a hanging chain *abA*, the
perpendicular *FbB* from the point *B* where tangents *aB* and *AB* meet crosses the chain *abA* in its
center of gravity *b*. The proof relies ultimately on a generalization to a continuous case of a theorem valid for a rod hanging from two nonparallel lines. Pardies first argued that the theorem is
valid if the chain becomes a rigid body. Then he claimed that equilibrium is not altered by replacing a portion of the string at arbitrary points *A* and *a*, for example, with weightless strings directed
along the tangent, such as *AE* and *ae*. Nothing changes if the chain is suspended from any point
along the straight line through *EA*, and similarly *ea*, in particular point *B*. Since the chain remains
in equilibrium if we suspend it from *B*, the vertical through *B* must meet its center of gravity *b*.

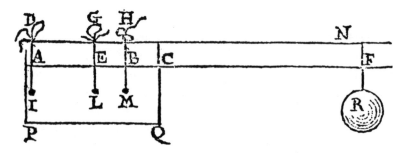

Figure 8.14. Mariotte's balance *(Traité du mouvement des eaux)*. A balance *AEBCF* is secured with three strings *ID*, *GL*, and *HM* to a block *PQ*, and a weight *R* is attached at *F. C* is the fulcrum, and *CF* = 12 feet, *CA* = 4 feet, *CE* = 2 feet, and *CB* = 1 foot. The strings break when they are stretched two lines. Taking them one by one, because of their respective distances, *DI* breaks if *R* = 4 pounds, *GL* breaks if *R* = 2 pounds, and *HM* breaks if *R* = 1 pound. If we take them together, however, *HM* can stretch only one-quarter of its maximum extension and *LG* only one-half, otherwise *DI* would break. Since the degree of stretching is proportional to the strings' load, *HM* supports only ¼ pound and *GL* only 1. Generalizing, the load each string supports varies as the square of the distance from *C*. A similar reasoning can be applied to a beam infixed in a wall, namely Galileo's case. After an extraordinarily contorted calculation, corresponding in the language of calculus to the integration $\int x^2 dx$, where x is the distance from the fulcrum, Mariotte was able to show that the coefficient he was seeking was not ½, as Galileo had thought, but ⅓.

8.14). They stretch proportionally to the weight to which they are subjected and break when they are stretched just over two lines, with the string furthest from the fulcrum clearly breaking first. Their contributions to equilibrium vary as the square of their distances, which enters the proportion twice: once for the law of the lever, and then again because the elongation of the inner strings is also proportional to their distance from the fulcrum, lest the outer one break. A similar reasoning applies to the cantilever, where performing a summation gives the factor ⅓. Mariotte performed experiments with wood, glass, and iron in order to confirm his newly found coefficient, and he reported some numerical data useful to architects.[67]

In 1684 Leibniz published *Demonstrationes novae de resistentia solidorum* in the prestigious *Acta eruditorum* of Leipzig, a journal that had begun appearing in 1682. In claiming that Blondel, Wurtz, Mariotte, and others had shown that Galileo's law about the moment of a force necessary to break a beam did not agree with experiments, Leibniz showed himself to have been aware of Mariotte's work two years before it was published. The correspondence with Mariotte provided Leibniz with the idea that the beam is elastic and bends before rupture. After some false starts, Leibniz showed that the coefficient was ⅓, a result Mariotte accepted (fig. 8.15).[68] Galileo's notion that the beam is perfectly rigid was deemed inadequate. By considering all bodies to be elastic to some degree, Leibniz combined Galileo's work with the law of elasticity stating that the forces are proportional to the extensions. He presented the "fundamental proposition" by referring to the expansion of air; he mentioned Hooke neither in his text nor in his correspondence with Mariotte.[69] In the same work Leibniz also stated that there is a connection between the elastic and the acoustic properties of bodies, a connection that had already been made by Marci, Hooke, and Mariotte, for example.

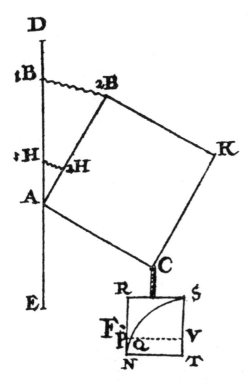

Figure 8.15. Leibniz's beam *(De resistentia solidorum).* Leibniz considered a beam shaped like a prism with a length equal to its height. The beam is elastic and is stretched from its rest position along *ED* by the weight *F.* Note the visual representation of the beam's elasticity by means of the wavy segments $_1B_2B$ and $_1H_2H$. Somewhat curiously, an auxiliary figure is drawn in weight *F.* Because of weight *F,* the prism is stretched from $_1B_1HA$ to $_2B_2HA$. The resistance of each fiber of the prism is proportional to the stretching force and to the distance from *A.* Since the stretching force is as the extension, which is proportional to the distance from *A,* the resistance is as the square of that distance. The auxiliary figure drawn in weight *F* shows a parabola where *PQ* is as PN^2 and *RS* is as RN^2 and distances from *N* along *RN* are proportional to distances from *A* along $_1BA$. The sum of all the resistances is as the area of *NRSQN,* namely ⅓*NRSTN.* Thus the auxiliary figure serves the purpose of performing an integration. Leibniz reached the conclusion that the resistance of a prism *ABCD* to being broken by pulling it straight is to its resistance to being broken by pulling it down (as in this figure) as *BC* is to ⅓*AB,* not ½*AB* as claimed by Galileo.

Only later refinements extended research in this area by considering not only the elastic tension but also the bending accompanying it, a topic Leibniz did not cover. His investigations, however, contributed to the understanding that solid bodies too, like air, could be seen as analogous to springs and were amenable to mathematical investigation by means of the new analytical tools.

CHAPTER NINE

A New World-System

Whether elastic fluids consist of particles that repel one another
is, however, a question for physics. We have mathematically
demonstrated a property of fluids consisting of particles of this
sort so as to provide natural philosophers with the means with
which to treat that question.

Newton

9.1 Teamwork and Anti-Cartesianism

This chapter is devoted to Newton's handling of falling, projected, oscillating, and especially orbiting bodies in *Philosophiae naturalis principia mathematica* (1687, 1713[2], 1726[3]), a work of extraordinary depth and complexity with a rich interpretive history. Although some of its propositions have always been considered crucial, others have had a varied history, and readers' perceptions have changed over time. For example, in the years immediately following its publication relatively few readers had the mathematical skills to tackle the work. Some of those who did, however, such as Huygens and Leibniz, rejected Newton's attractions and did not examine in great detail some of his most complex conclusions. In the eighteenth century, by contrast, scholars with different views on attraction and its causes, from Alexis Clairaut and Jean Le Rond d'Alembert to Leonhard Euler and Pierre-Simon Laplace, reveled in exploring and improving upon some of Newton's most challenging results. More recent historians too have emphasized different aspects of the work, such as the analyses of celestial motions, the brilliant mathematics, the conceptual systematization of mechanical notions, and Newton's concerns with alchemy and theology. Some commentators have quite legitimately read Newton's *Principia* from an eighteenth-century perspective as a first step toward the works of a number of mathematicians culminating with Laplace.[1] In this chapter, however, I focus on the *editio princeps* strictly within a seventeenth-century context and against the backdrop of the themes explored so far in this work.

"Teamwork" is not the first word to spring to mind in connection with Newton's *Principia*, yet there is no doubt that without the input of a few other English mathematicians and astronomers Newton's masterpiece would not be the work we are familiar with, and indeed it might never have been written. Despite the magnitude of Newton's individual achievement, there are good reasons to consider his work in the light of his exchanges with other mathematicians. As we saw in section 7.6, Newton's exchanges with Hooke and Flamsteed around 1680 played a significant role in the formulation of his ideas. Likewise, the exchanges among Hooke, Halley, and Wren in the early 1680s show that they were pursuing an approach in some respects similar to Newton's. Newton must have realized this during Halley's visit in August 1684. The exchanges with Flamsteed in the mid-1680s provided Newton with valuable astronomical data.

It has long been thought that the title of Newton's work may have been intended as a critical qualification on Descartes' *Principia philosophiae.* Newton wished to underline the solid mathematical foundations he had provided to his own work as opposed to Descartes' fancies. Newton was very forceful and effective in his criticisms because he had himself adhered to a large portion of those doctrines he was criticizing. His reaction against Descartes and his views about space and motion can be traced from a manuscript preceding *Principia mathematica, De gravitatione et aequipondio fluidorum,* whose dating is controversial and has been variously assigned from the mid-1660s to the mid-1680s. In *De gravitatione* we see the influence of the views of the Cambridge Platonists, especially Henry More, on God's role in the universe and with respect to space and time. More generally, the more one studies Newton's *Principia,* the more one discerns a far-reaching plan aimed at banishing Cartesian notions. Newton's mathematical propositions and experiments often reveal both a systematic construction effort and a ruthless demolition plan. My reading strategy in tackling Newton's *Principia* is based on my belief that the organization and style of the work were colored by the intention to create an alternative to Cartesianism. Recent interpretations of Newton's preface to the *Principia* support this reading by pointing out both his break with Cartesian mathematics and the significance of that break for a mathematical natural philosophy.[2]

Section 9.2 situates Newton's early efforts within a research tradition involving broadly similar questions and mathematical tools shared with Halley, Hooke, and Wren. I argue that what distinguished Newton from his fellow mathematicians at that early stage was not so much a new philosophical method or his views on alchemy and theology, but his mathematical skills. That situation rapidly changed, however, since in composing the *Principia* Newton soon began to enrich his construction with theological, philosophical, and alchemical views. These themes remained joined in a set of relationships that continued to evolve throughout his life.

Section 9.3 examines the structure and concepts employed by Newton in the opening of his work. I discuss briefly its intended audience, its definitions and laws, and the type of mathematics employed. Section 9.4 examines an area seldom studied in connection with the *Principia*, namely the role of experiments. I identify conceptual and material links among experiments described across the entire book and highlight their links to the third law of motion, the definition of mass, and the rejection of Cartesian fluids. I believe this exploration to be helpful in grasping that some portions of the book that had appeared to some commentators as asides were in fact integral parts of it.

Section 9.5 is devoted to the main bulk of Newton's treatise, including material from books I and II. I start from a proposition in book II in which Newton explored certain mathematical features of fluids, supposing them to consist of repelling particles and seeking that specific repulsion leading to Boyle's law. I then try to show analogies in the style of the two books, where Newton provided a general mathematical treatment of a problem as a grid for physical investigations. The Achilles' heel of this approach is already apparent from the study of Boyle's law, namely that it relies on attractive and repulsive forces whose existence was a highly contentious issue. Section 9.6 revisits the theme of teamwork, this time with respect to Astronomer Royal John Flamsteed, who willingly supplied Newton with a wealth of data he needed for book III on the system of the world. Moreover, relying on a rejected figure from a preliminary version of book III, I show how the approach based on objects was being replaced by one based on abstract principles and higher mathematics.

9.2 Halley, Wren, Hooke, and Newton

In August 1684 the astronomer Edmond Halley visited Newton at Cambridge and asked him which curve would be generated by the combination of a rectilinear uniform motion and an attraction toward a center inversely proportional to the square of the distance. This and other related problems had been the focus of discussions among Halley, Wren, and Hooke in previous years, but nobody could produce a proof. Newton replied that the curve would be an ellipse, but being unable to find the relevant papers, he promised to send the proof to London. Within a few months Newton had composed *De motu corporum in gyrum*, and he was able to send Halley the proof of the inverse, namely that if a body moves along an ellipse and is attracted toward a focus, the attraction is inversely proportional to the square of the distance from the focus, and much more. By March 1687, after several progressively larger drafts, Newton was able to put forward a stunning five-hundred-page work in three books, on the motion of bodies in nonresisting media, the motion of bodies in resisting media, and the system of the world.[3] The last book relied on the propositions of book I and applied them to an extraordinary range of phenomena, such as the motions of planets and satellites, especially the problem of lunar motion, tides, the shape of the Earth, the precession of the equinoxes, and comets. The work brought celestial bodies into the domain of the science of mechanics in a new way.

Newton's *De motu corporum* was remarkable in its own right, but the finished product surpassed it in dramatic fashion. Newton had achieved in just a few years what normally would have been the product of a few generations, and indeed that is how long it took readers to come to grips with it. Thus the mid-1680s were no less *anni mirabiles* than the mid-1660s, when Newton had developed the method of fluxions and began his investigations on motion, gravity, and the nature of light. Newton may have started out simply playing with mathematics following Hooke's suggestions and Halley's visits, but he must have soon recognized the extraordinary and previously unimaginable power of his construction. A huge range of phenomena in the heavens and on the Earth were coming under the compass of the idea that all portions of matter attract each other reciprocally as the square of the distance, an idea known as universal gravity.

Hooke, Halley, and Wren too were seeking to solve a mathematical problem of orbital motion having no immediate recourse to physical causes, let alone alchemical or theological ones. In a letter of 1686 Halley reported a revealing prehistory:

> I waited upon Sir Christopher Wren, to inquire of him, if he had the first notion of the reciprocall duplicate proportion from Mr Hook, his answer was, that he himself very many years since had had his thoughts upon making out the Planets motions by a composition of a Descent towards the sun, & an imprest motion; but that at length he gave over, not finding the means of doing it. Since which time Mr Hook had frequently told him that he had done it, and attempted to make it out to him, but that he never satisfied him, that his demonstrations were cogent. [A]nd this I know to be true, that in January 83/4, I, having from consideration of the sesquialter proportion of Kepler, concluded that the centripetall force decreased in the proportion of the squares of the distances reciprocally, came one Wednesday to town, where I met with Sir Christ. Wrenn and Mr Hook, and falling in discourse about it, Mr Hook affirmed that upon that principle all the Laws of the celestiall motions

were to be demonstrated, and that he himself had done it; I declared the ill success of my attempts; and Sir Christopher to encourage the inquiry said, that he would give Mr Hook or me 2 months time to bring him a convincing demonstration thereof, and besides the honour, he of us that did it, should have from him a present of a book of 40s. Mr Hook then said that he had it, but that he would conceale it for some time, that others triing and failing, might know how to value it, when he should make it publick; however I remember Sr Christopher was little satisfied that he could do it, and tho Mr Hook then promised to show it him, I do not yet find that in that particular he has been as good as his word.[4]

This remarkable account of the attempts to find planetary trajectories by combining a rectilinear uniform motion with a descent toward the sun due to an attraction inversely as the square of the distance teaches us which problems were at the center of attention of leading astronomers and mathematicians. We also learn that Wren, Halley, and Hooke were unable to solve the problem despite having tackled it in what retrospectively looks like a fruitful fashion. With the partial exception of Hooke, we do not have the details of their failed attempts, but it seems clear that neither the straightjacket of the mechanical philosophy nor the lack of a correct philosophical method was the source of the difficulty.[5] Rather, it appears that before Newton, attempts to solve the problem of planetary motion by combining Kepler's laws and a central attraction were unsuccessful because of the difficulties of joining mathematics with the science of motion. Finding a way to take into account Kepler's first law, whereby the areas swept out by the radii are used as a measure of time, was especially significant. The problem of orbital motion was not wholly exceptional in regard to physical causes. We have seen in previous chapters that in the *Discorsi* Galileo too had put forward a science of motion that refrained from providing a physical cause for gravity. Huygens too, in *Horologium oscillatorium*, hypothesis 2, book II, was prepared to treat gravity mathematically, temporarily setting aside the question of physical causes, and the same is true for the problem of impact, as we saw in his exchanges with Neile. Newton's case differed, however, because he soon became convinced, or at least he strongly suspected, that material fluids could not be filling the heavens; therefore the problem of a physical cause for gravity was not simply put aside but appeared, far more problematically, to be undermined by his theory. Although in the mid-1680s Wren was prepared to tackle the problem along lines similar to Newton's, after 1687 he parted ways with Newton, believing the cause of gravity to be mechanical (see sec. 10.4).

Historians have debated on how to read *Principia mathematica* with respect to Newton's well-known interests in alchemy and theology. The parameters of this study, focusing on mechanics and on texts either published or known to contemporaries, do not allow us to address this issue fully. However, we can draw some conclusions even from our perspective. Some historians have tried to identify the key to Newton's success in terms of his method of inquiry and views about nature and its operations, such as the role of active principles and attractions derived from chemical investigations. Newton was far from unique, however, since a number of mathematicians and natural philosophers, from Borelli in Italy to Roberval in France, believed in attractions, or the tendency of bodies to move toward other bodies, while a number of English scholars whose works were familiar to Newton, such as Charleton, Francis Glisson, and Boyle, also believed in active principles.[6]

Attempts to identify a unique method of inquiry developed by Newton in the first stages of his work are problematic. It seems likely that at the beginning Newton perceived his research as a mathematical investigation on motion, and what distinguished him from his fellow English mathematicians was primarily his mathematical prowess. Starting probably around the winter of 1684−85, however, Newton began to see his research in a more far-reaching light, with regard both to the growing complexity of his mathematical investigations and to his theological and alchemical themes. It may thus be more profitable to look at Newton's work from a different perspective, namely not at what methodological and philosophical presuppositions led him to, but at where unexpected mathematical findings led philosophical and theological speculations.[7]

In the mid-1680s Newton worked at his system with the fervor of a man possessed, convinced that his mathematical findings in natural philosophy were revealing a new world-system with profound philosophical and theological implications. The removal of aethereal fluids from the heavens opened the way to Newton's notion of space as a *sensorium Dei*, a means through which God could exert his power on the bodies in the universe. In the second edition Newton introduced a general scholium touching on theological issues, but in the first edition it is harder to find indications of his broader philosophical and theological concerns.

Although these concerns were left at the margins of the *editio princeps*, Newton continued to mull over them privately for the rest of his life, mostly in correspondence and only rarely in print. One exception occurs in the discussion on comets in book III. In a passage left unchanged in all three editions, Newton claimed:

> Comets seem to be required, so that from the condensation of their exhalations and
> vapors, there can be a continual supply and renewal of whatever liquid is consumed
> by vegetation and putrefaction and converted into dry earth. For all vegetables grow
> entirely from fluids and afterward, in great part, change into dry earth by putrefac-

tion, and slime is continually deposited from putrefied liquids. Hence the bulk of dry earth in increased from day to day, and fluids—if they did not have an outside source of increase—would have to decrease continually and finally to fail. Further, I suspect that the spirit which is the smallest but most subtle and most excellent part of our air, and which is required for the life of all things, comes chiefly from comets.[8]

Newton was trying to piece together a new cosmology wherein his computations about cometary paths went hand in hand with his views about processes on Earth and God's providence. According to his proposal, cometary tails would constantly supply the Earth with liquids consumed by vegetation and putrefaction, as well as the subtlest part of air required for life. Thus comets have a major role in God's providential plan, a role defined in terms of Newton's views about matter theory. His views on such topics were related to his vision about nature and emerge fleetingly in hypothesis 3, book III, of the first edition, which states that each body can be "transmuted" into another of a different kind.[9]

9.3 The *Principia's* Structure and Conceptual Framework

Before attempting to present some of the main results in Newton's masterpiece, I wish to discuss some issues such as its title, intended audience, the type of mathematics used, and the taxonomy of concepts employed. The title sounds like a qualification of Descartes' *Principia philosophiae,* and Newton probably intended it so.[10] It tells us much about the book's content and intended audience. The work did not present two new sciences or an in-depth study of an instrument like the pendulum clock but was an ambitious study of motion joined with a new mathematical method for studying natural philosophy, a new ontology of forces, and a partly new mathematics.

The subject matter was not amenable to easy presentation, but even so Newton made few concessions to his readers. Whereas Galileo's *Discorsi* occasionally lightened the weight of the mathematical proofs in days 3 and 4 with discussions among the three interlocutors, Newton compounded the complexity of an already difficult work by adding theorems and propositions that were not central to an understanding of it and that readers could in fact skip over. Indeed, at the opening of book III, and privately to Richard Bentley, Newton advised readers to focus on the first three sections of book I and then move to book III, going back to the propositions of the previous books referred to in the text. The only breathing space is provided by occasional natural philosophical scholia, which break the forbidding structure of the book.[11] By presenting his work thus, Newton was excluding all those who were untrained in higher mathematics. He had reasons to do so, since after his optical publications of the 1670s he had become involved in several exasperating disputes that made him vow to abandon philosophy.

Newton opened his preface with a reference to Pappus and situated his own work in the tradition of rational mechanics: "In this sense *rational mechanics* will be the science . . . of the motions that result from any forces whatever and of the forces that are required for any motions whatever."[12] Whereas mechanics had traditionally dealt with the manual arts, Newton argued that he was extending its scope by dealing with natural things and especially motion as it relates to philosophy. Thus he argued that the task of philosophy is to investigate the forces of nature from the phenomena, and then to demonstrate the other phenomena from those forces. The first two books are structured in axiomatic fashion, much like the Latin portions of Galileo's *Discorsi* and Huygens's *Horologium oscillatorium*. Newton proceeded by arguing that if certain forces are given, then bodies move in given ways. In book III the mathematical structure finds its application to nature in the explanation of the system of the world. Both Galileo and Newton would have loved to provide a convincing and presentable explanation for gravity, but for different reasons they were either unable to do so or refrained from doing so. Both argued that their mathematical proofs provided an excellent description of the phenomena they were studying.

Newton began with a set of definitions involving quantity of matter, quantity of motion, innate and impressed force, and centripetal force and various features of it. Terminological problems had plagued mechanics from the beginning, but the success of Newton's system helped establish a more widely shared nomenclature.

In the Cartesian world, body consisted in extension, at least at the abstract level. By contrast, Newton emphasized that he was considering not just the volume of a body but also its density, which could be determined by means of hydrostatics, as Ghetaldi had done at the beginning of the century. Newton opted to include this definition only in the final stages of composition; it does not appear in the preliminary *De motu* manuscripts. By means of experiments with pendulums Newton was able to establish that mass and weight were proportional (see sec. 9.4). Mass, however, is a constant feature of a body, whereas weight depends on its location; the weight of a body is greater at sea level than on a mountaintop. Consistent with the reigning mechanical philosophy, a number of actions were ascribed to vortices moving through the pores of bodies. Thus the notion a body's mass or bulk was seen in connection with the subtle matter pervading its pores. In his definition Newton did not take this hypothetical medium into account, a choice justified by further pendulum experiments in which he showed that the resistance to motion does not depend on the internal parts of a body.[13]

Innate force, or *vis insita* (also known as the force of inertia, or *vis inertiae*), "is the power of resisting by which every body, so far as it is able, perseveres in its state either of resting or of moving uniformly straight forward." One may wonder why Newton invoked the notion of force to account for rectilinear uniform motion. The reason is that "a body exerts this force only during a change in its state."[14] Thus one can conceive *vis insita* as a "latent" force in the body, manifesting itself when an external force endeavors to change its state. According to Newton, *vis insita* can be seen in two different ways, the first whereby the body strives against the impressed force, the other whereby the body strives to change the state of the source of the impressed force. Both appear to be related to the third law of motion and to the reciprocity of action and reaction, but arguably Newton could not say so at that stage because he had not yet introduced the laws of motion. In preliminary drafts of the definitions Newton had provided as an example of the former the *vis centrifuga gyrantium*, or the centrifugal force of orbiting bodies. Thus centrifugal force seems to be reinterpreted here as a reaction to the force bending a body's orbit.[15] Once again, though, Newton would have had difficulty making that statement at this stage because he had not yet introduced the notion of centripetal force. Centrifugal forces do not figure prominently in the *Principia*, so one is left to wonder what Newton made of them. The clearest statement of his views can be found in a memorandum of the mid-1710s in which, speaking in the third person, Newton stated that "the centrifugal endeavor is always equal to the force of gravity and is directed in the opposite direction because of the third law of motion in Newton's *Principia mathematica*." The context of the passage was orbital motion, and Newton was quite explicit about the trajectory being not circular but "eccentric."[16]

Centripetal force was defined as that force whereby bodies are impelled or tend toward a center, such as gravity, magnetic force, and the force keeping the planets in their orbits. It will become clear in book III that this third force can be identified with gravity, whereas magnetic force is different. Newton argued that magnetic attraction is not proportional to matter, because some bodies are strongly attracted, some less so, and many not at all. Moreover, the magnetic force of a body can be increased and decreased and is roughly inversely proportional to the third power of the distance. When Newton wished to provide examples of attractive forces, he often chose magnetism rather than alchemy. Newton further defined the absolute, accelerative, and motive quantities of centripetal force, proportional to the efficacy of its cause, the speed, and the quantity of motion it generates in a given time, respectively.[17]

The eight definitions are accompanied by a scholium wherein Newton discussed the nature of space and time and introduced the distinction between relative and absolute space, time, and motion. The former are the subjects of our sensations, whereas the latter are mathematical and true. According to Newton it is not possible to discern absolute uniform rectilinear motion, but in the case of circular motion a body's tendency to fly off along the tangent provides evidence of its absolute motion.[18]

Newton then proceeded to provide three laws of motion. Traditionally they have played a major role in expositions of Newtonian mechanics, but it appears that neither Newton nor his contemporaries believed them to be especially novel, with the partial exception of the third. The first law, now commonly referred to as the law of inertia, derived from Descartes and states, "Every body perseveres in its state of being at rest or of moving uniformly straight forward, except insofar as it is compelled to change its state by forces impressed."[19] The second states that the motive force impressed is proportional to the change in the quantity of motion, which occurs in the direction of the line of the force. This law applies both when the force is impressed all at once, as in impact, and when it is impressed gradually, as in centripetal force. Since mass was generally constant, impressed force could be proportional either to the change of velocity or to the change of velocity in a given time, namely acceleration. The action of a centripetal force could be conceptualized as a continuous series of impulses, thus the roots of the second law are to be found in the collision of bodies. The third law states that the reciprocal actions of two bodies are always equal and opposite. This law had important applications to phenomena of impact, which Newton discussed also in relation to experiments performed by Wren at the Royal Society, by Mariotte, and by himself. Newton emphasized that the third law is universal and the conservation of quantity of motion does not depend on bodies being perfectly hard or elastic, but rather occurs even with soft ones, such as balls of compressed wool. Moreover, Newton employed the third law to establish a fundamental result, namely the reciprocity of attraction.

Newton believed the first two laws to be sufficiently established to need no extensive justification. A large portion of the ensuing scholium, however, was devoted to the third law. His attempts to ground it followed a two-pronged approach, one for the collision among bodies and the other for attractions, as we shall see below. From these preliminary remarks it is easy to grasp the central importance of impact in the mechanics of this period and for Newton (see sec. 5.2). Phenomena of impact were enormously important in themselves, but they also provided key tools and principles that, with suitable generalizations, could be extended to broader domains. Impact led to the identification of the centrality of the notion of quantity of motion "in one direction" and its conservation, namely the building blocks used in different ways by Newton.

Ernst Mach claimed that Newton had stated all the principles of mechanics, and that after him no substantially new principle was formulated, but an examination of the issue shows this view to be untenable.[20] Newton's analysis of certain problems was defective because he lacked adequate laws and principles. Whereas in some cases those principles were formulated in an adequate fashion after his death, in other cases Newton failed to take seventeenth-century principles into account. Here I examine two cases instantiating my claim, namely the pierced cistern and the rotating Earth.

In book II Newton studied motion in resisting media and used the pierced cistern as a tool to investigate the effects of a continuous medium. Instead of considering the resistance experienced by a sphere moving through the medium, Newton tried to determine the force of a moving fluid on a sphere at rest. It was in the course of these researches that he investigated the speed of water flowing out of a pierced cistern. It is not clear whether Newton knew Torricelli's work, but he was certainly familiar with Mersenne's *Cogitata physico-mathematica*, wherein the problem was discussed in experimental terms.[21] Newton determined the speed of water spurting from a hole at the bottom of a cistern and the speed of water falling in a vacuum from the height of the water in the cistern, comparing the quantity of motion in the two cases. However, he did not treat them symmetrically. In the one case he considered the amount of water in the cistern and in the other the amount of water in a container with a height double that of the water in the cistern, because in the same time of fall the water covers with the speed acquired at the end of the fall twice the height of the water in the cistern; thus Newton introduced an error of a factor of two that led him to claim that if the motion of the water spurting from the bottom of a cistern is turned upward, the jet would reach only half the height of the water in the cistern. In the case of water falling in a vacuum, Newton did state that it would ascend to the height from which it had fallen, but he did not believe that the same was true for the water exiting the container. His erroneous conclusion proved controversial and led to debates among several mathematicians and revisions in the subsequent editions. After publication Newton slowly became convinced that his account was faulty by experiments performed by Halley at the Royal Society and Fatio de Duillier, rather than by conservation. Halley, for example, performed experiments at the Royal Society on 18 and 25 March 1691 and found that the jet reached well above half the height of the water in the cistern. He also pointed out a discrepancy in the estimation of the speed based on the quantity of water spurting out of the hole and the height it can reach.[22] Overall, Newton disliked conservation principles and saw decay rather than conservation in the world (see sec. 10.3).

Another revealing mistake emerged in Newton's study of the precession of the equinoxes, a phenomenon known since antiquity involving the very slow rotation of the Earth's axis. This phenomenon is due to the gravitational attractions of the sun and moon on a rotating Earth with a bulge at the equator. Although Newton's value of approximately 50″ per year agreed with astronomical observations, in the mid-eighteenth century Jean d'Alembert showed his calculations to be incorrect, and Euler later systematized this area of mechanics. Newton tackled the rotation of rigid bodies without adequate principles and applied the conservation of linear rather than angular momentum. It is revealing in this regard that in the brief explanation following the first law, Newton provided the example of the spinning top that continues to rotate unless it is retarded by friction. In fact, the spinning top is adequately described by the conservation of angular momentum, not by the first law of motion.[23]

It is not easy to characterize Newton's mathematics in *Principia mathematica* because Newton employed a variety of techniques, such as infinitesimal geometry (which he called the method of first and last ratios), infinite series, and occasionally calculus or theorems on quadratures. Some have found it surprising that Newton, who had invented the calculus of fluxions in 1665–66, did not make more use of it in his work. His method of first and last ratios, however, was a suitable and powerful tool for dealing with the geometrical objects he was investigating, such as orbital trajectories and their tangents, in a form closely tied to the calculus. Newton was highly conscious of mathematical style, and his concerns are visible in *Principia*. Yet his masterpiece was not a treatise on mathematics but a work of extraordinary complexity ultimately aimed at mechanics, natural philosophy, and astronomy; thus it is not surprising that he employed the entire range of mathematical tools at his disposal. Later in life, at the time when he was involved in an acrimonious dispute with Leibniz over the invention of calculus, Newton claimed that he had first formulated his masterpiece in the language of calculus, and then had translated its mathematics for publication. Such claims may apply to some portions of his work, but not to the main propositions about orbital motion.[24]

Rather than attempting a characterization of Newton's mathematical style in general, I shall point out certain features as we proceed. Here I wish to introduce what is probably the most characteristic lemma of his method of first and last ratios. Lemma 10 in section 1, book I, states that the spaces traversed by a body urged by a regular, or finite, force are at the very beginning of motion as the square of the times (fig. 9.1). Thus lemma 10 enabled Newton to generalize Galileo's result

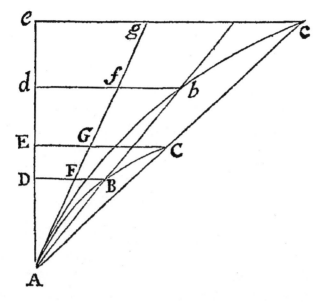

Figure 9.1. Principia, lemmas 9–10. Since lemma 10 is the application to motion of the purely geometrical lemma 9, I discuss them together. The straight line *AE* and the curve *ABC* are given. If
lines *AD* and *AE* represent time and *DB* and *EC* are the correspondent speeds, the areas *ABD* and
ACE are as the distances or spaces traversed. *Ad* and *Ae* are proportional to *AD* and *AE*, and the
curve *Abc* is similar to *ABC*. *AFGfg* touches both curves in *A*. If points *B* and *C* approach *A* while
Ae remains fixed, *Abc* approaches *Ag*. Thus *Ae* and *Ad* are finite and always proportional to the
vanishing *AE* and *AD*. When the angle *cAg* vanishes, the curvilinear areas *Abd* and *Ace* coincide
with *Afd* and *Age* and therefore are proportional to the square of *Ad* and of *Ae*. Because of the proportionality between the vanishing triangles *ABD* and *ACE* with the finite *Adb* and *Ace*, the areas
ABD and *ACE* are also proportional to the square of *AD* and of *AE*. Thus at the very beginning
of motion the spaces are as the square of the times. Notice the analogy with the diagram from
Galileo's *Discorsi* (see fig. 3.17), especially if it is rotated by 180 degrees. Galileo's diagram was frequently reproduced.

about uniformly accelerated motion under a constant force to the case of a variable force, at the beginning of motion. Together with the second law of motion,
lemma 10 plays an important role in Newton's investigations on motion.[25]

9.4 The Role of Experiments

Traditionally, the kernel of Newton's masterpiece has been located in its mathematical demonstrations and in their application to the system of the world in
book III. Commentators often followed Newton's suggestion about reading the
definitions, the laws, and the first three sections of book I before moving to the

system of the world. They often focused on mathematical proofs, and at times on astronomical data, while considering Newton's optical publications as the prime locus for studying his experimental research.

The *Principia* is a work of extraordinary complexity and originality that was composed in an exceedingly short amount of time. It is remarkable that in that short period of intense mathematical research Newton found the time to embark upon experimental investigations ranging from motion in a resisting medium to the determination of the speed of sound. He emphasized the important role of his experiments in such foundational areas as the definitions and laws of motion, and often stressed the empirical foundations of his work even in portions of *Principia mathematica*. Therefore it would seem worthwhile and perhaps overdue to attempt a reading strategy going against the grain of common practice and to investigate what role experiments played in his work. One may ask whether there were intellectual and material links in the apparatus among at least some of the experiments in *Principia mathematica*, or whether it is possible to identify some common strategy in them. I will not discuss Newton's narrative strategy or attempt an exhaustive investigation here, but even a selective survey will highlight that Newton used experiments largely in two ways, by no means mutually exclusive: to underpin some fundamental notion such as a definition or law, and to investigate motion in a resisting medium. The latter aimed to show that subtle fluids either did not exist or had no measurable effect on motion.

Newton became progressively more engaged in experiments while working on the *Principia*, but the initial tract drafted for Halley, *De motu corporum in gyrum*, contains no reference to them, thus suggesting that they did not have a significant heuristic role in the early stages of composition, unlike Galileo's case. Overall, the same is true for the immediately following preliminary draft, *De motu sphaericorum corporum in fluidis*, although there Newton mentioned the resistance to motion of mercury, water, and air, namely the three fluids on which he reports pendulum experiments in book II. The reference, however, is very brief and offers no firm indication that Newton had already performed the same experiments later described in book II.[26] In the published version, experimental reports can be found among the preliminary materials preceding book I and in book II. A renowned experiment in book III concerns the notion of mass and thus relates to the definitions. Later editions of the *Principia* differ so extensively over the issue of experiments that they can almost be seen as different books.

In the scholium to the definitions Newton argued in defense of absolute space, time, and motion. Here we find the famous rotating-bucket experiment, the first in the book. A very long cord to which an empty bucket is attached is twisted tightly and held still. Then the bucket is filled with water and the whole system is released. At the beginning the bucket rotates rapidly, but the water inside it has not yet acquired any rotation, and its surface is flat. After a while the water starts to rotate, and its surface becomes concave. Newton thus concluded that motion should be determined not relative to contiguous bodies but rather with respect to absolute space. Only absolute rotation engenders an endeavor to recede from the axis of rotation; relative rotation with respect to a contiguous body such as the bucket produces no effect. Therefore, at least in the case of rotation, Newton argued that absolute motion not only exists but is also detectable. This experiment was probably an attack on Descartes, who in *Principia philosophiae* had argued that motion is defined with respect to contiguous bodies.[27]

The scholium to the laws of motion contains two experiments in support of the third law, one for impacts and the other for attractions. Since it is problematic to test the latter for gravity, Newton used lodestone and iron: "If they are placed in separate vessels that touch each other and float side by side in still water, neither one will drive the other forward, but because of the equality of the attraction in both directions they will sustain their mutual endeavors toward each other, and at last, having attained equilibrium, they will be at rest."[28] Newton's aim here was to show that the two bodies' mutual attraction is exactly equal and opposite. If it were not so, the two bodies would move in the direction of the stronger attraction after they had come together. After having introduced magnetic force in definition 5 as an example of centripetal force, Newton had recourse to it at this juncture in connection with an experiment that would have been impossible with gravity. In the *Principia* Newton occasionally had recourse to magnetic force as an example of an attractive force with properties different from those of gravity. One such example occurs in the scholium to II, 23, where Newton argued that the attraction of magnetic bodies ends approximately in nearby bodies of the same kind.[29]

The other experiment is considerably more complex and relies on the pendulum, namely the single most significant tool used in the first edition of *Principia mathematica*. First Newton found an ingenious way to subtract the effect of air resistance so as to perform all the experiments as if they were taking place in a vacuum rather than in air. He marked the place to which the bob returns after an entire oscillation. The difference between the point where the bob was released and the point to which it returns represents the effect of air resistance over an entire oscillation, while a quarter of that arc represents the effect of air resistance while the bob falls to its lowest point, namely a quarter of the entire oscillation. The actual tests of the third law were performed with a ten-foot pendulum in a fashion similar to that pioneered by Huygens and discussed in section 8.3. Newton concluded, "As a result of the meeting and collision of bodies, the quantity of motion—determined by adding the motions in the same direction and subtracting the motions in opposite directions—was never changed." Newton was confirming here Huygens's result and refuting Descartes' conservation law. Newton confirmed also another important result established by Huygens, namely that the experiment and the law it embodies are valid for all sorts of bodies, whether hard or soft, elastic or not. In addition, Newton provided the ratio of the speeds before and after collision for a number of bodies. For tightly compressed balls of wool the ratio was about 5 to 9, for steel it was nearly 1, for cork it was just less than 1, whereas for glass it was approximately 15 to 16.[30] Experimental evidence in support of a law of nature was intertwined with the investigation of elastic properties of bodies. Here and elsewhere in his work Newton provided numerical data that could be read in a variety of ways, including as a guide to further theoretical and experimental investigations on the behavior of bodies.

In the scholium to the laws of motion Newton's concern was to subtract the effects of air resistance in order to establish the equality of action and reaction and to determine the bodies' elasticity. In the scholium to proposition II, 40, Newton again used pendulum experiments to study air resistance, this time the focus of attention being that resistance which had to be eliminated while studying the third law. As we saw in section 7.4, pendulum experiments on air resistance had already been performed at the Royal Society in 1669. The idea behind Newton's efforts was that total resistance could be represented as a polynomial in different powers of the maximum speed. Resistance may be rather complex to measure, but in proposition II, 31, and its corollaries Newton established a proportion enabling him to determine the resistance by measuring the decrease in the arcs described. He took the portion of the arc lost in relation to the pendulum's length and the resistance on the bob in relation to its weight. Thus the variables had no physical dimensions.[31]

Newton expressed the portion of arc lost in a given number of oscillations as a polynomial with different powers of the maximum speed. His investigations relied on algebraic techniques and would have been unfeasible within the framework of the theory of proportions, as we saw in section 6.6 with Boccabadati's experiments, where the theory of proportions hindered a formulation combining pressure and speed. Newton's innovative efforts were brilliant and at the same time highly problematic. The scholium can be divided into four parts. The first and longest involves experiments in air, the second in water, and the third in quicksilver, while the fourth attempted to determine whether any resistance arose from the internal parts of matter, as we shall examine later on in this section.[32]

The experiments in air were quite complex and involved attempts to determine both the coefficients of the terms with different powers of the maximum speed

Defcenfus Primus	2	4	8	16	32	64
Afcenfus ultimus	$1\frac{1}{2}$	3	6	12	24	48
Num. Ofcillat.	374	272	$162\frac{1}{2}$	$83\frac{1}{3}$	$41\frac{2}{3}$	$22\frac{2}{3}$

Figure 9.2. Results of pendulum experiments in air (*Principia*, scholium to prop. II, 40). The first row in the table shows the arc of descent for six pendulum experiments; the second row shows the last arc of descent, which is determined as to be three-quarters of the initial arc; the last row gives the number of oscillations required to have the reduction in the arc. For large arcs the data show a greater resistance.

and the dependence of resistance on the surface of spherical bobs (fig. 9.2). Newton used a wooden ball of $57\frac{7}{22}$ ounces, and the pendulum measured 10½ feet from the center of suspension to the center of oscillation. Newton relied on a horizontal ruler placed 5 inches higher than the center of oscillation to measure the bob's displacement. He performed two series of six experiments each with initial displacements of 2, 4, 8, 16, 32, and 64 inches. Since he could not determine exactly the portion lost in one oscillation, he measured the number of oscillations required by the pendulum to reduce its displacement first by one-eighth and then by one-quarter. The number of oscillations ranged in the first series from 164 for the smallest displacement to 9⅔ for the greatest, and in the second from 374 to 22⅔.

Without warning or justification, Newton used polynomials not simply in the powers of V and V^2 but also of $V^{3/2}$, where V is the maximum speed, setting the portion of arc lost $\delta_{arc} = AV + BV^{3/2} + CV^2$, where A, B, and C are arbitrary coefficients that need to be determined. Even with such arbitrary moves, his entire construction remained plagued with problems. Out of six experiments with different initial displacements giving six equations for the coefficients A, B, and C for V, $V^{3/2}$, and V^2, he arbitrarily chose three. The values for A, B, and C associated with different combinations of equations varied greatly and in some cases were even negative. At one point Newton surmised that part of the problem was due to the resistance of the thread, which needed to be minimized by making the balls comparatively larger. Those experiments led him to conclude, "Those parts of the resistances that are (the balls being equal) as the squares of the velocities are also (the velocities being equal) as the squares of the diameters of the balls."[33]

Newton's overall aim was to show that the term in V^2 was dominant and its coefficient was proportional to the fluid's density and, for a sphere, to the square of its diameter. This was a crucial result for Newton's philosophy of nature. In the corollaries to proposition II, 40, Newton had stated a link between the resistance of a fluid to the motion of a sphere being proportional to the square of its diameter and the size of the fluid particles:

—Corollary 1. Therefore, if from the increased magnitude of a spherical solid its resistance is increased in the duplicate ratio, the resistance of a given spherical solid will by no means be diminished as a result of a diminished magnitude of the particles of the fluid.

—Corollary 2. But if, by increasing the spherical solid, the resistance is increases in less than a duplicate ratio of the diameter, it will, by diminishing the particles of the fluid, be diminished in the ratio by which the increased resistance is wanting from the duplicate ratio of the diameter.[34]

This meant that by diminishing the size of the particles of the fluid, the resistance they offered to motion would decrease even faster; therefore the heavens could conceivably be filled with a fluid so subtle as to offer virtually no resistance to motion, yet they would not be empty. In a passage from the general scholium following proposition II, 40, Newton referred to the corollaries to that proposition claiming that the resistance to motion could not be much diminished by making the particles of air smaller. Therefore media less resistant than air had to be less dense than air. If the underlying implications had been missed by someone, in the following paragraph Newton expressed his wish that such experiments could be carried out with greater precision, because "the demonstration of a vacuum depended" on them. In *De motu sphaericorum corporum in fluidis,* written around late 1684 or early 1685, Newton still talked of the resistance of the aether being either nil or very small. Thus his "demonstratio vacui" plausibly dates from 1685 at the earliest. In his own copy of Newton's masterpiece Leibniz underlined the words "demonstratio vacui."[35] We see here the subtle correlations between Newton's immersion in the peculiarities of matter and his belief in the existence of a vacuum. Once again, the behavior of pendulums had far-reaching consequences.

Newton proceeded relentlessly experimenting with different fluids, such as fresh water and mercury. He used a 166⅙-ounce lead ball attached to a thread about 10½ feet long oscillating in a container filled with water and an iron pendulum about 3 feet long oscillating in mercury. In both cases he provided extensive data from experiments, in tabulated form for motion in water. Newton had even more ambitious projects of performing experiments with "molten metals and certain other liquids, hot as well as cold," but he believed he had established "that the resistance of bodies moving swiftly is very nearly proportional to the density of the fluids in which they move."[36] Since reducing the size of the fluid's particles could not diminish resistance, the regularity of celestial motions indicated that the density of heavenly fluids must be zero and therefore celestial bodies move through a vacuum.

Brilliant and innovative as Newton's method was, it was plagued by a number of problems both theoretical and experimental. For example, the components associated with different powers of V cannot be separated in the way he envisaged, and pendular motion generated an oscillatory motion in the fluid that affected the results.[37] Although it was already known that book II had an anti-Cartesian sting, Newton's concerns with a demonstration of the existence of the vacuum appear so deeply rooted in its structure as to affect its meaning as a whole. While displaying his mathematical and experimental skills in providing a comprehensive account of motion in a resisting medium, Newton was aiming from all angles at clearing the heavens of subtle fluids.

Whereas the experiments supporting the third law can be found in the appropriate scholium, those in support of the definition of mass or quantity of matter are announced in definition 1, but they are actually described hundreds of pages later. In the definition of mass Newton faced two problems. The first is that whereas the notion of weight could be determined experimentally, the notion of mass seemed to be entangled in the world of subtle fluids and aethereal particles penetrating the pores of bodies. Newton cut that Gordian knot by stating that he was "not taking into account any medium, if there should be any, freely pervading the interstices between the parts of bodies." The issue was whether such a choice made sense for the establishment of a science of motion, namely whether one could ignore subtle fluids with impunity. The second problem concerned the determination of mass. Newton argued, "It can always be known from a body's weight, for—by making very accurate experiments with pendulums—I have found it to be proportional to the weight, as will be shown below."[38] He addressed the first problem at the end of the scholium to proposition II, 40, and the second in proposition III, 6. In both cases Newton used eleven-foot pendulums whose bobs consisted of round wooden boxes, possibly the very same apparatus. Thus this material link strengthens the conceptual link between the two experiments and the definition of mass.

The idea of the first experiment was to compare the resistance to air offered by the outside of the pendulum to the hypothetical resistance to an aethereal medium offered by the internal parts inside the bob. In performing the experiment Newton employed techniques analogous to those he had used to justify the third law of motion, namely he marked the places reached by the pendulum with the empty wooden box he used as a bob after successive oscillations. He then filled the bob with lead and other heavy metals. The pendulum with the bob filled weighed 78 times more that the pendulum with an empty box. Newton added a plethora of details to highlight the great accuracy of his experiment, my favorite being that he calculated the weight of the air contained in the empty box. Letting the pendulum with the filled box oscillate between the places he had previously marked, Newton counted each time 77 oscillations. Since the *vis insita* of the pendulum with the filled bob was 78 times greater than that with the empty bob, he reasoned that the filled pendulum ought to have returned to the marked places in 78 oscillations, and the resistance of the full bob was not greater than the resistance of the empty bob than 78 is to 77. If A represents the resistance of the external surface and B that of the internal parts of the empty box, $A + B$ will be the total resistance of the pendulum with the empty bob and $A + 78B$ that with the filled bob. Their ratio should be 77 to 78, as determined above. From this proportion $A = 5,928B$, hence the resistance B of the internal parts would be utterly negligible compared with the resistance A of the external parts.

It is remarkable that Newton devoted so much time and effort to an experiment inherently so problematic. We notice again a correlation between the peculiar behavior of pendulums in very special circumstances and far-reaching conclusions about the role of a hypothetical aether for the science of motion. Newton concluded his report by claiming that he was relying on memory, since he had lost the relevant papers. This does not necessarily mean that he had performed the experiment a long time before, since in all likelihood he was able to rely on the same apparatus for another experiment described in relatively few words but of considerable moment.[39]

In proposition III, 6, Newton stated that "at any given distance from the center of any one planet the weight of any body whatever toward that planet is proportional to the quantity of matter which the body contains." Setting perturbations aside, Newton's claim amounted to stating that all bodies fall with the same acceleration regardless of their weight. Newton used pendulums rather than falling bodies and relied on proposition II, 24, establishing that in simple pendulums of equal length, the quantities of matter are as the weights and the squares of the periods combined. Thus if the periods are the same, the quantities of matter are as the weights. Newton used two eleven-foot pendulums with wooden round boxes as bobs. He filled one with wood and the other with the same weight of a long list of substances such as gold, silver, lead, glass, sand, common salt, wood, water, and wheat. The two pendulums oscillated isochronously for a very long time, thus the amount of matter each pendulum contained was exactly proportional to its weight. Newton claimed great precision for these experiments, arguing that they would have detected a difference in the amount of matter of less than a thousandth.

The conclusions were highly general and significant. Mass and weight, though conceptually distinct, were exactly proportional, and the mass in the expression for quantity of motion in the second law was exactly proportional to the mass responsible for gravity.[40] In corollary 3 to proposition III, 6, Newton defended again the existence of empty space with the words "Itaque Vacuum necessario datur," a claim duly noted by Leibniz in his reading of the *Principia*. Since his experiment had shown that mass was always proportional to weight, changing a body's shape could not alter its weight. Therefore, according to Newton, interstitial empty spaces must exist, otherwise air would be so dense that even bodies as dense as gold could not descend in it.[41] In *Principia philosophiae* Descartes had explicitly denied that gravity of bodies on the Earth was always proportional to their quantity of matter, because gravity depended on the variable speed and size of the particles flowing through them. Once again we find Newton targeting Descartes by means of an experimental refutation.[42]

The last experiment I examine here is Newton's determination of the speed of sound. Here too the pendulum was crucial to his work, both conceptually and experimentally. Some commentators consider Newton's result as a detour in a work chiefly aimed at celestial phenomena,[43] but even in this case Newton wished to undermine the role of the aether in the propagation of sound. In *New Experiments Physico-Mechanical Touching the Spring of the Air* (Oxford, 1660) Boyle had reported several experiments carried out in his air pump. Whereas a watch in an evacuated container seemed to produce no audible sound, a bell seemed to produce a sound as loud in the emptied container as when the container was full. This experiment led Boyle to hypothesize that there may be a fluid subtler than air responsible for sound propagation. There is no question that Newton was responding to those concerns, since in the scholium at the end of section 8 of book II he referred to those experiments. Moreover, the previous experiment in Boyle's treatise deals with the motion of a pendulum in a container exhausted of air. This experiment had a lasting appeal for Newton and probably inspired some of his experiments discussed above.

What Newton set out to prove by measuring the speed of sound was that sound consists exclusively in the motion of air, not a subtler fluid like the aether. The theoretical basis for his work was a remarkable combination of the elasticity of the air with the known laws for the motion of the pendulum (fig. 9.3). First Newton established in proposition II, 44, that the oscillations of water in a U-shaped tube are isochronous and correspond to those of a cycloidal pendulum whose length is half the height of the water in the tube. In proposition II, 48, Newton determined that the speed of pulses propagated in an elastic fluid is as the square root of the pressure of the fluid over its density, giving a speed of sound of 979 feet per second.[44]

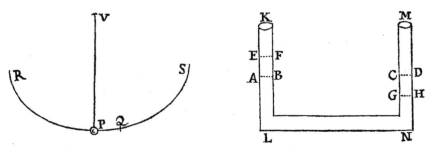

Figure 9.3. Pendulum and U-shaped pipe (*Principia*, prop. II, 44). In order to study wave motion, Newton established an analogy between the oscillations of water in a U-shaped tube and those of a cycloidal pendulum. In the pendulum the force is proportional to the displacement, whereas in the tube the force is proportional to the excess weight in one arm over the other. If the equilibrium position is *AB*, *CD*, the excess weight is proportional to *AE* plus *CG* or twice *AE*. Hence a pendulum with a length *VP* will oscillate with the same period as the water in a tube wherein the heights *KL* and *MN* are half *VP*.

Newton remarked that this value was considerably smaller than the one given by Mersenne of 1,474 feet per second and larger than the one given by Roberval of about 600 feet per second (the latter estimated at the siege of Thionville from the delay of the sound of cannons after the flash was seen). Newton performed his own experiment in the courtyard of Trinity College by measuring the time it took the sound generated at one end to produce a fourfold echo. In order to measure the time, Newton used a pendulum with a variable length. After several trials he determined that its length was greater than 5½ and less than 8 inches. Since it is the square root of the length that is proportional to the period, the difference in the periods is smaller, more than 920 and less than 1,085 English feet. As he put it in the scholium, "Accordingly, since this motion depends on the density of the whole air, it follows that sound consists not in the motion of the aether or of some more subtle air but in the agitation of the whole air." This was not the only stab at Descartes in section 8 of book II, on the propagation of motion through fluids. The first proposition of that section states that "pressure is not propagated through a fluid along straight lines, unless the particles of the fluid lie in a straight line." The target was probably Descartes' theory of propagation of light as pressure, studied by Newton in his early college years.[45]

The picture of Newton's experimental activity emerging from the *editio princeps* reveals the existence of a set of common experimental practices and tools as well as his commitment to refuting the notion of subtle fluids. Large portions of book II in particular were devoted to such efforts and appear as a *pars destruens*, a necessary clearing of the ground before moving to the "true" system of the world. From the standpoint of experimental practice, *Principia mathematica* appears to

be more coherent in its aims and selection of material than has thus far been recognized.

9.5 The Mathematical Principles of Natural Philosophy

Approaching the *Principia* by way of its experiments highlights the different
phases of its composition. The existence of absolute space, the notion of mass and
its proportionality to weight, the third law of motion, and the denial of the existence of the aether were all missing from *De motu corporum in gyrum* and other
early drafts. They probably emerged in 1685, while Newton was fully engrossed in
his work. In their basic structure and contents, sections 2 and 3 of book I cover material from the earliest formulations of the second half of 1684.

Newton's philosophical outlook and style of presentation changed as well. In
his quest for certain knowledge, Newton sought a new strategy, a move made all
the more necessary by his departure from the mechanical philosophy. In a wholly
unconventional move, I have selected proposition 23 from book II to illustrate
some features of his philosophical and rhetorical approach. The proposition
claims, "Particles that are repelled from one another by forces that are inversely
proportional to the distances between their centers constitute an elastic fluid
whose density is proportional to the compression."[46] Conversely, "if the density of
a fluid composed of particles that are repelled from one another is as the compression, the centrifugal forces [forces of repulsion] of the particles are inversely
proportional to the distances between the centers." By "centrifugal forces" Newton meant simply repulsive forces, as correctly interpolated by the translator. In
the accompanying scholium, Newton generalized the result to repulsive forces inversely as any power n of the distance, the compressing forces will be as the cubed
root of E^{n+2}, where E is the density; therefore if $n = 1$, the compressing forces are
as the density. This result—Boyle's law—is valid with some serious restrictions,
such as that the forces do not extend beyond the first set of particles.

The conclusion of the scholium is especially significant: "Whether elastic
fluids consist of particles that repel one another is, however, a question for physics.
We have mathematically demonstrated a property of fluids consisting of particles
of this sort so as to provide natural philosophers [*Philosophi*] with the means with
which to treat the question."[47] Newton provided a grid of mathematical options,
but the grid would be unhelpful if particles did not repel each other. Thus physics
or natural philosophy enters the picture twice: first to determine whether there
are indeed such repulsive forces; and second, if they exist, to determine their dependence on distance. By "physics" Newton intended an enterprise involving experiments but not necessarily restricted to them.

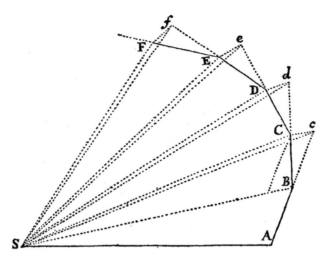

Figure 9.4. Motion under central forces (*Principia,* prop. I, 1). Newton's orbital path is a polygon
with vertices *A, B, C,* etc., where instantaneous impulses of centripetal force deflect the body toward
S. Dividing time into equal parts, Newton could show, using elementary mathematics, that the areas of triangles *SAB, SBC, SCD,* etc., are equal and are proportional to the times required to traverse
AB, BC, CD, etc. The same would be true for any composition of such triangles. At this point the
centripetal force is made to act continuously, and the width of the triangles is indefinitely diminished so that the polygon *ABCD*... becomes a continuous curve. Newton adopted the same mathematical approach while studying the motion of very small bodies acted upon by each of the particles of a large body or medium, as in light propagation. He first supposed the medium to consist of
a finite number of homogeneous strata and then diminished their thickness and increased their
number indefinitely, so as to have a continuous motion.

A not entirely dissimilar approach is visible in proposition I, 1, from section 2—
on the determination of centripetal forces—where Newton provided an extraordinarily powerful and elegant theorem proving that the areas swept out by radii
drawn from revolving bodies to an immovable center of force *S* are proportional
to the times in which the corresponding arcs are traversed (fig. 9.4). Of course, as
Leibniz immediately pointed out, Newton's theorem is valid for a central force in
general, whether attractive or repulsive. The dependence of the central force on
the distance is arbitrary. Here the curve described by the orbiting body is initially
represented as a polygon. This is a generalization of Kepler's area law, according
to which the areas swept out by the radii from the planets to the sun are proportional to the times. Newton also accounted for and modified within a far more
complex system Kepler's other two laws, stating that planets move in ellipses
where the sun is at one of the foci, and that the square of revolution times are proportional to the third power of the major axes of planetary ellipses. In *De motu
corporum in gyrum* Newton was rather hasty in moving from mathematical

propositions to astronomy and in a scholium claimed that "the major planets orbit, therefore, in ellipses having a focus at the centre of the Sun, and with their
radii drawn to the Sun describe areas proportional to the times, exactly as Kepler
supposed." At that stage Newton had not yet established universal gravity. Subsequent refinements showed that the matter was considerably more complicated because the mutual attractions of all particles of matter render the calculations enormously complex.[48]

The analogy with the case of Boyle's law becomes more evident if we consider
proposition I, 2, stating the converse, namely that if a body moves along a curve
where a radius drawn to a point at rest or moving in rectilinear uniform motion
describes equal areas in equal time, the body is acted upon by a centripetal force
to that point. But in the eyes of some contemporaries this result based on the assumption that centripetal forces exist looked like a peculiar coincidence of mathematics, since real bodies do not move through empty spaces attracted to a center,
and it seemed at least conceivable that a fluid carrying a planet may make it move
in accordance with Kepler's laws.[49]

The rest of section 2 shows how to find centripetal forces in different circumstances. We can get a sense of the power and intuitive nature of Newton's geometrical method from proposition I, 6, in which he found an expression for centripetal force for a body moving around a center of force (fig. 9.5). In proposition
I, 10 Newton studied a body moving along an ellipse and attracted toward its center. He found that the force is as the distance and all the periods of revolution in

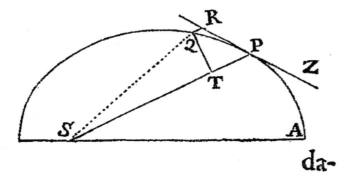

Figure 9.5. The determination of centripetal force (*Principia*, prop. I, 6). *P* revolves in any curve
APQ around a center of force *S*, where *RPZ* is the tangent to the curve in *P*, *QR* is parallel to the
radius *SP*, and *QT* is perpendicular to *SP*. In the limit when *P* and *Q* coincide, centripetal force is
proportional to the deviation *QR* and inversely as the square of the time, because of the generalization of Galileo's result in lemma 10. But since the time is as *SP* × *QT*, in the limit, centripetal force
is as the distance *QR* over the square of time $SP^2 \times QT^2$. For specific curves Newton was able to
find different forms of this expression.

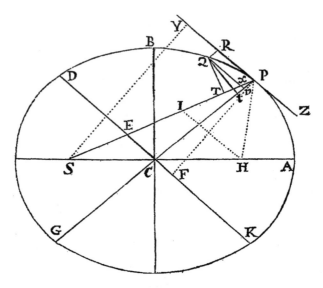

Figure 9.6. Centripetal force in an ellipse (*Principia*, prop. I, 11). Newton proved geometrically that if a body moves along an ellipse, the centripetal force toward the focus is inversely proportional to the square of the distance. In the ellipse *ABDGK*, *C* is the center and *S* the focus. *P* is the position of a body and *ZPR* the tangent to the curve. *Q* is another point on the curve, and *QR* is parallel to *PS*. From the corollary to proposition I, 6, when *Q* approaches P the force is proportional to $QR/(SP^2QT^2)$. For the ellipse Newton showed that QR/QT^2 is constant; therefore the force is inversely as SP^2 or inversely as the square of the distance.

all ellipses around the same center are the same. Newton proved this proposition, implying the isochronism of harmonic oscillations, first for similar ellipses and then for all.[50]

Section 3 of book I tackles the problem of finding the forces toward the focus for bodies moving along conic sections. In propositions I, 11–13, Newton proved that if bodies revolve along conic sections, the force toward the focus is inversely proportional to the square of the distance (fig. 9.6). In the first corollary to proposition I, 13, Newton stated that the inverse is also true, but mathematicians disputed whether he demonstrated his statement.[51]

Sections 4–5 deal with conic sections and are part of the *Principia's* scaffolding. Originally Newton inserted them for the sake of their role in the study of comets, but by the time he found a better way to tackle comets, book I had already been printed. Section 6 deals with the problem of finding the position of an orbiting body at any given time.[52]

Sections 7 and especially 8 have a long interpretative history. Section 7 deals with the rectilinear ascent or descent of bodies acted upon by centripetal forces.

Proposition I, 39, the last proposition in that section, is a springboard to the following section, on finding the orbits of bodies acted upon by arbitrary centripetal forces. The problem of determining the orbit given the force is one of daunting complexity, and Newton could offer only a general geometrical representation, providing a more detailed construction only for some specific cases that were not central to his chief endeavor, as in corollary 3 to proposition I, 41, when the force is inversely as the third power of the distance. His attempted solution of the inverse problem of central forces in the corollary to I, 13, follows a different path.

In section 9, on the motion of bodies in moving orbits and the motion of the apsides, Newton proceeded toward still greater complexity. Whereas section 8 provides a mathematical generalization, section 9 has implications for astronomy, notably lunar motion. Propositions I, 44 and 45, are to my mind among the most impressive of the entire book. In the specific case of orbits differing very little from circles, Newton found by means of series expansions a way to link deviations from the inverse-square law to the rate of rotation of the apsides. This is another instance in which Newton provided general results for centripetal forces varying with an arbitrary power of the distance A from the center in the form A^n, or even a polynomial $bA^n + cA^m$, where b and c are given numbers and n and m arbitrary exponents. Newton's procedure enabled him to find the centripetal force from the motion of the apsides and to retrieve from a more general standpoint some of the results found in sections 2 and 3, such as propositions I, 10 and 11.[53]

The results proved by Newton in the opening sections of book I consider a center of attraction that is either at rest or in uniform rectilinear motion. The reciprocal attraction of bodies, however, required a more elaborate approach whereby centers of attraction move in more complex ways. This is the topic of section 11. In its introduction Newton stated that although he was talking of attractions, perhaps physically one should talk of impulses. Attractions suggest continuous actions at a distance, whereas impulses suggest a discrete contact action, as in the impact between bodies. He also added that since he was dealing with mathematics, he could set aside debates regarding physical issues. Such comments, which can be found in the later editions as well, are peculiar because Newton also argued that the heavens are void of any resistance, thus leaving readers to wonder whence the impulses would be coming. This was precisely one of Leibniz's objections. At the end of the section Newton returned to the same issue and outlined a procedure of inquiry based on mathematics:

> Mathematics requires an investigation of those quantities of forces and their proportions that follow from any conditions that may be supposed. Then, coming down to physics, these proportions must be compared with the phenomena, so that it may

be found which conditions of forces apply to each kind of attracting bodies. And then, finally, it will be possible to argue more securely concerning the physical species, physical causes, and physical proportions of these forces.[54]

Here Newton put mathematics in the driving seat, but this programmatic statement clashed with the one quoted at the beginning of this section on Boyle's law. There he had left to physics the investigation of whether fluids do consist of particles endowed with repulsive forces; here he took for granted that attractive forces exist and left to physics the investigation of their nature. No wonder his contemporaries were baffled.

Proposition I, 66, deals with the problem of determining the trajectories of three bodies attracting each other with forces inversely as the square of the distances, known as the three-body problem, which presented one of the greatest mathematical challenges of Newton's system (fig. 9.7). This proposition and its many corollaries were the basis for studying lunar motion in book III. It is remarkable that although Newton tackled the problem geometrically, he managed to find some approximate solution.

Sections 12 and 13 deal with the attractive forces of spherical and nonspherical bodies where each point attracts all the others with a force inversely proportional

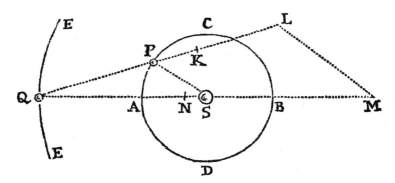

Figure 9.7. The three-body problem (*Principia*, prop. I, 66). Newton indicated the three bodies as *S, P,* and *Q,* of which *S* is the largest. The orbit of *P* is *ADBC,* and that of *Q* is *EE. QK* is the mean distance between *P* and *Q* and also represents the accelerative attraction of *P* versus *Q* at that distance. *QL* is constructed so that $QL:QK::QK^2:QP^2$ and represents the accelerative attraction of *P* versus *Q* at the distance *QP.* At this point Newton decomposed *QL* into *LM* and *MQ,* where *LM* is parallel to *PS* and *MQ* is along *SQ.* The effect of *LM* is to increase or decrease the attraction between *P* and *S,* whereby the radius *PS* still sweeps out areas proportional to the times, but the force is no longer inversely as the square of the distance. The component *QM* adds a force that is no longer directed along the radius *PS.* Newton proceeded to study different cases depending on the respective sizes of the bodies *S, P,* and *Q* and to estimate the effects of perturbations on their orbits.

to the distance. In propositions I, 73 and 74, Newton was able to prove that a particle located inside a spherical body is attracted by a force proportional to the distance from the center; one outside is attracted by a force inversely proportional to the square of the distance from the center. In the latter case the spherical body acts as if all the mass were concentrated in the center. Thus insofar as planets are perfect spheres, they can be treated as points. In some cases this is an adequate approximation, but in others it is necessary to take into account that planets have a bulge at the equator due to their rotation. In section 13 Newton extended his investigations to the case of nonspherical bodies, a case relevant to the shape of the Earth and lunar motion.

Book II, dealing with the motion of bodies in resisting media, has been considered far less appealing by historians and has consequently been studied less than book I. The first two books are presented in a similar mathematical style. In book I central forces vary with the distance, whereas in book II resistance forces vary with the speed. But book II differs in another important respect from book I in that it discusses a number of careful quantitative experiments, as we have seen.

Newton dealt with resistance proportional to simple velocity and the square of velocity, pendular motion and the resistance of fluids to the motion of projectiles, wave motion, and the motion of sound. Throughout the *Principia* Newton gave numerical values for experimental results that are in close, but not complete, agreement with theoretical predictions. This is an important feature of his work.

Since I provided a brief sketch of book II in the previous section, here I mention only one aspect of it, Newton's attempted refutation of interplanetary vortices. He attempted to refute them by means of rather dubious propositions on the rotation of a sphere in a fluid and by applying rules about the motion of fluids, trying in both cases to point to a disagreement with Kepler's laws of planetary motion. In proposition II, 53, Newton drew a simplified picture of the solar system according to Copernicus, with the sun in the center S, an outer circle CF concentric with S, whereas circles DA and EB are not centered in S (fig. 9.8). According to Kepler's second law, planets should move faster in E and D than in A and B, but according to the laws of mechanics, the vortices should be swifter in the narrower spaces A and B than in the wider ones E and D. Thus Newton was trying to show that the hypothesis of vortices contradicts either the laws of planetary motion or Castelli's law of water flow (discussed above in chaps. 3 and 6).[55]

9.6 A New World-System: Newton and Flamsteed

Book III differs in style from the previous ones in that it deals with astronomical phenomena on the basis of the theorems of book I. One could say that book

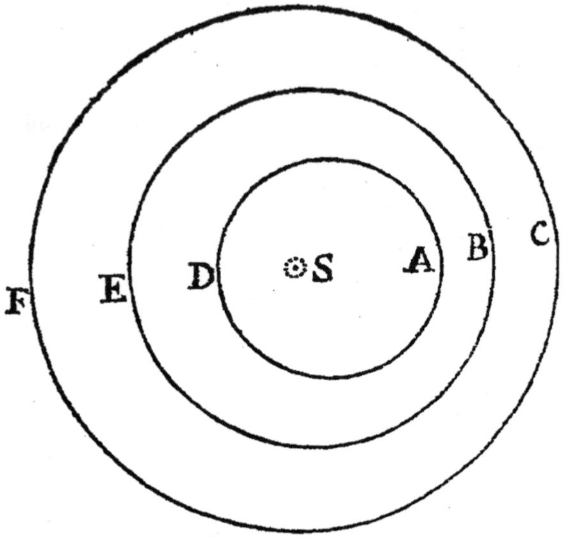

Figure 9.8. Refutation of interplanetary vortices (*Principia*, prop. II, 53). The figure shows the sun in *S*, a planetary orbit *FC* concentric to *S*, and two eccentric orbits *EB* and *DA*. *A* and *B* are the aphelia, whereas *D* and *E* are the perihelia. On the basis of Castelli's theorem, the fluid matter and a hypothetical planet floating in it would move faster in the space *AC* than in *FD*. This, however, would contradict Kepler's area law. Newton proceeded to discuss the orbits of Venus, the Earth, and Mars, claiming that the ratio between the distance of Mars and Venus at the beginning of Virgo to their distance at the beginning of Pisces is approximately as 3 to 2. The speed of the Earth, therefore, should be faster at the beginning of Pisces than at the beginning of Virgo in the same ratio. According the astronomical data, however, the Earth moves faster at the beginning of Virgo, when it is closer to the sun, in accordance with Kepler.

III of Newton's *Principia* corresponds to parts III and IV of Descartes' *Principia*, in that they represent the real-world application of the previous mathematical and philosophical principles. Other scholars had previously tackled the problem of orbital motion and other issues such as tides as part of mechanics, but Newton brought this area of study to a new level by joining astronomy and mechanics in a fashion that had not been seen before—and, indeed, was not to be seen for several decades after the first edition of his work.

In *Principia mathematica* we find for the first time in a treatise on mechanics and motion an extensive and detailed treatment of astronomical data. The issue was not just Kepler's laws; the work also included data about planets, satellites, comets, and tides. This feature of Newton's work is highlighted by his correspondence with Flamsteed. Following Halley's account at the Royal Society in December 1684 of his meetings with Newton and the subsequent registration of Newton's preliminary tract *De motu*, Flamsteed was fully aware of Newton's project and eagerly provided him with all the data he requested. As late as September 1686, Newton inquired whether Flamsteed could confirm Cassini's observation that Jupiter is flattened at the poles, adding that if this was so, it would help explain the precession of the equinoxes, something Flamsteed had tried to explain by means of a solar vortex. When Flamsteed reported his inability to observe any satellites of Saturn beyond the first, Newton excised the relevant references in the text. Although Flamsteed lacked a deep understanding of mathematics and the science of motion, book III of the *Principia* would not have been produced in the same time frame and in the form we know it without his help. Despite their later bitter dispute, their collaboration marked a new alliance between mechanics and astronomy foreshadowing similar collaborations in the eighteenth century.[56]

Newton had originally cast book III in a popular format, but he subsequently adopted a more rigorous style. The aborted version contains a figure highly relevant to my account (fig. 9.9). As in figure 3.18, where Galileo had showed the link between falling and projected bodies, Newton showed the link between orbiting

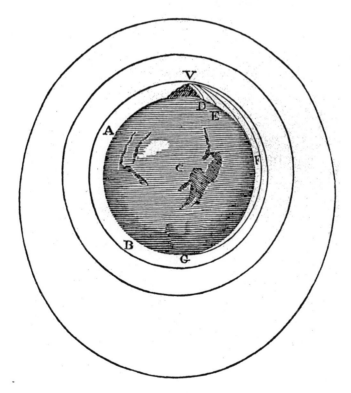

Figure 9.9. Orbiting bodies and projectiles *(The System of the World).* Projectiles shot at increasingly greater speeds from a high mountain *V* fall at increasingly greater distances *D, E, F,* and *G.* If the initial speed is sufficiently high, the projectile becomes an orbiting body.

bodies and projectiles that have been shot with a sufficiently high speed. Newton's figure resulted from a profound conceptual shift that had began in the 1660s with the moon test, in which he had tried to determine whether the force deflecting the moon from its path is the same gravity that makes heavy bodies fall on the Earth. The matter was not so straightforward, as Fabri's example from section 7.2 shows. Significantly, Newton's picture was not known in the seventeenth century; it was first published posthumously in 1728, when the aborted version appeared as *The System of the World*. In the meantime, Newton had settled the question by means of reflections and computations that to his mind rendered the figure redundant. This example is representative of broader shifts toward rigorous deductions from principles and laws, away from merely visual analogies among objects.[57]

Unlike the previous portions of Newton's work, book III is not divided into sections, although the material would have easily lent itself to such a partition. The first seventeen propositions deal with the system of planets and satellites, establishing a modified form of Kepler's laws and universal gravity. Propositions 18–20, examine the shape of the Earth and other rotating bodies. Propositions 21–39, deal with lunar motion and related problems, such as the precession of the equinoxes and tides.[58] Finally, lemma 4 and the extensive propositions 40–42 deal with comets.

Book III opens with a series of nine hypotheses, a singularly ill-chosen name for this heterogeneous collection of methodological statements and empirical observations. In later editions the first two came to be more appropriately called *Regulae philosophandi*.[59] The fourth one states that the center of the system of the world is at rest. The following five rely on astronomical observations and claim that both planets and satellites move according to Kepler's laws. For example, the radii from the sun or primary planet to the orbiting body sweep out equal areas in equal times, and the square of the periods of revolution are as the third power of the median distances from the center. The last hypothesis claimed that lunar motion too satisfies the area law, but immediately below Newton admitted that there is a tiny perturbation due to the solar force. Newton provided tables showing a remarkable agreement between data obtained by several astronomers of the distances of Jupiter's satellites from their planet and of the planets from the sun, and the values obtained from Kepler's third law of the periodic times.[60]

Proposition 7 establishes universal gravity, possibly the most famous result of Newton's treatise: "Gravity exists in all bodies universally and is proportional to the quantity of matter in each."[61] The fall of heavy bodies on the Earth, the motion of satellites around planets, and the motion of planets and comets around the sun all suggest that attraction is a pervasive feature of matter and is proportional to the inverse square of the distance. The proportionality constant, however, might differ, for example equal portions of mass on Jupiter, the Earth, and the Moon could attract differently. It is the third law of motion that enabled Newton to make the argument for universal gravity, since the equality of action and reaction requires all portions of matter to exert the same attractive force.

Concerning the figure of the Earth, Newton assumed that a rotating fluid body adopts a stable configuration as a result of the combination between gravity and centrifugal force, and then he proceeded to determine that configuration. He imagined an L-shaped canal filled with a fluid in equilibrium to go from the pole to the center and from there to the equator. While searching for the equilibrium conditions, Newton found that the radius of the Earth is seventeen miles longer at the equator than at the poles, supposing the Earth to have a uniform density. Measurements of the lengths of the seconds pendulum at different latitudes had been carried out a few years earlier in Paris, at Cayenne in French Guiana, and on Gorée in the Cape Verde Islands, off the western coast of Africa. Those measurements provided both a prima facie confirmation of his estimations and a tool to explore variations in the Earth's density from the surface to the center.[62]

Although many problems, such as lunar motion and the precession of the equinoxes, could not be fully solved by Newton, his overall achievements were unprecedented. In his study of comets, for example, he was able not only to show that they obey the same laws as planets and other celestial bodies but also to determine with graphical methods, to a stunning degree of accuracy, the parameters of the 1680–81 great comet. In other cases Newton was less successful. For example, as we have seen, in his attempt to provide a quantitative account of the precession of the equinoxes, Newton ought to have relied on laws for the rigid body involving the conservation of angular momentum, which was discovered around the mid-eighteenth century.

Whereas Newton could investigate the behavior of point-masses, the behavior of rigid, fluid, and elastic bodies was more successfully tackled with conceptual and mathematical tools of greater complexity after his death.

CHAPTER TEN

Causes, Conservation, and the New Mathematics

> There is nothing left, then, besides finding the expression of the centrifugal force with which the body tends the thread MC in order to reduce this matter to pure geometry.
>
> *Guillaume de L'Hôpital*

10.1 Mechanics at the Turn of the Century

Galileo's views on several aspects of the science of motion were a central matter of debate among his contemporaries soon after the publication of the *Dialogo* and *Discorsi*. His reflections shaped significant portions of the works of scholars from Torricelli and Gassendi to Mersenne and Riccioli. By contrast, a dozen years after its publication, Newton's *Principia* had attracted far fewer debates than Galileo's masterpieces. The main area of contention was the role of physical causes, but there was no sustained debate in any way comparable to what followed the *Discorsi* until the eighteenth century, in connection with the priority dispute over the invention of calculus. In the 1630s and 1640s Galileo was a highly visible and controversial figure, whereas in 1687 Newton was known to a restricted group of scholars as a mathematical genius and to a broader public through his optical papers of the 1670s. Probably the difficulty of his work was a major factor in limiting its initial readership. The private annotations of such scholars as Leibniz, Huygens, and David Gregory testify to how hard it was not only to grasp Newton's main achievements but also to muster some preliminary results. That the Padua professor of mathematics Stefano degli Angeli recommended *Principia mathematica* to his student Jacopo Riccati in 1695 seems to have been the exception rather than the rule.[1]

Soon after the original edition in Italian, the *Dialogo* was published in a Latin translation that went through several editions, and the *Discorsi* was paraphrased in French by Mersenne in 1639, thus reaching a wider European audience. Although Newton's *Principia* was initially published in Latin, it had a rather limited print run and had to wait over a quarter of a century for a new edition. Upon publication, however, the new literary genre of the book review in the recently established scholarly and scientific periodicals provided a means for advertising some of its contents. The review in the *Philosophical Transactions* was by Halley, who had seen the book through the press, and it was, not surprisingly, highly favorable. The Leipzig *Acta eruditorum* provided a positive and detailed account that we now know was due to the mathematics professor at Leipzig, Christoph Pfautz. Pfautz seems to have been especially impressed by how comprehensive the book was, but it is probably fair to say that he failed to grasp some of the major results. For example, he did not remark on proposition I, 1, and in general propositions I, 1–17, in sections 2 and 3 were treated very superficially, with no mention any of the fundamental results therein contained. Pfautz did not miss the attack on vortices at the end of book II and explained the contents of propositions II, 52 and 53, which argue that celestial motions occur in spaces free of vortices. The account of book III was more satisfactory. Pfautz mentioned Kepler's laws and some of the main results, devoting a large section to comets. The account in the *Bibliothèque universelle* was more a synopsis of the sections than a review, whereas that in the *Journal des sçavans* argued that Newton had provided a most perfect mechanics, based largely on arbitrary hypotheses. For this very reason, however, Newton would have failed to provide a true *physique*. Thus the author of that review understood mechanics as a geometrical construction based on abstract principles or hypotheses, whereas only *physique* was a true description of the world.[2]

Huygens and Leibniz attempted a limited response to Newton's far-reaching investigation on celestial motions, the cause of gravity, and motion in a resisting medium. Huygens, however, died in 1695, leaving Leibniz as the key scholar behind several research areas crossing from the late seventeenth century well into the eighteenth. The first was the development of the differential and integral calculus, initially put forward in a series of memoirs in the *Acta eruditorum* starting from 1684. Jakob Bernoulli and his brother Johann at Basel were the first to seize on the new methods, which were rapidly adopted from Paris to Padua, developed, and applied to a range of problems in mechanics, such as elasticity and the shape of the catenary. In the pages of the same journal Leibniz announced his findings on conservation principles and a new science he called dynamics. Unlike the modern namesake, Leibnizian dynamics had profound philosophical roots stretching to Aristotelian notions such as entelechy and prime matter. It was only at the beginning of the new century that theorems and problems in Newton's *Principia* started being investigated by means of the new analysis. Varignon was a leading figure in this process and published a series of memoirs in the Paris *Mémoires de l'Académie des Sciences*. Others, notably Johann Bernoulli, followed suit with more hostile intentions.

This chapter provides only a brief and largely nontechnical characterization of the state of mechanics at the turn of the century. Section 10.2 explores the new analysis and some of the peculiar features of the differential calculus. Section 10.3 takes a selective look at the debates on conservation principles originating in the 1680s and extending well into the eighteenth century. The last two sections deal with responses to Newton's *Principia*. Overall, the early responses addressed the issue of physical causes for gravity, whereas starting around 1700 the focus shifted to rendering Newton's results in the language of the calculus in a systematic way following established algorithms.

This chapter highlights the difference between Newton's work in mechanics and on the system of the world, on the one hand, and his contemporaries' perspectives, on the other. Newton was the preeminent mathematician of his age and he had profound interests in the foundations and methods of mathematics, but he brought his array of mathematical techniques to bear in the *Principia* because he wished to study the physical world in a mathematical fashion, providing quantitative data in exact form or, when this proved unfeasible, by devising clever estimation methods. This was a notable achievement of his work in a wide range of phenomena such as the three-body problem and motion in a resisting medium. Very few of his contemporaries, especially on the Continent—with the partial exception of Huygens—had a comparable desire and ability to combine mathematical acumen with empirical data. In the case of astronomy and celestial mechanics, for example, a program for joining mathematical research with observations capable of surpassing Newton's achievement in a substantive way emerged only in the 1730s and especially the 1740s with collaborations such as those between Tobias Meyer and Euler, Pierre Le Monnier and d'Alembert, and Joseph-Nicolas Delisle and Clairaut.

10.2 The New Analysis

In 1684 and 1686 Leibniz published two fundamental essays in the *Acta eruditorum*, "Nova methodus" and "De geometria recondita," with an explanation of the rudiments of the differential and integral calculus, which he had developed in Paris around 1675–76. Several methods for finding tangents and areas were already known at the time, via the method of indivisibles, for example. What Leibniz added was not only the recognition that the two operations were one the inverse of the other, but also a new way of conceptualizing and manipulating infinitesimal magnitudes by means of an algorithm and a convenient notation. The reciprocity of two operations is known as the fundamental theorem of calculus, a key result in the history of mathematics. The new developments had an impact on mechanics both because infinitesimal magnitudes are crucial to the investigation of motion and because the new algorithm enabled mathematicians to integrate differential equations in a variety of areas.[3]

Brief as Leibniz's essays were, they provided enough information to stimulate other mathematicians. The first followers were the brothers Jakob and Johann Bernoulli at Basel, who in their turn developed and promoted the new calculus in Paris and elsewhere. The term "integral" is due to Johann Bernoulli, who soon be- *true?* came the master of the new art. Basel and Paris were the main centers of the early practitioners of the Leibnizian calculus. Jakob Bernoulli held until his death in 1705 the chair of mathematics at Basel, where he was the teacher of Jakob Hermann, who later obtained Galileo's former chair at Padua, thus spreading the knowledge of the Leibnizian calculus in Italy. In Paris Johann taught the Marquis de L'Hôpital, under a complex legal and financial arrangement regarding intellectual property, and the Marquis published the first textbook of the new calculus, *Analyse des infiniment petits* (Paris, 1696). The early practitioners of the new calculus, such as L'Hôpital, were to be found among a group of scholars associated in different degrees to the philosopher Nicholas Malebranche, including Varignon, Bernard le Bovier de Fontenelle, and Joseph Saurin. They all became members of the Académie des Sciences. Malebranche, Fontenelle, and Saurin were avid supporters of Descartes and his system of vortices.[4]

Leibniz's rules for differentiation and summation or integration resemble those still common today, but there are some important differences especially significant to the study of motion and mechanics worth discussing here.[5] In the Leibnizian calculus the symbol dx stands for the differential of the variable x, meaning an incomparably small variable, while ddx is the second-order differential. The latter depends on the way dx is selected: it could be zero if dx is constant, or different from zero if it is variable. This choice exists only for second- and higher-order differentials. Integration and differentiation do not change the physical dimen- sions of the variable, but only its size. Thus the integral of an incomparably small distance ds, for example, is a finite distance s. Likewise, the integral of an incomparably small speed dv is a finite speed v. The differential of an incomparably small distance ds gives a distance dds that is twice incomparably smaller than a finite magnitude. For Leibniz speed was not the integral of acceleration over time, and accelerations do not figure prominently in his study of motion, although some of his followers used them in a different fashion. Time plays a peculiar role. According to Leibniz, the differential of time dt could be chosen so as to be either constant or variable. Leibniz often chose it to be constant, whereby it behaved like any other constant factor in differentiation, and consequently he represented it with a symbol for a constant factor rather than a variable. For example, the differential of speed dr/dt is ddr/dt, or an incomparably small speed, where dt could be indicated as a letter usually employed for a constant factor such as a or b, for example. Since dt is constant, Leibniz often stated that the speed is proportional to dr and its differential is proportional to ddr, neglecting to indicate constant factors, including time.

The use of mathematics in mechanics usually involved a range of problems including, but not limited to, finding a proportion between two or more magnitudes, identifying a curve, and setting and solving an algebraic equation. Of course, these problems are not mutually exclusive; a proportion could result in a curve, for example. Examples of the first kind include proportions between time, speed, and distance for a moving body, or length, width, and height of a beam, as we have seen in Galileo. Galileo tackled the second kind too with the search for the curve of swiftest descent, or the curve described by a chain hanging from two nails, which he incorrectly identified as the circle and parabola, respectively. Examples of the third kind are the equations expressing conservation in impact or Newton's polynomials in different powers of the speed in the study of resisted motion. While proportions became less prominent, the determination of complex curves grew in importance in the course of the seventeenth century. With Huygens the cycloid gained a major role in mechanics, but it was in the 1690s that the search for curves gained momentum. In the age of calculus, beginning about the turn of the century, a new mathematical object emerged as the single most important mathematical tool in the study of mechanics, namely differential equations. Differential equations required special techniques to be solved or integrated and special attention to aspects that were not always very prominent in other mathematical domains, such as signs and constant factors. In dealing with proportions, for example, constant factors are of no significance, whereas they can be crucial in differential equations, especially higher-order ones. Techniques such as separation and substitution of variables were crucial to differential equations.[6]

These transformations had profound implications for how mechanics was practiced and for the role of experiments. Within the scope of simple proportions, experiments could have a heuristic role in ascertaining how variables varied with respect to each other, such as distance and time in free fall. In the realm of equations and complex curves, experiments often took a different role with respect to the resolution of complex mathematical problems. Newton tried to use them to determine the coefficients of the terms in different powers of the speed in his study of motion in a resisting medium. Their role in investigating the catenary was limited to helping ascertain what the curve was not, namely it was not a parabola, but in the study of elasticity they provided Jakob Bernoulli with some initial information on the behavior of strings and beams.

Mathematicians challenged each other in the pages of the main journals of the time to find a number of curves relevant to mechanics. Some of them had been identified in the time of Galileo, as in the case of the catenary and brachistochrone, but the new analysis could treat them in a way that Galileo and his contemporaries could not. Leibniz tried to argue precisely on this ground for the superiority of his method against Viviani, who defended classical geometric methods. Other curves were introduced later in the century, such as the isochrone curve, namely the curve of descent wherein a heavy body falls with uniform speed, and the curve of equal pressure, wherein the sum of the relevant component of gravity of a heavy body and of its centrifugal force is a constant.[7] Continental mathematicians were able to focus on a number of curves, producing important results at the intersection between mathematics and mechanics. Although these curves were obviously relevant to mechanics, the methods used to deal with them did not involve equations of motion; rather, mathematicians found inspiration in mechanics to pose problems that were in the end chiefly mathematical. Their strategy consisted in using mechanics for setting the problem's conditions, so that the question could move on to the main part, or, be "reduced to pure geometry," an expression used by Leibniz in *De resistentia solidorum* and often repeated.[8] Following Pappus, dal Monte had used the same verb in the attempt to uncover levers in more complex machines. But whereas originally the reduction was a geometrical and visual affair and was also the solution, now the reduction "to pure geometry" was more abstract, and the mathematical task after the reduction could be quite major. Around 1700 "geometry" often meant mathematics broadly conceived, including methods we would not call "geometrical" today.

An exemplary case is that of the centrifugal curve, namely the curve on a vertical plane described by a body exerting on the curve a constant force equal to its weight (fig. 10.1). In 1695 Johann Bernoulli proposed the problem of determining such a curve to the Marquis de L'Hôpital, who candidly replied that he had no clear idea what centrifugal force was. Given the Marquis' status as one of preeminent mathematicians of his time, his lack of clear understanding of a notion Huygens had introduced over twenty years earlier is revealing of his interests and background.[9]

Elasticity was one of the most fertile areas of research for the new analysts, as Leibniz's 1684 *De resistentia solidorum* (discussed in sec. 8.5) stimulated Jakob Bernoulli to extend those investigations. Jakob's initial instigation came from an unnamed expert maker of balances, who had inquired about the problem of flex-

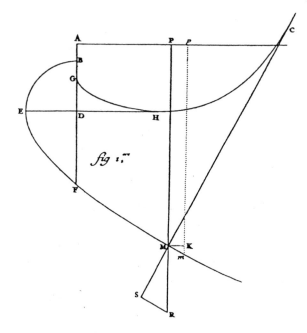

Figure 10.1. Centrifugal curve (de L'Hôpital, *Solution d'un problème physico-mathematique*). Johann Bernoulli's problem consists in determining a curve on a vertical plane such that a body falling along it exerts at each point the same amount of pressure equal to its own weight, resulting from the combination of the relevant component of its weight and centrifugal force. An equivalent way to look at the problem is to imagine the body attached to a string unrolling from a curve. In this case one seeks the curve described by the body exerting at each point the same amount of tension on the string. *BEFM* is the sought curve, *CHG* is the curve from which the string *CM* unrolls, and *MS* is the component of the weight *MR* at right angles to the curve. The equation determining the sought curve is $2PM \times MR/MC + MS = MR$, where *PM* is the height from which the body falls and *MR* its weight; therefore $2PM$ is as the square of the speed, *MC* is the radius, and $2PM \times MR/MC$ represents centrifugal force. The problem consists in writing the equation above as a differential equation and carrying out the integration.

ion of the arms. In 1687 Jakob consulted Leibniz's work and started experimenting and applying the new analysis, rapidly becoming Europe's leading authority on the topic. Initially he inquired about the empirical adequacy of the proportionality between force and elongation. He attached weights to a gut string from a musical instrument and found that the elongation did not increase proportionally to the weight, but rather less, as shown by an accompanying table displaying the difference between theoretical predictions and experimental results. He also claimed that the unnamed technician had observed that a loaded iron beam infixed in a wall began to bend not at the juncture with the wall but rather in the area one-third to one-half its distance from the wall.[10]

In the following years Jakob published a series of profound essays on elasticity, as did Varignon. In the last essay before his death Jakob introduced four lemmas that refined the understanding of the elastic behavior of bodies and sought the analytic form of the curvature of an elastic body. The first lemma states that the elongation or compression of elastic fibers of equal material and thickness subjected to the same force is proportional to their length, whereas the second states that if the length and elongation or compression are the same, the force is proportional to the thickness of the fibers. These two lemmas refine Hooke's law, which states that the force F is proportional to the displacement Δl; now the force is proportional to the cross-section S of the fibers times the ratio $\Delta l/l$ between their displacement and length, $F \propto S \times \Delta l/l$. The third lemma states the result Jakob had communicated to Leibniz in 1687 and reports the same data. Attaching weights of 2, 4, 6, and 8 pounds to a string 3 feet long, for example, the elongations are 9, 17, 23, and 27 lines rather than 9, 18, 27, and 36. Finally, the fourth lemma deals with the stretching and pressing of the fibers of a loaded beam or rod. As Leibniz observed, the discrepancies between expected values and experimental data increase as 1, 4, 9.[11]

Jakob Bernoulli was the chief actor in another contribution relevant to mechanics and to this work. Returning to the problem of the center of oscillation, he conceived the compound pendulum or oscillating rigid body as a lever with weights attached at different distances rotating around a fixed point of suspension and proved Huygens's result from the equilibrium of the moments of the weights, showing in addition the identity of the center of oscillation and percussion.[12]

10.3 Conservation

In the works of Galileo and Torricelli we often find stated that whatever falls can rise back to its original height. With Descartes and Huygens the notion of conservation started playing a major role, for Descartes in conjunction with theological concerns, for Huygens without such concerns. For example, in impact we have the idea that something is conserved before and after. As we have seen, the two cases are related because colliding bodies can be made to rise by means of a pendulum or an inclined plane.

In 1686 Leibniz published in the *Acta eruditorum* an essay devoted to conservation, "Brevis demonstratio erroris memorabilis Cartesii et aliorum," a brief polemical tract whose title appears to have been designed to generate controversy. Leibniz argued from commonly accepted axioms in order to refute the notion that Cartesian quantity of motion is conserved in nature. He stated that a body falling from a given height acquires enough force to return to its original height, and that the same amount of force is required to raise a body *A* of one pound to four yards as to raise a body *B* of four pounds to one yard. The latter can be easily grasped by imagining *B* as composed of four bodies of one pound where each is raised by one yard. Leibniz's choice was especially adroit because in *Explicatio machinarum,* which had just been published posthumously at Kiel in 1672, Descartes himself had stated the same principle:

> The invention of all these machines is founded on only one principle which affirms: with the same force with which we can lift a weight of (e.g.) 100 pounds to the height of two feet, we can also lift a weight of 200 pounds to the height of one foot. And this principle must necessarily be admitted, if we consider that there must always and necessarily be a proportion between action and the effect produced by it.[13]

From these premises Leibniz showed that if the force or the effect is measured by the Cartesian quantity of motion, body *A* in falling would acquire half the force of *B.* By contrast, if force is measured by the product of the body times the square of its speed, the two forces would be the same at the end of the fall. Leibniz's force was proportional either to the height of fall or to the square of the speed. He called the latter expression *vis viva,* or living force. From the end of the seventeenth century to the middle of the eighteenth, extensive debates about force and its conservation ranged from metaphysical to experimental arguments. Leibniz engaged in extended disputes with, among the others, the Abbé François de Catelan, an associate of Malebranche, and Denis Papin. In order to settle the matter Leibniz tried to reduce the terms of the debate to logical form, while later attempts turned to experiments.[14]

In 1695 Leibniz published in the *Acta eruditorum* an important essay, "Specimen dynamicum," in which he outlined the principles of a new science of dynamics. His new science was not simply devoted to the study of motion under the action of forces but was an attempt to link the new science of motion to traditional Aristotelian doctrines. Whereas the "Brevis demonstratio" was targeted at Descartes, in the "Specimen dynamicum" Leibniz was building bridges toward the Peripatetic school and his style was irenic rather than confrontational. Preliminary drafts of the work contain references to Newton, although his name does not appear in the final version. Leibniz introduced several distinctions between forces, such as between active and passive and between primitive and derivative. Whereas primitive forces pertain to metaphysics, derivative forces concern phenomena. Active derivative forces can be living, combined with actual motion and proportional to the square of the speed, or dead, such as gravity, centrifugal force, and elasticity. Passive derivative forces are connected with resistance to motion and also with a certain laziness or inertia. Leibniz's classification is partly philosophical, partly related to actual motion and infinitesimal mathematics. Unfortunately, in these cases Leibniz relied on verbal accounts rather than the more rigorous language of calculus.[15]

In a letter to the Leiden professor of mathematics and philosophy Burchard de Volder, Leibniz stated that "according to the analogy of geometry or of our analysis, solicitations are as dx, velocities are as x, forces as xx or as $\int x dx$." Thus it seems that solicitations are as dead forces, whereas living forces result from an integral of dead forces. In his published works and correspondence Leibniz did not express the relationship between dead and living force unequivocally, leading several interpreters to doubt that he had grasped what type of integration was involved. Both the letter to de Volder quoted above and manuscript sources dealing with motion, however, show that Leibniz intended the integral $\int dr \times ddr = \frac{1}{2} dr^2$, where dr is the differential of a distance, proportional to speed, and ddr is the second-order differential, proportional to solicitation. Leibniz's integration comes from a manuscript dealing with centrifugal and centripetal forces and is therefore related to the science of motion. It is also significant in this respect that he introduced the factor $\frac{1}{2}$, whose omission in other texts had suggested to some that Leibniz had never grasped the link between dead and living forces.[16] However, his rather elliptical treatment of time derived from a mathematical style based on proportions and makes it slightly awkward at first sight to interpret his expressions in terms of distances, speeds, and forces.

Leibniz's idea of conservation of force, meaning living force or another of its manifestations, seemed to be contradicted by some cases of impact, wherein if the colliding bodies are not elastic, living force is not conserved. Leibniz believed all bodies to be elastic and denied the existence of absolute hard bodies. Quantity of motion in one direction is conserved, but Leibniz disliked the association with a special direction and the minus or plus sign associated with it, and he preferred to focus on living force, which is always positive. In the "Essay de dynamique," presented to the Paris Académie in 1692, his answer to those cases where living force seemed to be missing after the impact was that it was absorbed by the individual components of the colliding bodies. This explanation may sound more convincing to us, raised with the notion of conservation of energy, than to his contemporaries, to whom it may have sounded ad hoc. Although Leibniz theorized extensively on elasticity, he did not provide a detailed account of how living force was transferred to the bodies' constituents. The conservation of force was established in a different context and then simply applied.[17]

By contrast, Newton disliked conservation principles and avoided attaching great significance to them in his system. In corollary 3 to the laws of motion he stated the conservation of quantity of motion in one direction, which is not changed by the mutual action of bodies. For Newton, however, this conservation descended from the third law of motion and did not have a particularly significant status, like the conservation of force for Leibniz. Curiously, both Newton and Leibniz seemed to downplay the association of quantity of motion with direction, possibly because of some seemingly paradoxical or surprising results. Two equal bodies colliding head-on with equal and opposite speeds and the same two equal bodies at rest seemed to be two very different situations, yet the bodies' common quantity of motion in one direction was the same in both cases. Although Newton did not accept Descartes' notion of conservation of quantity of motion, he apparently took—contrary to later views on mechanics—Cartesian quantity of motion to be a meaningful variable providing the measure of a body's motion and of the motion in the universe, which could increase or decrease but never remained constant. Even though both theoretical possibilities were open, in practice Newton was convinced that motion in the universe was on the decline: "But by reason of the Tenacity of Fluids, and Attrition of their Parts, and the Weakness of Elasticity in Solids, Motion is much more apt to be lost than got, and is always upon the Decay. For Bodies which are either absolutely hard, or so soft as to be void of Elasticity, will not rebound from one another." In query 31 of the *Opticks* Newton continued to talk of the collision of pendulums of lead or soft clay, whereby motion stops, and of the viscosity of fluids such as water, oil, and pitch, which will make pendulums stop pretty rapidly once they are set in a vortical motion. Of course, the reference to vortices was not accidental, and Newton proceeded to draw conclusions on the decay of motion in a hypothetical fluid filling the heavens. But even without such a fluid, motion would be decaying unless it was conserved and increased by active principles, such as gravity and fermentation.[18]

Thus for both Newton and Leibniz, views about laws of nature and conservation, or the lack thereof, were linked to theological and philosophical concerns. Some of these themes were debated in the dispute between Leibniz and Clarke, who tackled the issue of the decay of motion in the universe from their very first letters.[19]

10.4 Early Responses to Newton's *Principia*

Studies on mechanics and motion developed through the interplay of experiments and theoretical reflections. Much like Huygens, Newton was an outstanding experimenter as well as a mathematician, whereas Mariotte was almost exclusively an experimenter; his death in 1684 did not help experimental work in mechanics in the succeeding years. Unlike Galileo, Newton had no Mersenne or Riccioli who pursued the experimental side of his work. One of the remarkable features of the *Principia* was the attention it devoted to empirical data in a huge range of cases, from motion in a resisting medium to the speed of sound, from the shape of the Earth to the trajectory of comets. This feature was almost completely lacking in the work of his Continental readers. Not surprisingly, numerical tables of data do not figure prominently in the literature on the reception of *Principia mathematica*. Probably the main exceptions are Halley's 1705 remarkable study of the periodicity of comets, in which he pointed out similarities in their orbits and predicted the return of the 1682 comet in approximately seventy-five years, and the Anglican clergyman William Derham's tabulations of different values for the speed of sound, including Newton's, in relation to other variables. Apart from the experiment on the pierced cistern performed by Newton and others mentioned in section 9.3, the next substantive body of experimental research in mechanics was carried out again by Newton in preparation for the second 1713 edition, on topics such as motion in a resisting medium. As in the first edition, Newton sought to determine not the exponents of variables in a proportion but rather the coefficients of the terms in a polynomial wherein the variable assumes different powers. Even among Newton's critics, it is hard to find challenges to his experiments. Huygens, for example, accepted the proportionality between quantity of matter and weight proved in proposition III, 6, by means of pendulum experiments. Derham's experiments with cannonballs and pendulums investigated Newton's estimate of the speed of sound and other features of its propagation, but the research on acoustics by the engineer and professor of mathematics at the Collège Royal Joseph Sauveur was not undertaken in response to Newton.[20]

Despite Newton's efforts to shape the audience of his work in such a way as to exclude those untrained in mathematics, *Principia mathematica* was not received without criticisms. Moreover, those very mathematicians Newton would have considered to be receptive readers of his work and to whom he sent copies of the book, such as Huygens and Leibniz, were very critical of his approach to physical causes and were shocked at his apparent endorsement of attraction at a distance, or at least without a medium. Huygens objected to several features of Newton's work, chiefly his belief in action at a distance. The issue was not simply one of philosophical principles but also concerned Newton's mathematical style and the empirical adequacy of portions of his theory. Huygens was prepared to accept Newton's proofs that this gravity extended to interplanetary spaces and was inversely to the square of the distance. In some cases he found Newton's proofs baffling because of their reliance on the method of first and last ratios and their determination of magnitudes at a point rather than through a proportion. But regardless of mathematical style, a major issue was that the cause of gravity remained to be determined. What Huygens was not prepared to accept was that all particles of mat-

ter attract each other, or the principle of universal gravity. Huygens had no reason
to question the empirical adequacy of Newton's inverse-square law on a planetary
scale, but he thought that gravity on the Earth's surface behaved differently from
how Newton believed it behaved. Relying on data from clocks aboard the ship *Alc-
maer* on her 1687 return journey from the Cape of Good Hope to Texel in the
Netherlands, Huygens hoped to show that the centrifugal force due to the Earth's
rotation was the only correction needed. By contrast, Newton believed that in addition to centrifugal force, gravity at the equator was less than at the poles because
the Earth's radius decreased as latitude increased. Huygens dropped hints of his
views in the *Addition* to the *Discours de la cause de la pesanteur* (Leyden 1690,
published together with his *Traité de la lumière*) and his correspondence, and
probably explained the issue more fully in conversations during his journey to England in 1689, but did not publish a detailed account. Huygens's *Discours de la
cause de la pesanteur* focuses on the word "cause," much like Leibniz's "Tentamen
de motuum coelestium causis." Huygens considered the inverse-square law as a
genuine discovery and was prepared to jettison Cartesian vortices. However, he believed in an aether causing gravity, whose existence he needed as a support for the
propagation of light. He argued that light propagation was analogous to the propagation of waves through a medium; therefore the heavens could not be void of
matter. Huygens admitted that he had conceived the action of gravity on the surface of the Earth or the sun, not across huge distances such as those between the
sun and the planets or even between the Earth and the moon. In his *Discours*
Huygens also outlined his views on motion in resisting media where the resistance
is proportional to the speed, a topic he had researched while at the Paris Académie
in the 1660s.[21]

Leibniz replied with two articles in the *Acta eruditorum* for 1689, one on the
motion of bodies in a resisting medium, "Schediasma de resistentia medii," the
other on planetary motion, "Tentamen de motuum coelestium causis." In drafting both essays he made use of the differential calculus, but he used it in print only
in the latter. The "Schediasma" dealt with a subject Leibniz had already examined in the past, whereas the "Tentamen" was hastily and ingeniously put together on the basis of what Newton had proved in the *Principia.* The latter is particularly interesting for the way Leibniz managed to retrieve some, though by no
means all, of Newton's results about celestial motions with a system based on vortices. We have seen how in proposition I, 1, Newton was able to account for Kepler's area law in terms of central forces. It is hard to imagine how such a striking
and powerful proof could be transformed. Yet Leibniz managed to show that elliptical planetary motion could be decomposed into a circular motion and one
along the radius. This approach resembles Kepler's, whose work was known to

Leibniz, and Borelli's *Theoricae,* which he probably had not seen at that time. If the circular motion occurs following a simple law, namely that the velocity of rotation is inversely proportional to the distance from the center, whatever the radial motion, Leibniz could show that Kepler's area law followed. At this point he could state, in order to save a vortex theory of planetary motion, that there is indeed a vortex rotating around the sun following the required law and carrying the planets. Leibniz was also able to prove that motion along the radius resulted from an imbalance between an outward centrifugal force due to the body's rotation and a force toward the center due to the vortex inversely proportional to the square of the distance. Thus his different mathematical decomposition of motion had a straightforward alternative physical interpretation. While his approach was moderately successful in dealing with one planet, it was unable to account for the planets as a whole and for Kepler's third law, according to which the squares of the times of revolution are proportional to the third power of the major axes of planetary ellipses. Moreover, Leibniz could not account for several other phenomena, notably comets.[22]

Much like his theory of celestial motions, Leibniz's theory of motion in a resisting medium was also inspired by Descartes. As we saw in chapter 5, Descartes had developed a loose a priori theory about resistance depending on the surface and volume of a body. Early in 1675 Leibniz had composed an essay for the Paris Académie, "Du frottement." Those preliminary results were expanded in the "Schediasma" of 1688. Leibniz considered two types of resistance generally acting together, *absoluta* and *respectiva.* The former depends on the *glutinositas* or viscosity of the medium, or on the friction between surfaces. As such, *resistentia absoluta* between a body moving in a fluid and a fluid's particle does not depend on their respective speeds. However, since a greater speed means that the body would encounter more particles in the same amount of time, *resistentia absoluta* is proportional to simple speed. *Resistentia respectiva* depends on the density of the medium and the speed imparted to its particles by the moving body, thus it is proportional to the body's speed. But since in a given amount of time a faster body encounters more particles of the fluid, in the end *resistentia respectiva* is proportional to the square of the speed. Leibniz further claimed that *resistentia absoluta* depends on the body's surface and *resistentia respectiva* on its *soliditas.* We have already encountered the same notion in Descartes' *Principia,* when he was trying to account for the motion of a body through the celestial vortex.

Leibniz's essay consists of six propositions arranged in two symmetrical sets of three, each set dealing with the two types of resistance. The first propositions in each set (1 and 4, respectively) examine the effects of the medium on a rectilinear uniform motion; the second propositions (2 and 5) deal with the retardation of a motion uniformly accelerated by gravity; the third propositions (3 and 6) deal with projectile motion, resulting from the composition of uniform and accelerated motion discussed previously. Proposition 6 turned out to be problematic. If resistance is as the speed, it is possible to decompose projectile motion into separate directions. With a resistance as the square of the speed, however, composition of motion cannot be done in the usual way. Huygens pointed out the mistake to Leibniz, who duly corrected it in the *Acta eruditorum* for 1691.[23]

Besides the reviews, these works by Leibniz and Huygens were the main reactions to Newton's *Principia* during the seventeenth century. However, we know of another reaction to Newton's masterpiece. Although the report is brief and secondhand, it is of considerable interest in many respects. In a memorandum dated 20 February 1697/98, David Gregory reported Wren's views on the cause of gravity and on Newton's beliefs: "Mr C. Wren says that he is in possession of a method of explaining gravity mechanically. He smiles at Mr Newton's belief that it does not occur by mechanical means, but was introduced originally by the Creator." Wren's opinion is especially interesting both because he unambiguously identified God as the cause of gravity in Newton's system, and because in the past he had attempted to solve the problem of planetary motion by combining an "imprest" rectilinear uniform motion and a descent toward the sun. From conversations with Halley and Hooke in the early 1680s, Wren was also acquainted with the notion of an inverse-square relation. Thus his belief in a mechanical explanation for gravity coexisted with attempts—albeit unsuccessful—to tackle the problem of planetary motion in a mathematical fashion supposing that the planets have a tendency toward the sun. Wren seems to have been involved in trying out some mathematical solution, leaving a detailed mechanical explanation to a later stage.[24]

10.5 The New Analysis and Newton's *Principia*

In the years immediately following 1687 Leibniz put forward a response involving the differential calculus, as we have seen. Leibniz's early response was concerned with physical causes but was not representative of later developments. At the turn of the century Varignon began publishing in the Paris *Mémoires* a series of essays in which he applied the Leibnizian calculus to such topics as motion under central forces and in a resisting medium, a topic to which he devoted twelve memoirs. Although at an earlier stage Varignon had published *Nouvelles conjéctures sur la pesanteur* (Paris, 1690), a work addressing the cause of gravity still belonging to the pre-*Principia* universe, in the later memoirs he eschewed this issue to focus on a systematization of the science of motion by providing general rules for determining the forces, speeds, spaces, and times, as well as for solving a num-

ber of specific problems. In his contributions Varignon often appears more eager
to show the virtues and universality of calculus than to develop new areas of research or to solve challenging new problems. Varignon was a committed teacher
of mathematics, and his efforts toward generality, ease, and systematization may
reflect didactic concerns.[25]

Varignon had already shown similar concerns toward systematization in *Project d'une nouvelle mechanique* (Paris, 1687), a work published soon after his arrival
in Paris. Varignon started by mentioning Descartes' rejection of dal Monte's attempt to reduce the pulley to the lever (see sec. 5.2), then questioned the idea of
reducing the inclined plane to the lever, and finally challenged dal Monte's overall project of reducing all simple machines to the lever. Following previous examples by Descartes and Wallis, Varignon sought to reorganize mechanics starting
from a more abstract principle. He chose the composition of forces along the di-
agonal of a parallelogram, from which he analyzed weights suspended from cords,
pulleys with weights attached at different angles, inclined planes, levers, and the
screw. Varignon's systematic concerns and his search for a more abstract principle
able to account for all simple machines at once represent long-term trends in mechanics. Almost exactly a century later his work was praised by Lagrange, for
whom abstract principles and the analytic method were the hallmarks of his contributions. Yet from the perspective of this work it is especially significant that
Varignon's emphasis in questioning dal Monte's approach appeared in the same
year as *Principia mathematica*, namely the work wherein Newton discarded the
visual analogy between projected and orbiting bodies of figure 9.9 in favor of a
more abstract and rigorous approach.[26]

Returning to the early 1700s, we notice some differences in the ways Leibniz
and Varignon used the differential calculus in the study of motion. Leibniz, as we
saw earlier, generally chose the element of time dt to be constant, and as such he
did not pay much attention to it. Time was frequently grouped together with other
constants so that it is not easy to find it in Leibniz's equations. Varignon, by contrast, gave time a more prominent position even if it was a constant, so that it is
always visible in his equations. In a memoir presented at the Paris Académie in
January 1700, Varignon named a number of variables and posed relations among
them. For example, he set the space traversed as x, the times as t, and the speeds as
v, with dx for the space traversed with a uniform speed v in each instant, dv for the
increase of speed, ddx for the increase of space due to the increase of speed, and
dt for the time instant. From here Varignon set two basic equations as general rules
for motion in a straight line, namely $v = dx/dt$ and $y = dv/dt$, or $y = ddx/dt^2$,
where y is the force. Varignon did not take mass into account here.[27]

There is a certain ambiguity in the mathematical rendering of the expression "an instant of time" and, as a consequence, of the other variables. If the instant is taken to be like a point to a line, then change is continuous, but if the instant is an incomparably short time interval, change is not continuous. Rather, one would have a series of tiny uniform motions interrupted by sudden changes, in the style of proposition I, 1, in Newton's *Principia*. The difference is subtle, but it carries important consequences for dealing with acceleration in the tiny segments or elements of the curve. The problem came to a crunch in the correspondence between Leibniz and Varignon. Initially both made inaccurate statements, but eventually they realized that there are alternative representations that cannot be mixed. Unlike Leibniz, in the end Varignon chose to consider a continuous curve with acceleration at each point because he believed motion to be truly accelerated since centripetal and centrifugal forces act continuously. Leibniz considered a curve to be composed of incomparably small rectilinear segments. Initially he had not fully appreciated the implications of his construction and had taken motion to be accelerated in those segments. Later, while appreciating the legitimacy of Varignon's construction, he preferred to retain his polygonal view of a curve, requiring no acceleration in the elements, where motion is uniform. In the *Acta eruditorum* for 1706 he had to publish an emendation to his "Tentamen" (fig. 10.2). His final choice fits well with his belief that accelerated motion, both rectilinear and curvilinear, was due to microscopic impacts; therefore accelerations were a macroscopic phenomenon that did not truly exist at the micro-level but were only a convenient mathematical idealization. In a sense, rectilinear uniform segments were also ideal entities, but they corresponded at least in principle to the actual physical state of affairs. In the end the solution adopted by Varignon resembled that used by Newton in lemma 10 (see fig. 9.1).[28]

Varignon's essays and the new analysis' role have been differently judged by different historians. Those interested in new results have been reluctant to show much appreciation for his systematization and rendering of many results in Newton's *Principia* in the language of the differential calculus. Those interested in conceptual systematization and mathematical methods, notably Michel Blay, have considered him as key figure in the establishment of mathematical physics. His memoirs involved a conceptual reframing of Newton's work that made it more accessible on the Continent.

Questions of a different nature affect the historical appreciation of another early promoter of the differential calculus, Johann Bernoulli. Johann subjected to careful scrutiny Newton's study of the orbital trajectory given the force. His starting point was the same as Newton's, but he then translated Newton's equation into the language of the differential calculus and was able to show explicitly by inte-

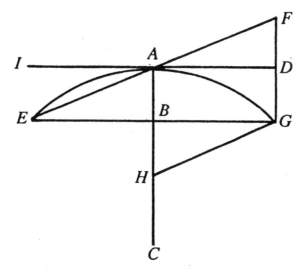

Figure 10.2. The analysis of curvilinear motion (based on "Excerptum ex epistola," in the *Acta eruditorum* for 1706). There are alternative ways of representing curvilinear motion, either as continuously curved, or as composed of incomparably small rectilinear segments. *EAG* is the arc of a circumference and *C* its center. The body moves with uniform circular motion, traversing the equal arcs *EA* and *AG*, or the equivalent subtended chords, in equal times. One can imagine a body to move along the arc *EA* where, if released, it would continue along the tangent *AD*. Here *DG* is the deviation from the tangent to the curve and represents the segment traversed with an accelerated motion, in Newtonian fashion. Alternatively, one can conceive the body to move along a series of incomparably short segments such as *EA, AG*, etc. In this case, if the body were released in *A*, it would continue along *AF.* This time the segment *FG* from the prolongation of the chord to the curve would be traversed with a uniform motion. Both accounts are feasible, but they cannot be mixed; if motion is curvilinear throughout, the deviation from the tangent to the curve is traversed with uniformly accelerated motion, whereas if the macroscopic curvilinear motion is composed of incomparably small rectilinear segments, the deviation from the prolongation of the chord to the curve is traversed with uniform motion. In the determination of the central force, the fact that *FG* = 2*DG* compensates for the factor ½ due to motion along *DG* being uniformly accelerated. Errors of a factor ½ result from mixing the two approaches.

gration that from the assumption that the force is inversely proportional to the square of the distance, the only possible orbits are conic sections. Other mathematicians too, such as Varignon and Hermann, tackled the same problem with similar if slightly less refined methods based on the calculus. Hermann and Johann Bernoulli used different integration strategies. Another point of contention concerned a series expansion in proposition II, 10, dealing with motion in a medium resisting as the square of the speed under the action of a constant gravity. Johann's nephew Niklaus pointed out the mistake to Newton, who hastened to prepare a correction for the forthcoming second edition of the *Principia*, at which he had been working with Cotes.[29]

Although the new analysis eventually overtook the more geometric style of *Principia mathematica*, the explicit solution of the inverse problem of central forces and the identification of a mistake in a coefficient of a series expansion in book II hardly justify the claim of the superiority of the new analytical methods in early-eighteenth-century mechanics. Moreover, for several decades after 1687 Newton's work remained unchallenged in dealing with applied problems ranging from motion in a resisting medium to the entire domain of book III, including tides, lunar motion, the shape of the Earth, and comets. Continental mathematicians needed several decades to advance significantly beyond Newton's masterpiece in the area that became known as celestial mechanics.[30]

Newton had a remarkable ability to extract numerical data and quantitative estimations from mathematics. This was one of his primary concerns in the *Principia*, where he adopted a malleable mathematics adaptable to the problem at hand. At the time of the priority dispute, Continental mathematicians tried to find fault in *Principia mathematica* and at the same time to prove the superiority of the Leibnizian calculus. By and large, they did not link mathematical investigations either to experimental philosophy or to astronomy. Whereas Newton was able, at least for some time, to team up with Flamsteed, Varignon's colleague Giandomenico Cassini did not accept Keplerian ellipses and proposed his own curve described by planets, the Cassinoid. In a memoir presented on 13 November 1700 at the Paris Académie, Varignon intended to demonstrate the versatility of his method by applying it to seven problems. He considered the ellipse, the Cassinoid, and the eccentric, each with the correct form of the area law and with a computational method that had commonly been used in the seventeenth century. The last example sought the central forces required to make the planets move along arbitrary orbits with a uniform motion. His examples had no astronomical significance and look like a rhetorical exercise and a display of diplomacy. Throughout his work, Varignon adopted noncommittal style aimed at avoiding offense to his fellow academicians rather than pursuing vigorous new avenues of research.[31] The reading of Newton's *Principia* at the Paris Académie tells us as much about the work as it does about the controversies and concerns internal to the Académie over the calculus, vortices, and a physical cause for gravity.

The debates on mechanics and on Newton's *Principia* in the early eighteenth century shifted progressively toward higher mathematics. Varignon's rejection of dal Monte's project and search for more abstract principles—appropriately, in the same year in which Newton's *Principia* was published—as well as, starting in

1700, his rendering of some of Newton's results in the language of the new analysis make him a natural endpoint of my narrative. Mechanics had changed quite dramatically from dal Monte's foundational concerns with unmasking levers in simple machines, becoming a far less geometrical discipline primarily concerned with increasingly abstract principles and differential equations.

Mapping the Transformations of Mechanics

Looking back at the ways in which mechanics was transformed in the course of the long seventeenth century, one detects a stepping-stone approach venturing from known and established areas to new problems and domains. This approach involved different strategies, such as the unmasking of levers or other devices in more complex objects, the search for and application of principles or laws, the use of experiments, and the straight solutions of mathematical problems. These different strategies were not mutually exclusive, since mathematicians and philosophers frequently relied on a combination of them.

The aim of dal Monte, following Pappus and other Greek scholars, was to explain systematically all simple machines in terms of the lever, the theory of which was firmly established. Galileo shared his patron's concerns, even though eventually he had to follow a different path. Starting from the lever, I have drawn a map of the main objects of seventeenth-century mechanics representing its chief transformations. The use of objects as experimental tools and as sources for visual and conceptual analogies in mechanics and the mathematical investigations of motion was so pervasive that it would be impossible to draw a complete map.[1] The one I have drawn *(opposite)*, however, covers a surprising amount of material.

Unlike the case of the simple machines discussed by dal Monte, I am obviously not arguing that all the objects on the map can be reduced directly or indirectly to the balance or the lever. I adopt here a broader perspective whereby an object can be explored and accounted for in terms of another one—not necessarily a lever—whose features are already known. As we shall see, the arrows represent not outright reduction but a wide range of practices and connections as understood from a seventeenth-century perspective, even if those connections are no longer accepted. More precisely, they indicate that an object was used in significant ways in a conceptual, mathematical, causal, or experimental fashion in the investigation of other objects. A two-way arrow indicates a reciprocal use of objects or the iden-

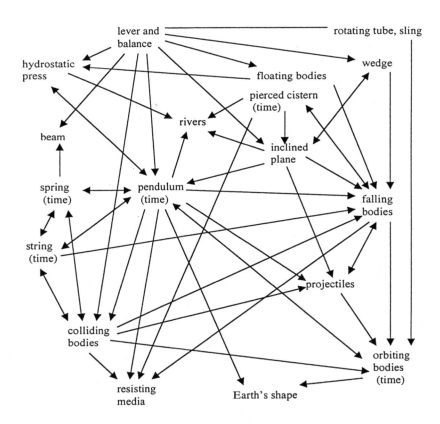

tification of profound similarities between them. For example, the pendulum, string, and spring all oscillate in similar ways and are nearly isochronous; analogies in their behavior were used to explore one object by means of the others. The line without arrow-points indicates objects used in conjunction with the lever (sling and rotating tube) to account for orbiting bodies. The length of the arrows and the proximity of some of the objects depend on drawing constraints rather than historical considerations.

Much like dal Monte's machines, the objects on the map exist in different forms and at different levels and are both theoretical or mathematical and practical or experimental. They exist as geometrical diagrams in a mathematical treatise and as time-measuring devices, as for example the pendulum, vibrating string and spring, and, in different ways, the pierced cistern and orbiting bodies, including the Earth itself. The objects can be built with the care and accuracy required by precision instruments, but they are also objects of common experience readers would have been familiar with, such as the inclined plane or vibrating string. I find their amphibious and ambiguous nature helpful because it reflects the status of mechanics as a mixed mathematical discipline, intermediate between mathematics and *physica,* partly abstract and partly dealing with nature. The vast majority of the objects on the map, especially those in the top half, from the lever to the string and from the spring to falling bodies and projectiles, were studied with the same mathematical tools, such as the theory of proportions, often joined with some version of the doctrine of indivisibles. This is an additional feature that made it easier to transfer knowledge among them. The intermediate position of many objects shows their double role as subjects of inquiry and tools for further research. Grasping some features of an object enabled scholars to deploy it in other investigations. Conversely, identifying in the subject of investigation features analogous to those of an object whose properties were known enabled scholars to deploy their knowledge to the new domain. This dual strategy bears a close resemblance to the methods of analysis and synthesis, as described at the opening of book VII in Pappus, *Collectiones.* In the method of analysis we assume that which is sought and move backward from there until we reach a known proposition. Conversely, in synthesis we move forward step by step, starting from known propositions, until we reach what was sought.[2]

The arrows should not be taken as representing automatic and unproblematic developments; rather, they often stand for research programs lasting decades, going through successively more refined stages and requiring the efforts of the best scholars of the time. Huygens's work with the pendulum led him to identify the analogies with a vibrating spring as a result of his investigation of the properties of cycloidal motion, and in particular the proportionality between force and displacement. His work involved a visual and conceptual shift but stemmed from a refined mathematical research. His determination of the center of oscillation of a compound pendulum relied on the astute use of suitable principles and computations involving quadratures.

It is worth recalling here the comparison established in chapter 7 between Borelli and Huygens. Borelli was educated in mathematics by Galileo's student Castelli and relied extensively on the theory of proportions. Huygens was educated in mathematics by Frans van Schooten the younger, the follower and translator of Descartes, who had a major role in introducing algebraic methods. Borelli relied on superficial and often unsatisfactory analogies with a range of tools from the lever to the pendulum, in an imitative style that represents a weakening of the best Galilean tradition. By contrast, Huygens's analogies were supported by deep mathematics and a creative use of principles, including some coming from the circle closest to Galileo, notably Torricelli. The gulf between them epitomizes major aspects of the transformations of mechanics, which was in a state of transition between the second and the last third of the century. Those transformations reached a high point with Newton's *Principia*, a complex work defying simple categorizations. On the one hand, as we have seen in chapter 9, Newton relied on pendulums to justify the third law and in the investigation of specific phenomena, such as resistance to motion. Moreover, a central result such as proposition I, 1, was framed in terms of colliding bodies, and the pierced cistern in book II was used to study motion in a resisting medium. Yet the mathematical solutions to a number of problems take the leading role in the book. Although objects remained a feature of mechanics beyond the period covered here, in later periods connections were more likely established between the types of differential equations involved or an abstract principle rather than visually through a geometrical diagrams. Thus Huygens's *Horologium oscillatorium* and Newton's *Principia* point to the eighteenth century, when mathematics took a more prominent role with the study of special curves by means of a geometry equipped to deal with the infinitely small, and later with differential equations. A map accounting for the transformations of eighteenth-century mechanics would look considerably different in its very constituents.

In order to show how the map works, I start from the lever, which was used both strictly in statics and in conjunction with motion. As we saw in figure 1.2, dal Monte identified a lever in the wheel and axle, and in figures 1.5 and 1.6 we see him doing the same for the pulley and the inclined plane, respectively. Galileo followed a similar approach and employed the lever to explain equilibrium conditions for the inclined plane. He imagined a lever or balance with a bent arm wherein one of the weights is partially supported by an inclined plane. He relied on the lever in the study of floating bodies and the siphon, as illustrated in figure 3.8. Figure 3.13 shows him ingeniously using a lever with a bent arm to explain the resistance of beams in what was the first rather primitive mathematical theory of the beam. He also tried to rely on the balance in the study of falling bodies, but those attempts were problematic, and in the end he had to rely on a combination of principles derived from the inclined plane and the pendulum. The investigation of falling bodies was not limited to the lever. In the mid-sixteenth century Tartaglia and, especially, Benedetti had sought to rely on floating bodies to account for falling bodies, something Galileo too attempted in *De motu antiquiora*, though in that case the extension of the theory of floating bodies was more conceptual than visual. Galileo further unmasked a falling body in the motion of a projectile, as in figure 3.18. In the mid-seventeenth century Torricelli tried to determine the speed of water spurting from a hole at the bottom of a container by means of parabolic trajectories, as in figure 4.8. Thus dal Monte and Galileo, while building on a tradition stretching back to antiquity, were providing influential exemplars for further investigations.

As we saw in section 6.3, Pascal referred to the hydrostatic press as a balance or lever wherein force is multiplied and equilibrium is determined by the height of water in the two containers. Several scholars, including Kepler, Descartes, Borelli, and Leibniz, used the lever in their accounts of circular and orbital motion. For example, Borelli imagined the planets to be pushed by the sun's light acting like a lever. Jakob Bernoulli used the lever to investigate the problem of the center of oscillation. The principle of the lever was central to early investigations on collision by Beeckman, Marci, and Descartes, who saw impact as a contest between bodies. The key notion of quantity of motion—in the form weight times speed—derived from the pseudo-Aristotelian treatment of the lever in *Quaestiones mechanicae*. Later Huygens and Wren presented their rules of impact in a form based visually and conceptually on the lever, as in figure 8.6. The speeds of the bodies after the impact can be determined by imagining them on a tilting lever. We witness here the crucial role played by the geometrical diagram; in this case its role seems to have been heuristic and at the same time didactic or mnemonic. Visualization via the geometrical diagram functioned in different ways and generally was a crucial ingredient making the objects on the map so amphibious and versatile. These reflections add to our understanding of the merits of geometrical representations in mechanics and the science of motion. If one might doubt whether the development of linear perspective in Renaissance art played a direct role, visual representations in general certainly proved crucial to the transformations of mechanics in the seventeenth century, as the figures in this work show.[3]

All these creative and fertile applications of the lever were conceptual and mathematical. However, this was not true in general for the objects on this map. It is very difficult, and I would argue impossible, to provide univocal characterizations of these objects precisely because they were so versatile.[4] Take the inclined plane, for example, a key object in the transition from equilibrium to motion. It was a classical mechanical tool, used for raising weights or as a wedge and studied as a mathematical problem in antiquity and in the Middle Ages. In Galileo's hands, it became in addition a practical tool for delaying falling bodies and for studying the trajectory of projectiles. In the latter case the geometrical diagram was not the result of theory, since it was drawn by a rolling ball in an experiment before any theory had been established or probably even conceived. In all probability, at the time of *De motu antiquiora* Galileo tried to make bodies with different specific gravity fall along planes with the same height and different inclinations in the same time, but failed. He used the inclined plane as a theoretical tool too in conjunction with the notion of *momento* and the theory of indivisibles. Galileo used the inclined plane with zero inclination to conceptualize the relations between force and motion. In addition, the inclined plane was used as a conceptual tool for studying pendular motion, since at each instant the bob can be conceived as moving along a different inclined plane, and for stating a principle whereby the speed acquired by a falling body depends only on the vertical fall, not the inclination and length of the plane. Finally, the inclined plane was employed to conceptualize the flow of rivers, seen as bodies sliding or rolling down a slope, as in figure 6.10.

Analogously, the pendulum was used as a time-measuring device in many areas of inquiry, from astronomy to falling bodies. It served to measure the periods of orbiting bodies, and, conversely, orbiting bodies were used to check its reliability as a time-measuring device. Its experimental and conceptual versatility made it probably the most useful and representative mechanical tool of the time, as evidenced in the map by the number of arrows starting from it. Around the mid-seventeenth century many scholars, including Mersenne and Riccioli, used pendular oscillations to find the distance a body covers in the first second of free fall. Huygens partly transformed the problem into a theoretical one and used the conical pendulum as a tool to compare gravity and centrifugal force. By this means he found an exceedingly accurate value for the length of the seconds pendulum. Huygens and others such as Wren and Mariotte also used the pendulum to perform experiments with colliding bodies, since it allowed the speeds of the bobs before and after impact to be measured indirectly by means of their height. Huygens again adapted cycloidal cheeks to the pendulum in order to make it perfectly isochronous. At first he adapted the cheeks on an ad hoc basis, and only later was he able to determine their exact shape with mathematical rigor. The Royal Society in the 1670s and Newton in *Principia mathematica* employed the pendulum to measure resistance to motion. The oscillations of the pendulum were also seen by Mersenne, Hooke, and Huygens, for example, as conceptually analogous to the vibrations of a string or spring, since both follow the same mathematical law. Hooke stated that the law now bearing his name, whereby force is proportional to the displacement, applies to the spring, the string, and the pendulum, for small oscillations. Jeremiah Horrocks, Hooke, and Wallis used the conical pendulum as a conceptual tool to represent important features of orbital motion, namely the idea that the bob tends to move in a straight line and is retained in its orbit by a central force. The pendulum was employed by Newton to justify the third law and to establish the proportionality of inertial mass, offering resistance to a force, and gravitational mass, responsible for gravity interactions. The pendulum was also used to measure the shape of the Earth and potentially its density.

principles

The search for and application of principles was a major concern during the entire period I have covered, especially, though not exclusively, among those mathematicians who sought an axiomatic formulation in the Archimedean style as well as those with theological concerns such as Descartes and Leibniz. Seventeenth-century scholars frequently introduced and referred to principles and laws in terms of specific objects that instantiated or embodied them, such as springs with Hooke's law of elasticity, a term often rendered in English as "spring." We have seen that Stevin, in his study of the inclined plane and in his hydrostatics, adopted the principle that perpetual motion cannot exist. Throughout the century many scholars provided different versions of the principles of equilibrium of the lever based on speed. Galileo anxiously sought a suitable axiom or principle for his science of motion, and found one stating that the degrees of speed acquired by a body falling along inclined planes with different inclinations are equal when the heights of those planes are equal. He also had recourse to the pendulum to argue that a falling body is capable of rising back to its original height. Torricelli too started from a postulate, stating that two connected bodies cannot move by themselves unless their common center of gravity descends. His postulate was explained in terms of balances, pulleys, and other simple machines. In *Principia philosophiae* Descartes redefined the boundaries and contents of mechanics, yet his laws of motion were all instantiated in objects, projectiles for the continuation of motion, the sling for its being rectilinear, and colliding bodies for the conservation of quantity of motion. Colliding bodies instantiated conservation of quantity of motion in one direction and, if they are perfectly hard, conservation of *vis viva*. Huygens was a master at working with suitable principles, and he did so throughout his career. He started with the notion of the invariance of unit of time for a law of falling bodies, and then in his work on collision he used a range of principles, from the notion of relativity of motion[5] to a more abstract form of Torricelli's principle applied to colliding bodies that are not connected. Leibniz too based his system on laws, most notably conservation of force, which much like Descartes' laws were often rooted in theological and philosophical concerns.

I am sure I have missed some roles of the pendulum and of other objects as well, but my aim is not completeness. I believe that this map is representative of how interrelated theoretical and experimental practices based on a small number of objects operated in seventeenth-century works on mechanics and motion. Starting from laws, and, following Koyré, especially the law of inertia, we exclude large areas of research and often have to descend several steps in the scale of abstraction to reach the middle of the action. By contrast, objects provide a more representative account of the areas of mechanics and motion, including problems unrelated or only vaguely related to inertia, such as the loaded beam, the pierced cistern, and the hydraulic press. Objects are located close to the center of the action and enable us to revisit issues such as the role of diverse mathematical representations, quantitative experiment, and the social status and boundaries of mechanics. They also enable us to grasp the fine-grained investigations whereby within a few decades the practitioners of mechanics dramatically expanded their domain. In this respect my account diverges from Gabbey's view of a plurality of "mutually independent revolutions, or minirevolutions—in collision theory, in statics and the theory of machines, in hydrodynamics, in vibration theory, in the theory of central forces and of rigid-body motion—each with its own new principles and procedures." I believe I have abundantly shown that these transformations were not mutually independent but rather profoundly linked. I see the interconnectedness of the objects at the center of this study and the creativity and ingenuity of the scholars working with them as going hand in hand in accounting for a remarkable phase in the history of mechanics. If my reading is correct, however, Gabbey's emphasis was less on the degree of connection among those transformations than on the need to offer a broader picture going beyond the law of inertia and the motion of projectiles and celestial bodies in the axis Galileo-Descartes-Newton; in this regard I would fully subscribe to his views.[6]

By focusing on objects we can grasp yet another feature of the history of seventeenth-century mechanics. Despite the many diatribes across the century on the continuity of motion and the cause of gravity, for example, allegiance with one or the other side was compatible with use of the pendulum and experiments with falling bodies. Thus the objects on the map could be interpreted and explained in different or partly contrasting ways, but to a large extent could also move across rival camps. Colliding bodies were the basis for causal explanations of phenomena according to strict mechanists such as Descartes and Huygens, but those who, like Newton, denied such a central role to impact accepted similar rules.

The map highlights the interplay between statics and the science of motion. Moreover, objects at the top of the map are generally closer to the classical and engineering traditions than objects toward the bottom. Some objects, such as celestial bodies, are of the sort we cannot handle, or they may consist literally of air, as for resisting media. These objects are at the margins of the map and generally at the receiving end of the arrows, meaning that they were investigated with the help of the objects of common experience. Orbiting bodies were studied by means of observational astronomy too, as well as with geometrical diagrams and equations. Thus the map also represents the rising status of mechanics from its practical and engineering roots to an area of inquiry rich with philosophical and cosmological implications, a transition in which Descartes was a pivotal figure. If the world was seen more and more as a machine, machines were becoming more and more significant in understanding the world, and objects gained an important cognitive role. They became tools for understanding the world by means of analogies, such as the conical pendulum with respect to orbiting bodies. The theme of mathematization too can be profitably revisited from the perspective of objects. Focusing on objects enables us to trace how scholars developed wider and farther-reaching mathematical descriptions of phenomena and how those mathematical descriptions progressively replaced traditional scholastic explanations based on causes.[7]

I have identified a few areas in which engineers, technicians, and instrument makers interacted with the scholars mentioned here. They include makers of rulers, compasses, and other precision instruments such as Barocci and Genga at Urbino; Contarini and his associate Savorgnan, both interested in military machines and shipbuilding; the *proti* or skilled workers at the Venice Arsenal referred to by Galileo in the opening of the *Discorsi;* hydraulic engineers, such as Bartolotti and Barattieri; military engineers and gunners; clockmakers who collaborated with Huygens and Hooke in transforming pendulums and springs into precision instruments, such as Thuret and Tompion; sailors in charge of regulating Huygens's clocks aboard ships; and, more generally, all those who were involved in experiments requiring more than one person, performing such tasks as counting pendular oscillations, as reported by Galileo and Riccioli. Other relevant practitioners include astronomers, whose role was crucial in the study of orbital motion. Engineers dealing with all types of construction, warfare, and water management interacted, and at times argued, with the scholars of mechanics while providing them with objects of investigation, such as the beam, the cannonball, and the riverbed.

Although the objects we have been investigating were more mundane and less glamorous than the telescope, the microscope, the air pump, and other "philosophical" instruments, the practices associated with them were of major importance. The inclined plane, the pendulum, the beam, the spring, and falling, projected, and orbiting bodies were central to seventeenth-century efforts to study nature and offer us a powerful tool to grasp the nature of those investigations.

Introduction

1. Koyré, *Galileo Studies*, 2–3; *From the Closed World.* Lenoble, *Mersenne*, 461. On broader differences between Koyré and Cassirer, see Friedman, "Kuhn," 29–35. Bertoloni Meli, "Inertia," in Applebaum, *Encyclopedia.* On laws, see Steinle, "Regularities."

2. Westfall, *Force*, ix–xii, 1. Westfall is especially concerned with the obstacles to the development of dynamics (his term), such as the lingering role of medieval notions, the legacy of statics, the complexities of circular motion and the notion of mass, and especially the mechanical philosophy. Westfall, "Floods"; "Science and Patronage."

3. Dugas, *Mécanique.* See also Dugas, *Mechanics.* Szabó, *Geschichte.*

4. Gabbey, "The Case of Mechanics," 521n6. For some thoughtful reflections on the notion of transformation, see Cohen, *Newtonian Revolution*, part 2.

5. Galileo, *Two New Sciences*, 185 (I have revised Drake's translation). Newton, *Principia*, 427. Mahoney, "Mathematical Realm," sec. 2.2. On the relationship between theory and experiment in more recent times, see Warwick, *Masters of Theory*, and Galison, *Image and Logic.*

6. Notable exceptions are Westfall, "Floods"; Benvenuto, *Structural Mechanics;* and Maffioli, *Out of Galileo.*

7. Ariotti, "Pendulum." Settle, "Experiment." Benvenuto, *Structural Mechanics*, 3–7. A classic philosophical work is the postscript to the second edition of Kuhn, *Structure*, reproduced also in the third edition, esp. 187–91.

8. Bennett, "The Mechanics' Philosophy," list of instruments at 3. Bennett challenges Kuhn's distinction between Baconian and mathematical disciplines. Bennett, "Practical Geometry"; "Challenge"; "Shopping for Instruments." Warner, "Scientific Instrument." Van Helden, "Birth." Machamer, "Galileo's Machines." Rossi, *I filosofi e le macchine*, 145–46. Gregory, *Scetticismo*, 71–77, 160–62. Wise, "Mediating Machines."

9. These methods adopted by Mersenne and Galileo are discussed in chaps. 3 and 4.

10. Bennett, "Practical Geometry." Lenoble, *Mersenne*, 23–24, 386–69.

11. Compare the historical account in Lagrange, *Mécanique analitique*, 1–12, 122–30, 158–89, 428–37. Montucla, *Histoire des mathématiques*, vol. 1. Duhem, *The Origins of Statics.*

12. In the Aristotelian sense, motion in general was a much broader notion encompassing all types of change in nature. Local motion corresponds more closely to our understanding of motion *tout court.* Sylla and Murdoch, "Science of Motion."

13. Wallis, *Mechanica*. Pardies, *Statique*, preface. Dechales, *Cursus*, vol. 2 (compare the indices of the section on mechanics in the 1674 and the posthumous 1690² editions). Gabbey has produced several articles on this issue: "Newton's *Mathematical Principles*" (compare the reference to Boyle at 313–14); "Descartes's Physics and Descartes's Mechanics"; "Between *Ars* and *Philosophia Naturalis*"; and "What Was 'Mechanical'?"

14. A classic study on Renaissance mathematics is Rose, *Renaissance*. Long, *Openness*.

15. McKirahan, "Subordinate Sciences." Wallace, *Sources*, 132–36, 141–46, 212–14. Pappus, *Collection*, book 8, 810. Newton, *New Translation*, 381–82, 418, 430. Gabbey, "Newton's *Mathematical Principles*," 315–22. On causal knowledge in *physica* and mathematics, see Jardine, "Epistemology and the Sciences."

16. See chap. 10 for usage of the term "dynamics." An example of this older historiography is Clagett, *Mechanics*, in which two of the four parts are devoted to medieval kinematics and dynamics. For the origin of the terms "dynamics" and "kinematics," see my entries in Applebaum, *Encyclopedia*, 196–97, 348–49.

17. Aiton, *Vortex*. Mahoney, *Fermat*. Bos, "Differentials." Mancosu, *Mathematics*. Giusti, *Euclides*. Andersen, "Cavalieri." Blay, *Naissance*. De Gandt, *Force and Geometry*. Guicciardini, *Newtonian Calculus; Reading*.

18. Classic studies include Dear, *Discipline*, and Daston, "Baconian Facts."

19. Alchemy is an interesting area in this regard, where mass balances, namely the weighing of the reagents and products of a reaction, required quantitative measurement. I thank William Newman for helpful information in this area.

20. Drake, "Renaissance Music." Palisca, *Musical Thought*.

21. Westfall, "Fudge Factor." Shapiro, "Experiments on Diffraction." Smith, "Newtonian Style."

22. Kuhn, "Mathematical versus Experimental Traditions"; "Measurement." Bennett, "The Mechanics' Philosophy."

23. Bertoloni Meli, "Numerical Tables."

24. Picolet, "Mariotte." Westman, "Astronomer's Role," is the *locus classicus*.

25. Laird, *Moletti*, 72–74.

26. Drake and Rose, "Pseudo-Aristotelian 'Questions,'" 92–96, 99.

27. Boas, "Hero's Pneumatica."

28. Bertoloni Meli, "Mechanics." On dedication copies, see Westman, "Reception." Howard, *Christiaan Huygens*.

One • Machines in the Field, in the Book, and in the Study

Epigraph: Guidobaldo dal Monte, *Mechanicorum liber*, 120r.

1. *Mechanics in Italy*, 245.

2. See Rose and Drake, "Pseudo-Aristotelian," esp. 96–98, for its dissemination among engineers and technicians. The humanist Fausto was connected with the Venice Arsenal. Concina, *Arsenale*. Laird, "Scope." Micheli, *Origini*, 137. On Biringucci, see *Dizionario Biografico degli Italiani*. Other engineers familiar with *Quaestiones* include Giuseppe Ceredi, Bonaiuto Lorini, Vittorio Zonca, and Giovanni Branca.

3. Keller, "Mathematicians, Mechanics," 23. On dal Monte's involvement with the transla-

tion, see Gamba and Montebelli, *Urbino*, 64n55. Manno, "Giulio Savorgnan." Galluzzi, *Mechanical Marvels*, 85.

4. Title page: "A beneficio di chi si diletta di questa nobilissima scienza; & massimamente di capitani di guerra, ingegnieri, architetti, & d'ogni artefice, che intenda pe via di machine far opre maravigliose, e quasi sopra naturali. Et si dichiarano I vocaboli, & luoghi più difficili." On previous manuscript variants, see Henninger-Voss, "Working Machines," 245.

5. Dal Monte, *Mechanicorum liber*, 1615. *Le mechaniche*, 1615. The occasion for the reprints was the publication in the same year by the same publisher of dal Monte's posthumous *De cochlea*, in folio. Popplow, "Why Draw," 43–45.

6. The English edition of *Quaestiones mechanicae* is *Mechanical Problems*, 331. Dal Monte, *Mechanicorum liber*, preface addressed to Francesco Maria II, Duke of Urbino (pages are not numbered). *Mechanics in Italy*, 241. Pappus too talks of mechanics as acting against nature, meaning against gravity, according to ver Eecke in Pappus, *Collection*, 809–10.

7. Hydrostatics is discussed in the following chapter. An Archimedean hydraulic machine operating "against nature" is the water screw, but it does not depend on hydrostatic principles. In the seventeenth century we encounter a machine operating "against nature" on hydrostatic principles, namely the hydraulic press (see sec. 6.3 below). Dal Monte, *In duos Archimedis aequiponderantium libros paraphrasis*, 2–3, "Immitando naturam (paradoxum id forte videbitur, cum tamen verissimum sit) prater naturae ordinem dicatur," quoted in Micheli, *Origini*, 146–47. On Moletti, see ibid., 145. Micheli's book, esp. chap. 2, is relevant to this theme and challenges Krafft, "Die Anfänge." Shiefsky, "Art."

8. Aristotle, *Mechanical Problems*, 335.

9. Several scholars have seen in this treatment of the lever the roots of the principle of virtual velocities, whereby a system is in equilibrium if the sum of the products of the forces and the displacements is nil. Benvenuto, *Structural Mechanics*, 28–29, 77–115.

10. Other questions relating to navigation are numbers 5–7. *Mechanical Problems*, 335 (on the balance and lever), 407–11. On the Arsenal, see Renn and Valleriani, "Arsenal."

11. *Mechanics in Italy*, 104–10, 179–96. Rose and Drabkin, "Pseudo-Aristotelian," 103–4, provide a list of commentaries on *Quaestiones mechanicae*.

12. *GOF*, 8:68–73. *Mechanical Problems*, 375.

13. Clagett, *Archimedes in the Middle Ages*. Rose, *Italian Renaissance of Mathematics*.

14. *Mechanics in Italy*, 259. Galluzzi, *Momento*, 59–62. Dal Monte, *Archimedis aequeponderantium paraphrasis*, is devoted to the determination of centers of gravity. Modern definitions do not require the center of gravity to be within the body, as, for example, in a horseshoe.

15. *Mechanics in Italy*, 244. Drake's introduction, 48, claims that dal Monte worked in the tradition of Hero and Pappus, but does not make much of it. Schmitt, "A Fresh Look," 165–66, highlights the importance of Hero's work for debates on the void.

16. Pappus, *Collectiones*, 8, sec. 31, 329. See also 8, sec. 10. The Greek form of the verb rendered by Commandino as "reducere" is αγονται. Ver Eecke, *Collection*, 873, translates φύσις as "structure."

17. *Mechanicorum liber*, 120r. I have slightly modified the translation appearing in *Mechanics in Italy*, 324.

18. *Mechanics in Italy*, 246. The term for "foundations" is *substructio*.

19. Ibid.

20. *Mechanicorum liber*, 106v. *Mechanics in Italy*, 318. Galluzzi, *Mechanical Marvels*.

21. Cuomo, *Pappus*, chap. 3. Long, "Picturing the Machine," 122–25. Passalacqua, "Le *Collezioni*."

22. *Mechanicorum liber*, 118r–v. On fol. 115r, dal Monte reports Pappus's opinion that the screw is a wedge without percussion. I have slightly modified the translation appearing in *Mechanics in Italy*, 322. Pappus, *Collectiones*, sec. 31. Laird, "Patronage of Mechanics," deals with the force of percussion.

23. *Mechanics in Italy*, 74–78, at 74. Tartaglia considers the distance from the beginning of fall alongside the distance from its end: "Every uniformly heavy body in natural motion will go more swiftly the more it shall depart from its beginning or the more it shall approach its end" (217). Benedetti: "The further distant [a body] is from its starting point the swifter it is." Koyré, *Galileo Studies*, 69.

24. *Mechanics in Italy*, 245.

25. Baldi, *Exercitationes*, 1, 3–4. On 36 he rejects using speed in analyzing statics. On the spinning top, see ibid., 177–85. Gabbey, "The Case of Mechanics," 499–502. Laird, "Scope," 55–58. Laird provides a thorough discussion of the nature of mechanics to the first quarter of the seventeenth century. In 1627, for example, Giovanni di Guevara argued that motion pertained to mechanics insofar as it was artificially produced by machines toward human ends. Ibid., 67. Murdoch and Sylla, "The Science of Motion."

26. Aristotle, *Mechanical Problems*, 347–51.

27. *Mechanics in Italy*, 261, 265.

28. Moody and Clagett, *Medieval*, 6, 16, 123–24.

29. *Mechanics in Italy*, 84.

30. Ibid., 261, 265, 275–76, 279. Montucla, *Histoire des mathématiques*, 1:691, perceptively argued that if the lines of gravity converge to the center of the Earth, the center of gravity varies both if the body approaches the center of the Earth and if it rotates. Thus the equilibrium of the balance is indifferent if the lines of gravity are parallel, but it is not so if they converge to the center of the Earth. This issue was the subject of a debate in the seventeenth century. *Mechanics in Italy*, 47. Gamba and Montebelli, *Urbino*, 241–42. Roux, "Cartesian Mechanics," 36–52.

31. Duhem, *Statics*, 151–52, is especially harsh with dal Monte, failing to grasp the structure of his argument. On 151 he states, "Because of his punctilious mind, Guido Ubaldo takes trifles for grave errors. He severely takes his predecessors to task for always having treated the verticals extending from different points on a lever or balance as lines parallel to each other." Rose, *Italian Renaissance*, 233: "Guidobaldo denounced Jordanus, Cardano and Tartaglia for assuming that the lines of descent of heavy bodies were parallel rather than convergent to the center of the earth," a demand for "unreasonable exactness." In *Mechanics in Italy*, 46, Drake states, "Guido's exaggerated concern for rigor appears in his insistence that, strictly speaking, the lines of descent of suspended weights are never mathematically parallel, but converge toward the center of the earth." Wallace, *Galileo and His Sources*, 204–5, 241, claims that Jordanus, Tartaglia, and Galileo were more willing than dal Monte and Benedetti to accept small deviations between theory and practice.

32. *Mechanics in Italy*, 284–85.

33. Ibid., 295. Micheli, "Guidobaldo del Monte," 100–104. The other balance is mentioned

in Gamba and Montebelli, *Urbino*, 246–47, 259. Dal Monte mentioned one of those balances in a letter to Giacomo Contarini of 9 October 1580. Favaro, "Due lettere inedite," 307. On the Genga workshop, see Henninger-Voss, "Working Machines," 242. Gamba, "Documenti di Muzio Oddi"; "Guidobaldo dal Monte."

34. *Mechanics in Italy*, 168–69, at 169; 174–78, at 176–77.

35. *SWD*, 3–13. Van Berkel et al., *History of Science in the Netherlands*, chap. 1, 12–36 (by van Berkel). Cook, "The New Philosophy."

36. *SWD*, 224–25, 242–43, 386–87. Dijksterhuis, *Simon Stevin*.

37. *SWD*, *De Beghinselen der Weeghconst*, 35–285.

38. *SWD*, *De Weeghdaet*, 287–373.

39. *SWD*, *De Behinselen des Waterwichts*, 375–483.

40. *SWD*, *Anvang der Waterwichtdaet*, 484–501. *Anhang van de Weeghconst*, 503–21.

41. *SWD*, 112–15.

42. Rose, "Proportional Compass"; "Jacomo Contarini." In the latter, Rose published excerpts of dal Monte's letters to Contarini. Rosen, "Reduction Compass."

43. *Mechanics in Italy*, 305–6.

44. Keller, "Mathematicians, Mechanics," 28. Favaro, "Due lettere inedite."

45. Favaro, "Due lettere inedite," 307–10, dal Monte to Contarini, Pesaro, 9 October 1580. Kuhn, "The Function of Measurement." *Mechanics in Italy*, 308.

46. *GOF*, 10:97–100, Galileo to dal Monte, Padua, 29 November 1602. I have modified the inaccurate translation of this passage that appears in Drake, *Galileo at Work*, 71.

47. *Mechanics in Italy*, 325–26. Pigafetta had access to a Greek manuscript of Pappus, *Collectiones*, in the library of Gian Vincenzo Pinelli. Ibid., 321. Cuomo, *Pappus*, 113–16.

48. *Mechanics in Italy*, 141–43. Cuomo, *Pappus*, 113–15, claims that according to Hero of Alexandria only a force smaller than any known force is required to move a body along a horizontal plane.

49. *Mechanics in Italy*, 55, 119, 141–43, at 142n51.

50. Ibid., 299–300, 316, 318. Galileo, *On Motion and Mechanics*, 166–67. Rose, *Italian Renaissance*, 233–34. Dear, *Mersenne*, 121. Benvenuto, *Structural Mechanics*, 80–81, while criticizing Lagrange's exaggerated praise, found merit in dal Monte's formulation of a principle of mechanics linking the spaces described by the moving power and the moved weight.

51. This work was reprinted in *Wisconstighe Ghedachtenissen* (Leyden, 1608) and translated into French and Latin.

52. *SWD*, 507–9. Baldi too in *Exercitationes* had objected to the use of motion in statics. Laird, "Scope," 58.

Two • Floating Bodies and a Mathematical Science of Motion

Epigraph: Galileo, *On Motion and Mechanics*, 69.

1. These works include book 9 of Vitruvius, *De architectura; Carmen de ponderibus;* the *Liber de ponderibus Archimedis*, wherein the notion of *gravitas in specie* was introduced; and the *Liber Euclidis de ponderoso et levi et comparatione corporum at invicem.* Moody and Clagett, *Science of Weights*. Napolitani, "Geometrizzazione," 165–71.

2. Dijksterhuis, *Archimedes*, 373–98, at 374. For the practical sides of Archimedes' work,

though not specifically related to *Floating Bodies*, see Schneider, *Archimedes*. For a classification of traditions in mechanics, see Drake's introduction to *Mechanics in Italy*. Schmitt, "A Fresh Look."

3. Clagett, *Archimedes*, 3:574. Details of Tartaglia's editions can be found in ibid., 1394, 1410, and *Mechanics in Italy*, 392, 400–401.

4. Clagett, *Archimedes*, 3:125ff., 316–18, 574–75, 583–84. De Pace, "Mazzoni," 169–70. Napolitani, "Geometrizzazione," 167.

5. In a printed marginal note Stevin mentions Archimedes' work, using the title of Commandino's translation. *SWD*, 384.

6. *SWD*, 397–401.

7. *SWD*, 409, theorem 7.

8. *SWD*, 415, theorem 8.

9. *SWD*, 421–61, props. 11–17; 487. Dijksterhuis, *Stevin*, 64–69.

10. Ghetaldi, *Promotus Archimedes*, 9, 24, 34, 72. Napolitani, "Geometrizzazione," 155–56, argues that all the ratios were calculated for water. Bertoloni Meli, "Numerical Tables," 168.

11. Benedetti's early works are reproduced and commented on in Maccagni, *Speculazioni*, and Bordiga, *Benedetti*. The volume *Cultura, scienze e tecniche nella Venezia del Cinquecento* provides a survey of Venetian culture in Benedetti's times.

12. Maccagni, *Speculazioni*, 10–13.

13. Ibid., 7, 13.

14. Ibid., 7–13. However, see 13 for a reference to *gravitas in specie*.

15. Ibid., 9–10, 13–14.

16. Ibid., xxxi, 19–61, esp. 24–42. Eden, *Navigation*.

17. Maccagni, *Speculazioni*, 73–74. *Mechanics in Italy*, 160–61.

18. *Mechanics in Italy*, 198–99, 211–14, 216–17.

19. Galileo, *Motion and Mechanics*, 84. Koertge, "Accidents." On composed proportions, see Giusti, *Euclides*. Guicciardini, *Reading*, 125–28.

20. *GOF*, 1:211ff., 225–28. In his geometric and military compass Galileo added data about the diameters of spheres of different materials. *GOF*, vol. 2 between 342 and 343. Bertoloni Meli, "Numerical Tables," sec. 2.

21. Galileo's correspondence with dal Monte and Clavius in early 1588 is in *GOF*, 10:22–25, 27–30, 34–35; Caverni, *Metodo*, 4:101–12. Galileo's work is in *GOF*, 1:179–208; Drake, *Galileo at Work*, 13–14; Wallace, *Sources*, 223–25. I do not discuss here Galileo's derivative notes from lectures delivered at the Collegio Romano, for which see ibid., 91–92. At the time, manuscript lecture notes often circulated without the professor's involvement, as was to happen to Galileo at Padua with the manuscript of *Le mecaniche*, known to Baliani and Mersenne. An incomplete list of manuscript sources is in *GOF*, 2:149–50. Galileo may have embarked on a study of the lecture notes from the Collegio Romano in preparation for the academic disputations known as *circoli*. Dollo, "Galilei."

22. The manuscript consists of a number of texts to be found in Ms 71 of the Galilean Collection at the National Library in Florence. In addition, a short memorandum on motion in Ms 46 belongs to the same period and deals with the very same problems. Several factors are traditionally invoked in dating the texts. In a portion of the manuscript in dialogue form Galileo refers to *La bilancetta*, a text dated by Favaro to 1586. The protagonists of the dialogue wish to have their conversation at the estuary of the Arno, and watermarks indicate Florentine and

Pisan paper. Camerota, *Gli scritti*. The views Galileo expressed in these manuscripts, especially the views about projectile motion, seem more primitive than those of the end of his Pisa period.

23. Viviani says that while at Pisa, Galileo expressed anti-Aristotelian views at the "dispute di conclusioni naturali." *GOF*, 19:603. More than forty years later Galileo referred to the "dispute circolari" in the *Dialogo*. *GOF*, 7:63, 138–39, 488. For references to "my adversaries" and "my opponents," see Galileo, *Motion and Mechanics*, 77–81. Near-contemporary references to the *circoli* are in Marcucci, *Lettere*, 7–9, dated November 1570. On the role of the *circoli* for Giovanni Alfonso Borelli (professor of mathematics at Pisa 1656–67) and his colleagues, see *MCA*, 1:23–25, Borelli to Malpighi, 11 December 1659, esp. 25n9 and *ad indicem* (under Pisa University). Camerota and Helbing, "Pisan Aristotelianism," 358–63, examine *inter alia* the notion of *motor conjunctus*. Helbing, *Buonamici*. Schmitt, "Pisa." On the Collegio Romano, see Wallace, *Sources*, parts 1 and 2. Baldini, *Legem impone subactis*.

24. Fredette, "*De motu.*" Galluzzi, *Momento*, 166–68. Giusti, "Relative Chronology."

25. Moody, "Avempace." Schmitt, "Pisa." Camerota and Helbing, "Pisan Aristotelianism." *Mechanics in Italy*, 379.

26. Galileo, *Motion and Mechanics*, 109. Koyré, *Galileo Studies*, 33, translated as "less and less slowed down."

27. *Mechanics in Italy*, 374, 375.

28. Fredette, "*De motu*," first established that E23 is complete. Galileo, *Motion and Mechanics*, 38–41.

29. A similar approach was used in *Elementa Jordani super demonstrationem ponderum* (in Moody and Clagett, *Medieval*, 136–37), but Jordanus applied it in a different fashion, having recourse to the angle of contact. Renn et al., "White Elephant," 69–70. *GOF*, 1:298. Galileo, *Motion and Mechanics*, 65.

30. Galileo, *Motion and Mechanics*, quotations on 65–66 and 68; see also 67. Drake, *Galileo at Work*, 25–27; *Mechanics in Italy*, 338. Galileo's approach makes it difficult to deal with oblique projection, namely when the body is shot at an angle with the horizon. Renn has analyzed the problem of oblique projection in Damerow et al., *Exploring*, 251–52. The problem of projection is discussed in chap. 3. Cardano had attained similar results on the inclined plane, arguing that the only force necessary to move a body along a horizontal plane was that to overcome the resistance of the air. *Opus novum*, in *Opera*, 4:480, prop. 40, in Caverni, *Metodo*, 4:386.

31. *GOF*, 1:301. Galileo, *Motion and Mechanics*, 68. He is talking about one body, since different bodies would behave differently depending on their specific gravity.

32. *GOF*, 1:301–2. Galileo, *Motion and Mechanics*, 69.

33. Galileo, *Motion and Mechanics*, 37–38. On the expression "periculum facere," see Schmitt, "Zabarella," 115–23.

34. Galileo, *Motion and Mechanics*, 107, 31–32n12. Laird, *Moletti*, 146–59. Camerota and Helbing, "Pisan Aristotelianism."

35. Galileo, *Motion and Mechanics*, 108. See also Galileo's *Memoranda on Motion*, ca. 1590, in *Mechanics in Italy*, 383: "That a contrary force is more strongly impressed in heavier bodies is clear from objects which are suspended by a thread and allowed to move back and forth. For the heavier these objects are, the longer will they move." Settle, "Tool of Investigation," 324. Cardano, *De subtilitate*, in *Opera*, 3:394, discusses some features of pendular motion. Caverni, *Metodo*, 4:386.

36. *Mechanics in Italy*, 74–81, 84–85, suppositions 1 and 2; 100–101. On Tartaglia, see

Cuomo, "Shooting." Cardano, *De subtilitate*, in *Opera*, 3:394. Caverni, *Metodo*, 4:512. The picture in *Opera* differs from Caverni's, quoting the 1580 Lyon edition. Baldi, *Exercitationes*, 4. Renn, "White Elephant," 46–48. Büttner et al., "Challenging Images."

37. Biagioli, "Social Status," 49. Renn et al., "White Elephant," 82, provide a dating before 1593. *GOF*, 14:386–87, Galileo to Cesare Marsili, Florence, 11 September 1632. Sarpi, *Pensieri*, 398–400.

38. Tartaglia, *Nova scientia*, in *Mechanics in Italy*, 80.

39. Rose, *Renaissance*, 228, somewhat inappropriately, also considers motion in the case of the hanging chain. A relevant passage is in Aristotle, *Metaphysics*, 1015a. Bertoloni Meli, "Archimedean Revival," 24. For Galileo's discussion of the analogy between projectile trajectories and a hanging rope in the fourth day of the *Discorsi*, see sec. 3.6 below. The manuscripts by dal Monte were first made known by Libri, *Histoire*, 4:397–98.

40. For example, on the occasion of the nova of 1604, dal Monte was reluctant to accept that the heavens were corruptible. Bertoloni Meli, "Archimedean Revival," 26–31.

41. Renn et al., "White Elephant," aptly argue for a more nuanced attitude toward the notion of discovery. However, they seem to believe that the law of falling bodies follows rather directly from the parabolic trajectory, whereas to me that step appears to be less direct.

42. Galluzzi, *Momento*, 188–97.

43. Mazzoni, *Praeludia*, 190–93: "dum tamen in medio eandem habeant resistentiam." Purnell, "Mazzoni and Galileo." De Pace, "Archimede."

44. Maccagni, "Mechanics and Hydrostatics."

45. *SWD*, 509–13. Maccagni, "Mechanics and Hydrostatics," 92–93.

46. *SWD*, 511–13.

47. Benedetti, *Diversarum speculationum liber*, letter to Lord Cambiano, 258–59; *Mechanics in Italy*, 225. Galileo, *De motu antiquiora*, *GOF*, 1:338–39, in Galileo, *Motion and Mechanics*, 111–12. Libri was apparently the first to identify a similarity between Benedetti and Galileo. Camerota, *Gli scritti*, 29–30n30. Caverni, *Metodo*, 4:275–6, 513. Drabkin, "Benedetti."

48. Galluzzi, *Momento*, 177–79, n94.

49. Galileo, *Motion and Mechanics*, 49.

50. Bertoloni Meli, "Archimedean Revival," 12–13. Galluzzi, *Momento*, 99, 172, 176–80.

51. Koyré missed this important distinction. Galluzzi, *Momento*, 178. Murdoch and Sylla, "Motion," 212–13. *Mechanics in Italy*, 379. On impetus see Clagett, *Mechanics*, parts III–IV.

52. Benedetti, *Diversarum speculationum liber*, 184, 194–95; *Mechanics in Italy*, 217, 220–21. *GOF*, 1:302–4. Quotation from Galileo, *Motion and Mechanics*, 71.

53. Laird, *Moletti*, 142–43. Tartaglia, *Quesiti*, 20v–21r. Biringuccio, *Pirotechnia*, 158v.

54. Renn et al., "White Elephant," 65.

Three • The Formulation of New Mathematical Sciences

Epigraph: GOF, 10:115, Galileo to Sarpi, 16 October 1604.

1. Wisan, "Method," 5, 44.

2. *GOF*, 2:535. Mazzoleni was a talented worker who made dozens of compasses. After his death of the plague around 1630 there was no one in Padua able to make compasses. *GOF*, 16:355–56, Micanzio to Galileo, Venice, 1 December 1635; 19:131–49 (mentioning several in-

struments fabricated by Mazzoleni). Drake, *Galileo at Work*, 55–56. Bedini, "Galileo and Scientific Instrumentation." Valleriani, "Galileo Practitioner in Padua."

3. Favaro, "Due lettere inedite," 304–6. Rose, "Contarini." Renn and Valleriani, "Arsenal."

4. Galluzzi, *Momento*, 199–201. Galilei, *Mecaniche*.

5. *GOF*, 2:161–3; 8:152–4. Galileo, *Motion and Mechanics*, 153–54. Mach, *Mechanik*, chap. 1, sec. 1, esp. par. 4, acutely objected that Galileo's proof assumes a linear dependence of the moment on the distance. This account differs from Dear, *Discipline*, 90, 127, 143. It is important to distinguish Galileo's private or informal investigations from the formal presentation of a new science in Archimedean fashion. See my "Axiomatic Tradition."

6. Galileo, *Motion and Mechanics*, 172.

7. Ibid., 179–82. Galileo, *Two New Sciences*, 281–306. Drake, *Galileo at Work*, 72.

8. Galileo, *Dialogue*, 230, 450–51 (*GOF*, 7:256, 475), and *Two New Sciences*, 87–88, 97, 99, 226–27. Galileo's experiments were problematic, and his proof (later presented in ibid., 211–13) could not work because circle arcs are not traversed in equal times. Hill, "Pendulums."

9. This is a contentious issue. Galluzzi, *Momento*, 215–17, 262–69, on 269, argues that by the time of his letter to dal Monte Galileo had accepted that acceleration was necessary and natural. Giusti, "Aspetti matematici," 14, dates the transition from uniform to accelerated motion to the end of 1604, but in "Leggi del moto," xxv, he claims that pendular motion is unequivocally accelerated. An opposite opinion is in Drake, *Galileo at Work*, 66–71, on 71. See also Wisan, "New Science," 161–65, and Sylla, "Oxford," 103–4. Early proofs of the chord theorem are National Library, Florence, Ms. Gal., vol. 72, fol. 147v (*GOF*, 8:377), fol. 151r (*GOF*, 8:378), and fol. 160r (reconstructed in Wisan, "New Science," 164).

10. Drake, *Galileo at Work*, 95–96. Galluzzi, *Momento*, 273–76. Naylor, "Theory of Motion." Sylla, "Oxford." Galileo may have performed an experiment similar to the one later described in day 3 of the *Discorsi*. In a letter to Leopold, *GOF*, 19:649, Viviani argued that Galileo had discovered that neither absolute nor specific gravity affects the speed of fall by experimenting with pendulums of different materials.

11. *GOF*, 10:115. Drake, "Galileo's 1604 Fragment," 340–41. The second quotation is from Sarpi's letter to Galileo of 9 October 1604, *GOF*, 10:114.

12. See esp. fols. 128 and 85. Galluzzi, *Momento*, 269ff. Drake, *Galileo at Work*, 98–104. Koyré, *Galileo Studies*, 65–78. On the role of space and time in Koyré and Meyerson's influence on him, see Friedman, "Kuhn and Logical Empiricism." Koyré cites Meyerson precisely in those pages.

13. *GOF*, 8:373. Ms 72, fol. 128, has the same watermark and ink as the letter to Sarpi (communication from Wallie Hooper). I have slightly modified the translation in Damerow et al., *Exploring*, 354–55.

14. Naylor, "Theory of Motion," 382, states that Galileo's "belief in his basic concept that *grado di velocità* increased with distance travelled was firmly grounded in his observation of percussive effects." Giusti, "Leggi del moto," xxxi–xxxiii; "Aspetti matematici," 15. *De motu accelerato*, in *GOF*, 2:263.

15. Sylla, "Oxford," 90–91n100: "Thus this physical argument about impetus carrying a heavy ball upward lies behind the proposal that in free fall velocities grow with distances."

16. Galluzzi, *Momento*, 294–98, based on an examination of fol. 179v. Sylla, "Oxford," 68: "Galileo approached the question of free fall via a study of the inclined plane and other ques-

tions of statics and hence was led to consider the variation of velocity over distance rather than time." See also ibid., 100.

17. Murdoch and Sylla, "Science of Motion," 222–51, at 235. Sylla, "Oxford," esp. 57–68. Drake, *Galileo at Work*, 370–71, makes much of Galileo's reformulation of the mean-speed theorem.

18. Galluzzi, *Momento*, 273–74, esp. n36. Sylla, "Oxford." Wallace, *Prelude to Galileo*.

19. Camerota, *Galileo*, 146–47, based on a letter of 1611 by Galileo to Daniello Antonini.

20. Sylla, "Oxford," 72–73. Giusti, "Leggi del moto," xxxvi n2, xlix–lii, lviii. Borelli, *De vi*, 149–59, at 153. Palmerino, "Infinite Degrees," 282–84 (see sec. 5.3). Mancosu, *Philosophy*, 38–50. Andersen, "Cavalieri." Maffioli, *Out of Galileo*, 126–28. Compare the case of Hooke in sec. 8.4.

21. The key manuscript is fol. 91v, first discussed by Wisan, "Science of Motion," 227–28. Giusti, "Leggi del moto," xliv–xlvi. *GOF*, 8:214, 281–83. In a stunning analysis Renn, in Damerow et al., *Exploring*, 180–97, identified the links between fols. 152r and 91v and reconstructed Galileo's path to the proportion between degrees of velocity and time.

22. The key text is fol. 164v, known as the *Mirandum* paradox, on which see Wisan, "Science of Motion," 201–4; Galluzzi, *Momento*, 292–93; and Damerow et al., *Exploring*, 200–202, 256–58. The same problem reappears in *Dialogo, GOF*, 7:47–50.

23. Galluzzi, *Momento*, 302–7.

24. *GOF*, 10:228–30, Galileo to Antonio de' Medici, 11 February 1609. Among the related manuscripts are fols. 152r, 91v, 81r, 114, 106v, 87v, and 116v, showing parabolic trajectories and a comparison of experimental and theoretical values of where the falling ball hits, and should hit, the ground. Galileo marked this difference with the word *doveria*. Its dating is toward the end of his Padua period, but its precise role in Galileo's complex path toward a new science of motion is unclear.

25. Naylor, "Theory of Motion." Damerow et al., *Exploring*, 180–85.

26. Galluzzi, *Momento*, 273, esp. the extended notes on 273-75.

27. Drake, *Galileo at Work*, 19. Renn, "White Elephant," esp. 92–113.

28. The classic work here is Westman, "The Astronomer's Role."

29. All the primary material is collected in *GOF*, 4:65. Extensive analyses of this dispute are in Shea, *Intellectual Revolution*, chap. 2; Biagioli, *Galileo Courtier*, chaps. 3 and 4; and Galluzzi, *Momento*, 227–46. Schmitt, "Experimental Evidence," 357–59, showed that the idea that water will decrease in volume upon freezing was widespread during the sixteenth century.

30. Shea, *Intellectual Revolution*, sec. 2.3. Biagioli, *Galileo Courtier*, 193–202.

31. *GOF*, 4:67–69. I have translated "pesi assolutamente eguali" as "equal weights." The qualification "assolutamente" is introduced to distinguish gravity from specific gravity.

32. *GOF*, 4:77–78: "il momento della velocità del moto in un mobile compensa quello della gravità." Galluzzi, *Momento*, 227–46. Shea, *Intellectual Revolution*, sec. 2.2. Maffioli, *Out of Galileo*, 73. Mach, *Mechanik*, chap. 1, sec. 6, pars. 5 and 6, argues that the analogy with machines is not perfect, because in the latter, equilibrium is indifferent, whereas in the hydraulic case any change in the level of the liquid in the vases raises their common center of gravity. Although Galileo's construction is analogous to the hydraulic press, he seems more concerned with the argument that a ship floats better in a small amount of water than in the ocean than with the idea of a press. In *Diversarum speculationum liber* Benedetti had already introduced the hydraulic

press. *Mechanics in Italy*, 231–32. Archimedes had treated floating without considering the water's raising, as in the open sea.

33. *GOF*, 4:165, 385. Shea, *Revolution*, 33–36. Biagioli, *Galileo Courtier*, 205–6.

34. Castelli discussed the key proposition of his treatise with Galileo at the end of 1625. *GOF*, 13:284–85, 289, 291, 293–94. Castelli, *Della misura*, 4–5. Castelli sent Galileo over fifty copies of his work. *GOF*, 13:464–65, 29 December 1628.

35. Castelli, *Della misura*, 19. Maffioli, *Out of Galileo*, 37–51. For experiments on the compression of water in the Galileo school, see ibid., 93–99. A second slightly enlarged edition of Castelli's work was published in 1639.

36. Bucciantini, "Trattato," 130n89. Riccioli, *Geographia*, 1672^2, 190, states that Ricci and Giovanni Battista Aleotti were consulted by Pope Clemens VIII in 1598.

37. Westfall, "Bisenzio." Bucciantini, "Trattato," 130–34n89. Maffioli, *Out of Galileo*, 54–57. "Annotazione alla lettera del Galileo," in Florence *Raccolta*, 1:231–32.

38. Castelli, *Carteggio*, 122–37, at 133–34 and 136. The exchange between Castelli and Arrighetti occurred in the summer of 1634. Maffioli, *Out of Galileo*, 57–59, emphasizes the lack of the notion of pressure for dealing with water forced through closed conduits.

39. In the vast literature on the *Dialogo*, see Jardine, "Galileo's *Dialogue*"; Clavelin, "*La philosophie naturelle*"; Shea, "*Intellectual Revolution*"; and the *Edizione critica* edited by Besomi and Helbing.

40. *GOF*, 7:49–52; 8:199–202; 5:133–35. Drake, *Galileo at Work*, 185–86.

41. Especially relevant passages are in *Dialogo*, 182–85, 214, 225, 233, 236.

42. Ibid., 248–55 and 269–71, where Galileo argues that given a fixed deviation from a tangent, the extrusion is greater for a smaller wheel than for a larger one. For the analogy with the projectile, see the end of the fourth day of the *Discorsi*. Clavelin, "*La philosophie naturelle*," 232–68.

43. Huygens's figure in *De vi centrifuga* (reproduced in the present work as fig. 7.13), appears as a corrected version of Galileo's and was probably inspired by him. *HOC*, 16:258–67. The corresponding statement is in prop. 3, *HOC*, 16:268–71.

44. In the first day of the *Dialogo*, 31–32, Galileo presented a strong defense of the primacy of circular motion. See chaps. 4 and 7.

45. Bertoloni Meli, "Numerical Tables," 171–75. Hon, "Error," 65–72. Galileo had already dealt with a problem vaguely related to error theory in the late 1620s. The question he posed then was, If a horse is worth 100, and one person is prepared to pay 1,000 and another to pay 10, who would commit the greater error? His answer was that they would be equally wrong. *GOF*, 6:565–612.

46. *GOF*, 7:47, 253, 256–57 (where Galileo discusses some features of the compound pendulum), 454, 475–77. Crombie, "Mathematics, Music," 299–300, and Dear, *Mersenne*, 165, point out that addition 7 to Mersenne, *Les mechaniques*, 77, contains the somewhat garbled enunciation. Although it is unclear whether by 1632 Galileo had grasped the relation between period and length, the *Dialogo* may have provided Mersenne with a qualitative relation on which he was able to elaborate. Galileo's own enunciation can be found in his letter to Lorenzo Realio, 5 June 1637, *GOF*, 17:96–105, at 102.

47. Palmieri, "Tides." Hooper, "Tides," 206–35.

48. Chaps. 39–40 of Cavalieri's *Specchio ustorio* are translated in Damerow et al., *Explor-*

ing, 380–82. Giusti, "Leggi del moto," xlix–lii, argues that instantaneous speeds are peculiar entities that cannot be linked to space and time, precisely because they are instantaneous. Thus it is their aggregate that allows that link to be built and a proper science of motion to be formulated. According to Giusti, Cavalieri considered the aggregate of speeds as a speed, rather than as a magnitude proportional to the space traversed, and hence he divined, rather than proved, the result $s \propto t^2$. Blay and Festa challenge that interpretation in "Mouvement," 76–85, at 82n30. Andersen, "Cavalieri."

49. Camerota, *Galileo*, provides an excellent account.

50. Drake, *Galileo at Work*, 386. On gunnery, see Hall, *Ballistics*. *GOF*, 10:228–30, Galileo to Antonio de' Medici. Galluzzi, *Momento*, 246–59.

51. Galileo, *Two New Sciences*, 12–13.

52. Renn and Valleriani, "The Challenge of the Arsenal."

53. *GOF*, 8:145–50. Dostrovsky, "Vibration Theory," 188. Camerota, *Galileo*, 30, argues that Galileo's father Vincenzio was already familiar with some of those proportions.

54. Similar comments can be found in the fourth day. *GOF*, 8:18–21, 132–34, 275–76. Westfall, *Force*, 33–34, argues that terminal velocity is incompatible with this analysis.

55. *GOF*, 8:63–65, 67–68, 93, 151. Galileo, *Two New Sciences*, 109 (I have slightly altered Drake's translation). See *GOF*, 16:138, Cavalieri to Galileo, October 1634, quoted in Palmerino, "Una nuova scienza," 317. On Galileo's matter theory, see Palmerino, "Galileo's and Gassendi's Solutions." But whereas in matter theory interstitial *vacua* are crucial, in the study of motion Galileo did not have recourse to infinitesimal moments of rest interspersed throughout the time of fall of a body. Biener, "Theory of Matter."

56. *GOF*, 8:156–58. Galileo, *Discorsi*, is especially valuable for Viviani's annotations. His own treatise on the matter, *Trattato delle resistenze*, appeared posthumously (Florence, 1718). For an excellent account of the discipline, see Benvenuto, *Structural Mechanics*. Truesdell, *Flexible and Elastic Bodies*. The pseudo-Aristotelian *Quaestiones* 14 and 16 contain relevant material, as does the commentary by Baldi, *Exercitationes*. I am grateful to Antonio Becchi for this information. A modern treatment would involve the moment of inertia of the beam.

57. *GOF*, 8:172. Galileo, *Two New Sciences*, 130 (I have modified Drake's translation). Viviani observed that not all Galileo's theorems are correct. For example, prop. 6 about the ratio between the moments of the weights to the moments of the resistances in similar cylinders should be in the ⅔ power of the side, not 3⁄2. Galileo, *Discorsi*, 133–35. Sylla, "Oxford," 60–61, points out that Galileo does not adopt the scholastic distinction between the maximum load before a beam breaks and the minimum weight with which it breaks.

58. *GOF*, 8:176. Galileo, *Two New Sciences*, 133 (I have slightly altered Drake's translation): "If two places are taken in the length of a cylinder at which the cylinder is to be fractured, the resistances at those two places have to each other the inverse ratio of the rectangle made from the distances of those two places," meaning distances from the two ends. By "resistances" and "forces" Galileo meant their respective moments. For a beam standing on a support C, the total moment of the resistance would be as the sum of $1/AC$ and $1/CB$, namely as $AB/(AC \times CB)$.

59. Benvenuto, *Structural Mechanics*, 196–97.

60. *GOF*, 8:161–62, 185–86. Libri, *Histoire*, 4:398.

61. *GOF*, 8:186–89.

62. Giusti, "Ricerche galileiane." Damerow et al., *Exploring*, 248–50.

63. Galluzzi, *Momento*, 309–29, esp. 310–12, suggests that the difficulty was that all bodies

have the same acceleration, regardless of their weight; therefore there are problems in linking acceleration to the weight or the moment of gravity of a body. At 311n4 Galluzzi challenges neo-Kantian interpretations that see Galileo's shift as a prelude to Kant. Cassirer, *Erkenntnisproblem*, 377–418, at 396 and 402ff. *GOF*, 4:139–40; 8:201–3. In the opening of the second day (*GOF*, 8:151), however, Salviati argues that all that matters for the forthcoming investigations is that bodies offer resistance to rupture. Wallace, *Prelude*, 284–88.

64. *GOF*, 8:193 (theorem 2), 203–4. Galileo, *Two New Sciences*, 160: "When speeds have the same ratio as the spaces passed or to be passed, those spaces come to be passed in equal times." The statement that speeds are proportional to distances is ambiguous. If speed is uniform and, given any distance, it is as that distance, then time is constant. If speed is not uniform, however, time has to be calculated with a definite integral. If $v = ks$, where v is speed, s is the distance, and k is a constant, then $t = \int ds/v$ or $\int ds/ks$, going as the logarithm with a singularity for $s = 0$. For analyses of Galileo's proof, see Sylla, "Oxford," 78–83, and Damerow et al., *Exploring*, 244–46.

65. *GOF*, 8:205–8. Giusti, "Leggi del moto," 191–94. Galileo, *Two New Sciences*, 164.

66. *GOF*, 8:210–14. I have marginally modified Drake's translation. The plane has a very small inclination to slow the motion sufficiently and to prevent the ball from sliding and jumping rather than rolling. Dear, *Discipline*, 127, argues that the odd-number rule was "the basic assumption" of Galileo's science of motion. In his formal presentation, however, Galileo defined uniformly accelerated motion as that motion where speed is as the time. Subsequently he proved the odd-number rule and relied on experiment to show that his mathematical theory is followed by nature.

67. Drake, in *Two New Sciences*, 265n23, makes much of this reformulation of the mean-speed theorem because the final speed can be measured.

68. For *De compositione continui* and the *Postils to Rocco*, see Galluzzi, *Momento*, 334, 345–46; Drake, *Galileo at Work*, 359, 361–67; and *GOF*, 7:569–750. Interpreters are divided over the role of mathematics in Galileo's science of motion. Wisan, "Science of Motion," 116, argues, "In *De motu locali* we do not find any new developments in mathematics, but rather a new use of Greek mathematics to study motion along inclined planes" (but see also 235). Drake, *Galileo at Work*, 94–95, 97, 116, 125–26, claims that Galileo's success in formulating a mathematical science of motion was largely due to the recognition of the legitimacy of the notion of speed at a point. For Galluzzi, *Momento*, 331–62, the central issues were the nature of the continuum, the presence of infinitesimal magnitudes, direct and oblique transit, and the comparison between infinite aggregates of lines—all topics where Galileo's foundations remained shaky and threatened, rather than furthered, his science of motion. According to Giusti, "Aspetti matematici," 3, 41–42, and "Leggi del moto," liv–lx, by 1604 Galileo had attained several results, such as the law of fall $s \propto t^2$, parabolic trajectories, the chord law, the law of fall along inclined planes, and also the false principle stated in the letter to Sarpi. Giusti compares Galileo's results to an archipelago of islands of certainty among which he was trying to construct mathematical bridges, an arduous task within the straightjacket of the theory of proportions. Giusti devotes great attention to paradoxes arising from the aggregate of speeds.

69. *GOF*, 8:209–10. Galileo relied on theorem 4 of book 1. Damerow et al., *Exploring*, 241, argues that theorem 2 is not restricted to uniform motion.

70. *GOF*, 8:215–19. Galileo relied on an extension of book 1, theorem 1. Wisan, "New Science" 215–22. Galluzzi, *Momento*, 359–62. Damerow et al., *Exploring*, 247–51. My reading is

based on the published text. Souffrin, "Mouvement uniforme," argued not implausibly that Galileo may have had in mind a previous version of the proof based on fol. 138, whereby Galileo would take the limit of a stepwise dependence of speed and time.

71. *GOF,* 8:261–64. Galileo had reached the correct result in his 1602 letter to dal Monte, in which he was also seeking to prove that the circle is the isochronous curve. We shall return to this problem in the following chapters.

72. Giusti, "Leggi del moto," xxix–xxx. Damerow et al., *Exploring,* 12–17, 197–202, 216–35, 237–53, 256–58, 263–66. Palmerino, "Galileo's Theories," 146–47.

73. *GOF,* 8:303–9, esp. 308–9.

74. Huygens proved this in the late 1640s, but the issue was taken up again at the end of the century. Gabbey, "Huygens and Mechanics," 168.

75. Damerow et al., *Exploring,* 266–68. Renn, "White Elephant," 113–26. Bertoloni Meli, "Archimedean Revival," 23–24.

76. Settle, "An Experiment in the History of Science."

77. *GOF,* 8:109. I have slightly altered Drake's translation. The words "on making the experiment" are missing in the Leyden text.

78. Galileo, *Two New Sciences,* 77–80, at 80. Galileo's views are derived from *De motu antiquiora.* Damerow et al., *Exploring,* 269–70.

Four • Novel Reflections and Quantitative Experiments

Epigraph: Mersenne, *Harmonie,* book 1, 111.

1. Before *Harmonie universelle,* Mersenne published a different version in Latin, *Harmonicorum libri XII* (Paris, 1635), reissued with additions in 1648. Descartes and the mechanical philosophy are discussed in the following chapter.

2. Lenoble, *Mersenne,* 393–94. Feingold, "A Friend of Hobbes," 271–72, reports the existence of early unpublished English translations of Galileo's *Dialogo* and *Discorsi.* See also Jesseph, "Galileo," 194. Galileo's work was also translated into English by Thomas Salusbury, *Mathematical Collections and Translations.* Carli and Favaro, *Bibliografia Galileiana.* On Galileo and university curricula, see Schmitt "Text-book Tradition." Westman, "The Reception of Galileo's *Dialogue.* Garber, "Frontlines."

3. Riccioli, *Almagestum,* part 2, 360b–81b, surveys Galileo, his critics, and his supporters. Roberval, *Aristarchi de mundi systemate,* 31–33. Mersenne, *Cogitata, Hydraulica,* 259–60. Palmerino, "Gassendi's Reinterpretation." Hooper, "Tides."

4. Mersenne, *Quaestiones in Genesim,* cols. 85–86. Dear, *Mersenne,* 119. Palmerino, "Jesuit Responses."

5. Drake, *Galileo at Work,* 84, 294. Palmerino, "Infinite Degrees," 298–99; "Galileo's Theories," 140–45. For a similar trial, see Camerota, *Galileo,* 404–5. *GOF,* 18:102–3, Baliani to Galileo, 16 September 1939. Galluzzi, "*Affaire,*" 239–42.

6. *MMC,* 2:257, 280; 3:188. Dear, *Mersenne,* 128–32. Mersenne, *Traité des mouvemens,* 46–47. On Mersenne, see Beaulieu, *Mersenne.*

7. Mersenne, *Harmonie,* vii–viii.

8. Floris Cohen, *Quantifying Music.* Dostrovski, "Vibration." Camerota, *Galileo,* 30.

9. Mersenne, *Harmonie,* book 1, 74.

10. Dear, *Mersenne*, chaps. 5 and 6. Mersenne, *Harmonie, Traité des instrumens a chordes*, 125. Auger, *Roberval*, 78–103, at 78–87. Dugas, *Mécanique*, 155–57. Dostrovsky, "Vibration," 186–87. Bertoloni Meli, "Numerical Tables," 175–88.

11. *GOF*, 7:54, 250; 14:342, Baliani to Galileo, 23 April 1632; 18:75–79, Galileo to Baliani, 1 August 1639. Castelli, *Carteggio*, Baliani to Castelli, Savona, 20 February 1627, 95–97, at 96. Drake, *Galileo at Work*, 306, 336, 399.

12. Mersenne, *Harmonie*, 85–88, esp. the corollary at 88; *Novarum observationum tomus III*, 192. Galileo's result means that a body falls 92 inches in the first second. Since Galileo had established his result for a sphere rolling down an inclined plane, we need to introduce a correction of a factor of ⁵⁄₇ due to the sphere's rotation, leading to nearly 11 feet, or 129 inches, per second.

13. Mersenne, *Harmonie*, 89–90; *Cogitata, Ballistica*, 35–38. Andersen, "Method," 22–23. I am grateful to Niccolò Guicciardini for the reference to Roberval.

14. Mersenne, *Harmonie*, 112, 125–26. Dear, *Discipline*, 132. Palmerino, "Infinite Degrees," 274.

15. If resistance halves the speed, for example, the speed increases as ½, ¼, ⅛, etc., in each time interval, while the increase from rest is as ½, ¾, ⅞, ¹⁵⁄₁₆, etc. Mersenne, *Harmonie*, 128, 206–7. Hall, *Ballistics*, 110.

16. For a body moving with a speed equal to ωR along a semicircle whose diameter is equal to the Earth's radius R, the time of fall to the center of the Earth is $t = \pi R / 2\omega R = T/4$, where $\omega = 2\pi/T$ and T is the Earth's period of rotation. Hence the time of fall is about six hours.

17. Mersenne, *Harmonie*, 96, 98; *Cogitata, Ballistica*, 50, 57–58. See the classics by Koyré, "Fall," esp. 340–41, and Galluzzi, "Galileo contro Copernico." See Mahoney, *Fermat*, 226–28, 384–87 on problems associated with $v \propto s$. Fermat, *Oeuvres*, 5:1–43. Palmerino, "Una nuova scienza," 298–301.

18. Mersenne, *Harmonie*, 137–50, at 145. Mersenne argued that in a time of 4‴ (¹⁄₁₅ of a second) a body would fall ¹⁶⁄₂₅ of an inch. In 4″″, he calculated, the body would fall ¹⁄₂₀₂₅₀₀₀₀ of an inch, namely ¹⁶⁄₂₅ divided by the square of the time, or 3,600. This calculation shows that Mersenne adopted the correct law of fall. See fig. 3.11. MacLachlan, "Mersenne's Solution."

19. Cabeo, *Commentaria*, 1:95b and 101–3, at 101. Borgato, "Cabeo," 384–85.

20. Mersenne, *Harmonie*, book 2, prop. 7, esp. 111–12, translation in Dear, *Discipline*, 131. Koyré, "Experiment in Measurement," 113. This discrepancy was due to the fact that a sphere rolling on an inclined plane behaves like a rigid body, not a mass-point, and follows more complex laws because it rotates, but neither Galileo nor Mersenne would have been able to take this phenomenon into account mathematically. Since a sphere rolling down along an inclined plane covers only five-sevenths of the distance Galileo and Mersenne had envisaged, Mersenne's results were not unreasonable. At high inclinations the ball bounces, thus making accurate predictions impossible.

21. Mersenne, *Harmonie*, 108–12. The issue of whether acceleration is continuous is discussed in Palmerino, "Infinite Degrees," esp. 274–82, taking into account also Mersenne's 1634 and 1639 translations of Galileo.

22. *MMC*, 2:231. Dear, *Mersenne*, 152–53. The entire section by Dear (139–60) is relevant to the present discussion. Mersenne, *Harmonie, Traité des instrumens a chordes*, 123. Dostrovsky, "Vibration," 185–88, at 187, suggests that Mersenne may have used "strings passing over pul-

leys; in which case friction in the pulleys could have made such corrections necessary." See ibid., 197–98, for Mersenne's experiments with long strings; also ibid., 190–92. Dear, *Mersenne*, 157–60.

23. The latter was discussed in correspondence with Beeckman and Descartes. *MMC*, 2:205–6, 317–19, 346; 5:126. Dear, *Mersenne*, 163–69.

24. Mersenne, *Harmonie*, 158, 165. Dear, *Mersenne*, 154, 156.

25. For Beeckman, see *MMC*, 2:206, 317–19, 346. For Descartes, see *MMC*, 5:126. Dear, *Mersenne*, 163–68.

26. Mersenne, *Harmonie*, props. 13–16, esp. 130–35. See also prop. 21 (erroneously numbered 20), 209–13. The relation between vertical components and times is not correct. In an inclined plane the component of gravity is the same at all points, whereas in an oscillating pendulum it varies at each instant. The relation Mersenne sought was more complex than he had envisaged.

27. Mersenne, *Harmonie*, *Traité des instrumens a chordes*, props. 18–20. In this portion of the book the pagination is in disarray, with two pages numbered 45, two numbered 46, and some having no number at all. Mersenne, *Novarum observationum tomus III*, 140–41. In the general preface to *Cogitata*, sec. v, Mersenne claimed that the lack of isochronism could be attributed to air resistance.

28. Gabbey, "The Case of Mechanics," 499–514. Costabel, "Controverse." Dubarle, "Théorie." Dugas, *Mécanique*, 164–73. *Discepoli*, 1:330–32, Mersenne to Torricelli, 6 November 1646; 338–39, Mersenne to Torricelli, 8 December 1646. *DOAT*, 4:362–71, 380–89, 420–33, 502–10, 597–600, 621–23. *MMC*, 14:47–50, 107–9, 116–24, 159–75, 235–36, 286–95, 480–86, 725–35. Mersenne, *Novarum observationum tomus III*, 114–26. Mousnerius [Fabri], *Tractatus*, 420–37. Lagrange, *Méchanique*, 168–71.

29. *GOF*, 16:214, Micanzio to Galileo, Venice, 17 February 1635; 218–20, 3 March 1635. Aproino objects mainly to the recourse to the infinitely small. *GOF*, 16:221–28. On 214 Micanzio reports that De Ville was regarded as a great mathematician, though his objections do not confirm that opinion. Galluzzi, *Momento*, 357–58.

30. *GOF*, 18:68–71, at 70, Baliani to Galileo, 1 July 1639; 75–9, at 77, Galileo to Baliani, 1 August 1639; 86–88, at 87, Baliani to Galileo, 19 August 1639. See sec. 3.4 above. Galluzzi, *Momento*, 249–50. Benvenuto, *Structural Mechanics*, 180.

31. *GOF*, 8:214–19. Galluzzi, *Momento*, 312–13. De Gandt, *Force*, 105–7.

32. *GOF*, 8:221–22, at 222. Torricelli, *De motu, Opere*, 2:105, 108–9. Giusti, "Master."

33. Torricelli, *De motu, Opere*, 2:108. Mersenne, *Novarum observationum tomus III*, 96. *Discepoli*, 1:179–81, at 180, Mersenne to Torricelli, Rome, 10 January 1645; 181–3, Torricelli to Mersenne, Florence, mid-January 1645; 187–91, Mersenne to Torricelli, Rome, 4 February 1645; 201–4, Torricelli to Mersenne, Florence, mid-February 1645. Galluzzi, "Torricelli," 73–82. Galluzzi, *Momento*, 323–26. Maffioli, *Out of Galileo*, 104–9.

34. Gassendi, *De motu impresso*, in *Opera*, vol. 3 (this later edition contains an additional letter), 497b; 498a: "Concipe rursus intervallum uniformiter crescens repraesentare velocitatem uniformiter increscentem, & quos pareis triangulos vides, totidem gradus velocitatis, & consequenter parteis spatij, quod grave decidens percurrit." Pamerino, "Infinite Degrees," 304–7; "Jesuit Responses," 211.

35. Gassendi, *Opera*, 3:566a–67b. Palmerino, "Infinite Degrees," 299–312; "Jesuit Responses," 204–14; "Galileo's Theories," 157–63. Festa, "Gassendi lecteur," 363–64.

36. Baliani, *De motu*, 182–85. Moscovici, *Expérience*, 68. Galluzzi, "*Affaire*," 265–70, argued that Fabri was Baliani's source.

37. Palmerino, "Jesuit Responses," 187–89, 192–204.

38. Drake, "Free Fall," esp. 362–63. Molland, "Atomisation." Moscovici, *Expérience*, 61–78. Galluzzi, "*Affaire*."

39. Mousnerius, *De motu locali*, 105. There were other issues relevant to mechanics and motion treated by Fabri. For example, he argued that in the pendulum (*funependulum*) the ascent does not reach the same height from which it had fallen (318–19), and he claimed that greater oscillations take longer; hence the pendulum is not isochronous (102–3). He also discussed the center of percussion (420–37). Subsequently Fabri argued that his time units were variable. Palmerino, "Jesuit Responses," 192–204.

40. Cabeo, *Commentaria*, 2:76, quoted in Dear, *Discipline*, 142. Dear, *Discipline*, 138–44. However, in the *Discorsi* Galileo did present the odd-number rule as a corollary to theorem 2, thus strictly speaking his rule was not taken from experience. *Commentaria*, 1:94a; 2:76. Borgato, "Cabeo," 373–74, attempts to save Cabeo's interpretation by arguing in terms of uniform average speeds during each time interval, and those speeds are as the distances.

41. *GOF*, 18:68–71, at 68, Baliani to Galileo, 1 July 1639. Dear, *Discipline*, 143. Moscovici, *L'expérience*, 140–47.

42. Baliani, *De motu*, 98, translated in Dear, *Discipline*, 144.

43. Yoder, *Time*, chap. 2.

44. *HOC*, 1:24–28, Huygens to Mersenne, Leyden, 28 October 1646. *HOC*, 11:68–72. Castelli's letter to Baliani of 20 February 1627, *GOF*, 13:348–49. Spatial invariance suggests an engineering origin, i.e., towers and bridges stand or fall independently of the units used to measure them. Bertoloni Meli, "Axiomatic," sec. 3.

45. *HOC*, 1:26. Calling $AB = a$ and $BC = b$, in a geometric progression $CD = bb/a$ and $DE = bbb/aa$. Applying Huygens's condition, we find that the only solution for a and b positive is $a = b$, meaning that motion is uniform and there is no acceleration. Torricelli had attacked the problem along the same lines as Huygens. Torricelli, *Opere*, 3:155.

46. *HOC*, 1:27. The odd-number rule is preserved even if we choose different units of time, such as one, two, or three units, for example, $1:3::(1 + 3):(5 + 7):(1 + 3 + 5):(7 + 9 + 11)$. Huygens has a body falling in equal times through spaces LM, MN, NO, OP, etc. Calling $LM = a$, in an arithmetic progression we have $MN = a + x$, $NO = a + 2x$, $OP = a + 3x$. From his condition $LM:MN::LN:NP$, we have $a:(a + x)::(2a + x):(2a + 5x)$, leading to $x = 2a$. Thus we have the odd-number rule.

47. Palmerino, "Jesuit Responses," 198–99, 215.

48. Palmerino, "Infinite Degrees," 295–97, 319–24. Lenoble, *Mersenne*, 423–25. *CMM*, 12:351, Deschamps to Mersenne, 1 November 1643. *HOC*, 1:558–59, Mersenne to Christiaan Huygens, 12 October 1646. Torricelli, *Opere*, 3:326–28, at 326–27, Torricelli to Mersenne, June 1645; 461–66, at 461–62, Torricelli to Renieri, August 1647. Torricelli's claim that no other progression apart from Galileo's works is incorrect. For example, if the spaces traversed from rest are proportional to the cube of the times, they would be as 1, 8, 27, 64, 125, 216, etc. The spaces traversed in successive equal intervals would be as 1, 7, 19, 37, 61, 91, etc., and $1:7::(1 + 7):(19 + 37)::(1 + 7 + 19):(37 + 61 + 91)$. That is, the same proportion holds if one chooses a unit of time that is twice or three times the original time interval. Dear, *Mersenne*, 208–22. Galluzzi, "*Affaire*." Mersenne, *Novarum observationum tomus III*, chaps. 15 and 16, 131–41.

49. Mersenne, *Cogitata, Ballistica*, 44–45; *Harmonie*, 206–7. On Mersenne's work, see Hall, *Ballistics*, 106–10. Roberval and Descartes also held that the initial speed of falling bodies is finite. *Discepoli*, 1:207–9, at 208, Mersenne to Torricelli, Rome, February 1645. Cabeo, *Commentaria*, 1:93.

50. Mersenne, *Novarum observationum tomus III*, chap. 9, 111–12, 324–28. His changing views are analyzed in Palmerino, "Infinite Degrees."

51. Torricelli, *Opere*, 3:349–56, Roberval to Torricelli, 1 January 1646. *CMM*, 6:373–75. Mersenne shared Roberval's views. Mersenne, *Cogitata, Ballistica*, 24–25, 30–34, 83–84.

52. Damerow et al., *Exploring*, 274–76. It is unclear whether Torricelli would have extended his views on rectilinear uniform motion from projectile trajectories to celestial motions.

53. Curiously, in a note in the margin to the letter Torricelli says, "As far as I recall I did not state this." *Discepoli*, 1:388. But Renieri was right: it is in *De motu*, *Opere*, 2:210.

54. *Discepoli*, 1:388–90, Renieri to Torricelli, Genoa, 2 August 1647. Baliani, *De motu*, book 3, 141–45, and the introduction by Baroncelli, 36. Moscovici, *L'expérience*, 24, 192–200. Maffioli, *Out of Galileo*, 108.

55. Cabeo, *Commentaria*, 3:42–44. Dear, *Discipline*, 128–29. *Discepoli*, 1:391–94, Torricelli to Renieri, Florence, mid-August 1647. Segre, *Wake*, 94–97.

56. *Discepoli*, 1:399–400, Renieri to Torricelli, Genoa, 24 August 1647; 405–7, Torricelli to Renieri, Florence, early September 1647.

57. Torricelli, *Opere*, 3:273–77, at 276, Torricelli to Mersenne, February 1645. Maffioli, *Out of Galileo*, 82–83. Blay, "La loi d'écoulement." On Torricelli's mathematics, see Galluzzi, "Torricelli."

58. Mersenne, *Cogitata, Hydraulica*, preface, twelfth unnumbered page, 45, 51–61. Maffioli, *Out of Galileo*, 85–89. Mersenne showed familiarity with Castelli's treatise and referred to its fundamental propositions. Ibid., 55, 176–77. In *Novarum observationum tomus III* Mersenne did not pursue his objections about the first drops of water.

59. Mersenne, *Cogitata, Hydraulico-pneumatica*, 149–55. Boas Hall, "L'élasticité de l'air," 55–56, reports that David Rivault attributed the invention to "sieur Marin." The wind gun, or *archibuso a vento*, was also mentioned by Castelli, *Della misura*, 17–18, who attributed its invention to Vincenzo Vincenti from Urbino.

60. Mersenne, *Cogitata, Mechanica*, 62–72; *Novarum observationum tomus III*, chap. 18, 148–51. Benvenuto, *Structural Mechanics*, 203–5. See also Mersenne, *Harmonie*, book 3, prop. 16, 193–97.

61. *GOF*, 8:319–46. Galileo, *Two New Sciences*, 281–306. Borelli, *De vi*, chap. 33, 249. Westfall's analysis of percussion in Torricelli in *Force*, esp. at 131 and 136, is unsatisfactory in its dimensional treatment of indivisibles. Westfall, *Force*, 129–38, 228–30. Andersen, "Method."

62. Riccioli, *Almagestum*, part 1, 84–91; part 2, 381–97. It is worth noting the disciplinary affiliation with music and astronomy of Mersenne's and Riccioli's works, areas in which numerical tables were common practice. Borgato, "Riccioli."

63. Dear, *Discipline*, 71–85. Koyré, *Metaphysics and Measurement*, 104–5, is concerned with Riccioli's failure to obtain a perfect seconds pendulum. This failure, however, in no way prevented him from using the pendulum. Riccioli also tested the heartbeat, finding it inaccurate. Mersenne had tentatively used it in *Cogitata, Ballistica*, 61. Riccioli, *Almagestum*, part 1, 88b, 89–90.

64. Arriaga, *Cursus* (1632), 582a–84a; (1647⁴), 515a–17b. Cabeo, *Commentaria*, 1:97a–98a, and 2:79b, argued that all bodies fall with the same speed, regardless of air resistance. Dear, *Discipline*, 67–76, esp. 74, discusses Riccioli's disagreement with Arriaga and Cabeo, with special emphasis on experimental narratives, but he does not mention that according to Arriaga falling bodies do not accelerate. Borgato, "Riccioli," 85–87. Cabeo experimented at Ferrara in 1634 with Riccioli. In 1641 Galileo's successor on the chair of mathematics at Pisa, Vincenzo Renieri, experimented with bodies made of different materials dropped from the leaning tower, finding that a lead ball fell faster than a wooden one. *GOF*, 18:305–6, Pisa, 13 March 1641; 8:106–16. Drake, *Galileo at Work*, 414–6. Baliani found that spheres of iron and stone dropped from the fortress at Savona reached the ground at the same time. *De motu*, 51–53, 263–64.

65. Riccioli, *Almagestum*, part 2, 391, 394a. In the *Dialogo* Galileo had discussed the motion of wooden balls falling in water. *GOF*, 7:262–63. *Dialogo: Edizione critica*, 2:570, points out that Galileo's passage is a paraphrase of Buonamici, *De motu*, 223.

66. Riccioli, *Almagestum*, part 2, 392b–393, and part 1, 68b–71a, where he criticizes Castelli's method for measuring running waters.

67. Ibid., part 2, 394a–96a. In his crucial conclusion on 396a Riccioli tries to argue that the *principale movens* is the substance of the body, and gravity becomes a mere instrumental cause. The force of a substantial principle would make a light body move faster than mere gravity.

68. Ibid., 398a–406b. Koyré, "Fall," 348–54.

69. Galluzzi, "Galileo contro Copernico," 93–94.

Five • *The Motion and Collision of Particles*

Epigraph: Descartes, *Principes de la philosophie*, IV, 203.

1. According to this view, a wound inflicted by a weapon could be healed by anointing the weapon rather than the wound. The claim about the weapon salve appears only in the French edition, Descartes, *Principes de la philosophie*, IV, 187. I refer to Descartes' work, whether in Latin or in French translation, according to part (I–IV) and paragraph. The two versions are in *DOAT*, vols. 8.1 and 9.2.

2. Descartes, *Principia philosophiae*, IV, 23, 133–83.

3. Plato, *Timaeus*, secs. 22–30. Dijksterhuis, *Mechanization*. Although Plato is quite specific about the relative sizes of his solids, commentators often ignore this detail and show them as being all of a similar size.

4. *GOF*, 6:197–372. Galileo, *Il saggiatore*, chap. 48. Descartes, *Principia*, IV, 191–95 (see 191 for tickling). Crombie, "Proprietá primarie." Rodis-Lewis, "Descartes' Life," 33.

5. Gaukroger, *Descartes*, chap. 3. Historians have debated the extent to which Beeckman influenced Descartes. Berkel, "Beeckman." Garber, *Metaphysical Physics*, 9–12, 231. Gaukroger and Schuster, "Hydrostatic Paradox."

6. Descartes, *Principes*, IV, 203.

7. This term was used in different ways in the seventeenth century. Different accounts are provided by Dear, *Discipline*, and Gaukroger and Schuster, "Hydrostatic Paradox." *DOAT*, 4:694–700, at 696.

8. *DOAT*, 1:431–32, 435–48 (the original title was *Explication des engins*), Descartes to Constantyn Huygens, 5 October 1637. His treatise was first published in Paris in 1668. *DOAT*,

4:200; 9.2:17. Insightful comments can be found in Gabbey, "Descartes's Physics" and "What Was 'Mechanical'?" *Quaestiones mechanicae*, questions 12, 15, and 35. I rely on the important work by Hattab, "From Mechanics to Mechanism." Garber, "Descartes, Mechanics,"

9. Dugas, *Mécanique*, 293, highlights the key role of the impact rules. Westfall, *Force*, 47, 188, and passim, presents the mechanical philosophy as a problem for mathematization and overall underestimates the role of the impact rules. Gaukroger, "Descartes' Project." Garber, "Different Descartes."

10. A classic and in some respect problematic account is Koyré, *Galileo Studies*, 79–94. Damerow et al., *Exploring*, 1–7, 24–69.

11. Beeckman, *Journal*, 1:263–65. Damerow et al., *Exploring*, 22–67, esp. 25, 30, 37–38, 41, 48, 55, 290–92.

12. *DOAT*, 2:379–405, Descartes to Mersenne, 11 October 1638. A partial translation is in Drake, *Galileo at Work*, 387–92, at 390.

13. Drake, *Galileo at Work*, 391.

14. *DOAT*, 3:807–8, Descartes to Huygens, 18 February 1643. Damerow et al., *Exploring*, 280–81, 377. The claim on 281 that Descartes "was the first to conceptualize the relations of time, space, and velocity in a manner consistent with the conceptual framework of classical mechanics" seems rather too clear-cut; in 1636 Mersenne had already associated the areas of the triangles of speeds to the distances traversed. Moreover, Descartes seems to have accepted an improved version of Galileo's rule only as an empirically adequate approximation.

15. An older classic is Mouy, *Développement*. Dugas, *Mécanique*, 251–64. Brockliss, *Higher Education*. Verbeek, *Descartes and the Dutch*.

16. *DOAT*, 9.2:11–12, 22–23.

17. *Principia*, II, 36–42. Garber, *Metaphysical Physics*, chap. 9, esp. 280–305. Bertoloni Meli, "Axiomatic and Experimental Traditions." Steinle, "Principles."

18. Garber, *Metaphysical Physics*, 203, 253–4. Bertoloni Meli, "Inertia."

19. Gassendi, *De motu impresso*, in *Opera*, 3:495, 489a. Koyré, *Galileo Studies*, 244–51; *Newtonian Studies*, appendices G and I. Westfall, *Force*, 99–109. The new interpretation of Gassendi is due to Palmerino, "Galileo's Theories," 150–53.

20. Gabbey, "Force and Inertia," 248–61. Garber, *Metaphysical Physics*, chap. 6, esp. 156, 172–75. Damerow et al., *Exploring*, 71–133, passim.

21. Koyré, *Closed World*, 143. Garber, *Metaphysical Physics*, chap. 6, esp. 162–72; "Descartes' Physics," at 330–31nn59 and 66. Gaukroger, *Descartes*, 371–77.

22. Descartes, *Principia*, II, 37, 38; *Principles*, 59–60. Murdoch and Sylla, "Motion," 212–13, 252n18. Damerow et al., *Exploring*, 22–24.

23. Descartes, *Principia*, III, 63.

24. Ibid., II, 39. Descartes, *Principles*, 60–61.

25. Bertoloni Meli, *Equivalence*, 29. Koyré, *Galileo Studies*, 79n61, 257. It is perfectly legitimate to have recourse to an outward tendency along the radius, later called centrifugal force by Huygens. What is problematic is to consider at the same time a tendency along the tangent and one along the radius. Bertoloni Meli, "Relativization." Descartes, *Principia*, II, 39; III, 55–59. In III, 59, he claims about the stone in the sling, "Idemque etiam experimus in fundâ: quò celeriùs enim lapis in eâ rotatur, eò magis funis intenditur; atque ista tensio, à solâ vi quâ lapis recedere conatur à centro sui motûs exorta, exhibet nobis istius vis quantitatem."

26. Descartes, *Principia*, III, 55–59, at 59. In those paragraphs Descartes tried to explain

the nature of light, but he also employed the notion of *conatus* in his general remarks about circular motion. The notion of *conatus* figures prominently in Hobbes and Leibniz, who use it in the sense of a tendency toward motion. Herivel, *Background*, 54–64. Westfall, *Force*, 78–81, highlights the analogy in Descartes' language between the endeavor due to circular motion and gravity. Gabbey, "Force and Inertia," 293–95. Bernstein, "*Conatus*."

27. Descartes, *Principia*, II, 40; *Principles*, 61–62.

28. Descartes, *Monde*, chap. 7; *Principia*, II, 4. Gaukroger, *Descartes*, 240–49.

29. Jammer, *Mass*, chaps. 5–6. Following a modern view, in all impacts the quantity of motion in one direction—understood as a vector—is conserved. In elastic impacts, *vis viva* or mv^2 is conserved; therefore elasticity is the key notion, and all bodies can be classified according to their degree of elasticity. I restrict my investigation to direct impact, excluding oblique collision, on which see Gabbey, "Force and Inertia," 256–7; Damerow et al., *Exploring*, 127–30; and Szabó, *Geschichte*, 435–36.

30. Beeckman, *Journal*, 1:266: "id enim in omnibus machinis animadvertitur, ut *duplex pondus, aequali vi sublatum, etiam duplò tardiùs ascendat quàm prius pondus.*" See also ibid., 2:45, spring 1620.

31. Ibid., 3:133–34. The term used for weight was "corporeitas seu pondus."

32. See Szabó, *Geschichte*, 429–36. Sørensen, "Study." Westfall, *Force*, 117–25.

33. Pappus, *Book VII of the Collection*, 94–104, 547–72.

34. Sørensen, "Study," 54, and Aiton, "Marci," 157, highlight the dimensional problems with Marci's definition of *impulsus*. Marci, *De proportione*, C and Mv–M2r.

35. Marci, *De proportione*, M4v. Szabó, *Geschichte*, 434–35.

36. Marci explains that that portion of *B*'s impulse equal to *A* is transferred to *A*, so that *A* is set in motion and *B* stops. The excess portion of *B*'s impulse can either make *B* rebound or *A* move faster. The latter case is explained thus: "Quod si ob motum velociorem [of *A*] nulla a percusso [namely *A*] inducitur plaga, minor expulso majori quiescit." Marci, *De proportione*, M3v.

37. Sørensen, "Study," 264.

38. Marci, *De proportione*, N4v–O3r. Sørensen, "Study," 265–66, reports other studies of this problem in the seventeenth century by Descartes, Mariotte, and Huygens; see also ibid., 267–68. Szabó, *Geschichte*, 436–37.

39. Marci, *De proportione*, N2r, puts it thus: "Minorem quidem globum *[B]* a majori *[A]* reflecti constat, propterea quod ex huius plaga impulsus quidem aequalis, major autem velocitas in minori consequator."

40. Ibid., Ov, O3v. Sørensen, "Study," 267–68.

41. *GOF*, 18:267–68. *HOC*, 1:307–8, Huygens to Kinner von Löwenthurm, 26 November 1654.

42. Gabbey, "Force and Inertia," 243–72. Garber, *Metaphysical Physics*, 231–62. Des Chene, *Physiologia*, 290–312. Damerow et al., *Exploring*, 71–133.

43. Garber, *Metaphysical Physics*, 240.

44. Ibid., 261. Damerow et al., *Exploring*, 99–100. Gabbey, "Force," 263–72.

45. Garber, *Metaphysical Physics*, 258–59.

46. A brief account can be found in Aiton, *Vortex*, chap. 3.

47. A relevant reference can be found in *Monde*, chap. 12, on tides. I am grateful to Mike Mahoney for having pointed this out to me. As we saw in chap. 3, the intuitive idea behind Castelli's formalized proposition was known before 1628 and was not difficult to arrive at.

48. Descartes, *Principia*, III, 51, 98; IV, 49. A more detailed account can be found in Shea, *Magic*, 289–90. Aiton, "Tides." According to Descartes, low tide occurs directly under the moon and on the opposite side.

49. *Principia*, III, 48–52; IV, 3. The French edition is explicit on this point.

50. Ibid., III, 121. The opening of Descartes' paragraph reads very much like a passage in Newton's definition of mass: "Per soliditatem hic intelligo quantitatem materiae tertii elementi."

51. Ibid., 121–23. For an earlier version of Descartes' views, see chap. 9 of *Le monde*, where Descartes discussed similar issues with the example of bodies floating in two rivers joined at one point. The example of barges and other bodies carried by the stream was taken up later in the century. Beeckman had expressed similar concerns in *Journal*, 2:77–78.

52. Descartes, *Principia*, III, 119.

53. Ibid., IV, 23–25. Newton's experiment is discussed in chap. 9 below.

54. Descartes, *Principia* and *Principes*, III, 119–22, esp. 121, where the tentative *peu-estre* is introduced into the French translation. By "agitation" Descartes seems to mean motion with no regard to its determination. Westfall, *Force*, 61–62.

55. Descartes, *Principia* and *Principes*, IV, 47.

56. There are linguistic differences in IV, 132, between the Latin *Principia* and French *Principes*. In the Latin version of the heading of par. 132, Descartes emphasized the notion of rigidity, whereas in the French he used the expressions *faire ressort* as equivalent to *estre roide*. In the French version of the text of par. 132 he used the term *ressort* more systematically. Breger, "Elastizität," 113. In chap. 8 we shall see that later in the century others sought to combine a physical and experimental study of elasticity with a more thoroughly mathematical account. Occasional references to elasticity in the study of impact can be found in the correspondence with Mersenne, for example *DOAT*, 3:631–37, at 635, Descartes to Mersenne, 23 February 1643; 648–55, at 652, Descartes to Mersenne, 26 April 1643.

Intermezzo • Generational and Institutional Changes

1. Jesseph, *Squaring*, chap. 3, 102–12; "Galileo." Westfall, *Force*, 99–114.

2. But see Torrini, *Dopo Galileo*.

3. Gascoigne, "Reappraisal."

4. Full documentation can be found in Howard, *Huygens*.

5. In the extensive literature on this theme, see Daston, "Baconian Facts," and Biagioli, "Etiquette."

6. The genre of pamphlets and broadsheets, generally called ephemera, is still awaiting an adequate study by historians of science. On water management, see chap. 6.

7. Schmitt, "Text-book tradition." The latter point is a key thesis in Damerow et al., *Exploring*.

8. Jesseph, *Squaring*, 40–47, 173–88. Andersen, "Changing Understanding." Malet, "Barrow." Hooke, *De potentia restitutiva*, in Gunther, *Early Science*, 8:351–53.

Six • *The Equilibrium and Motion of Liquids*

Epigraph: Guglielmini, *Aquarum fluentium mensura*, book 2, prop. 2. I have slightly modified the translation in Maffioli, *Out of Galileo*, 199.

1. Magiotti, *Renitenza*. [Magalotti], *Saggi*, 196–204. Michelini, *Trattato*, 11. Maffioli, *Out of Galileo*, 95–99.

2. The chief work on the science of waters in Italy is Maffioli, *Out of Galileo*, which identifies an Italian tradition following Galileo and a northern European tradition focusing on pressure. In this chapter I follow a division between scholars focusing on motion and speed, on the one hand, and scholars focusing on equilibrium and pressure, on the other, crossing to some extent national boundaries.

3. Barattieri, *Architettura*, vol. 1. The copy of Barattieri's work in the Burndy Library was owned and annotated by an unnamed hydraulic engineer who questioned some of his statements. See ibid., 1:153.

4. A specific instance of manuscript circulation is ibid., 1:182, quoting the works of Alessandro Betinzoli. For the cardinals' visitation, see 237. Collections of texts published at the time include Castelli, *Della misura;* Barattieri, *Architettura*, 2:150–248; and *Raccolta di varie scritture.*

5. *Discepoli*, 1:26–27. Bucciantini, "Trattato," 132–34 (the text of the quotation on 134 is incorrect). Maffioli, *Out of Galileo*, 75–76.

6. For Borelli's early involvement, see his letters to Malpighi, *MCA*, 1:17–18, Florence, 4 October 1659, and 21–23, Pisa, 7 November 1659. Late in 1659 Borelli had tried to provide a proof that turned out to be erroneous and was therefore withdrawn. Bucciantini, "Trattato," 137–38. Borelli's own copy of Castelli's *Della misura* is in the Lilly Library, Indiana University, Bloomington.

7. Cabeo, *Commentaria*, 1:332–63, at 335a–36b. Maffioli, *Out of Galileo*, 54. Bucciantini, "Atomi," 185–87, at 185n40, states that the Ferrarese Cabeo was a member of the committee dealing with the waters of the Reno and Po in 1624. Fiocca, "Ferrara," 344–50.

8. Castelli, *Carteggio*, 235–38, at 236–37, Castelli to Cavalieri, Rome, 1 January 1642. Maffioli, *Out of Galileo*, 65, suggests that the experiment was performed at Castelli's residence in the Monastery of St. Callisto in Rome.

9. Castelli, *Della misura*, 92–94.

10. Barattieri, *Architettura*, 1:182. Maffioli, *Out of Galileo*, 64, claims that Betinzoli had died in 1612. See also 56n23 on Gabriele Bertazzolo, who had put forward the proportionality between weight and speed. Bucciantini, "Trattato," 129.

11. Barattieri, *Architettura*, 2:66–70. Maffioli, *Out of Galileo*, 166–67n12.

12. Cassini's publications appeared in 1682 and were reprinted in the Florence *Raccolta*, vol. 4, esp. 403–7 and 410, with the facing table showing how much the Po would rise as an effect of the waters from the Reno.

13. Michelini, *Trattato*, 9 and 13, "piccolissima forza"; the comparison with the crystal is at 11. Maffioli, *Out of Galileo*, 100–101. Ibid., 102–4, discusses degli Angeli's letter to Leopold criticizing Michelini's idea that there is no lateral pressure. Michelini's work was reprinted at Bologna in 1700 in the wake of Guglielmini's *Della natura.*

14. Borelli, *De motionibus*, 329. Borelli was concerned with the micro-structure of fluids, atomism, and the existence of the void. Maffioli, *Out of Galileo*, 123–24.

15. Riccioli, *Geographia*, 234–40.

16. Torricelli's barometric experiment seems to belong more to the history of experimental philosophy than to that of mechanics and is therefore not discussed here. Middleton, *Barometer*.

17. Pascal, *Traitez de l'equilibre*, in *The Physical Treatises of Pascal*. Pascal, *Oeuvres*, 2:478–95, Pascal to Ribeyre, 12 July 1651; 3:145–55, introduction to the *Traitez*. Dugas, *Mécanique*, chap. 8, esp. 238–41. Dear, *Discipline*, chap. 7, esp. 205–6 on Pascal's tables. Benedetti discussed the hydraulic press in *Diversarum speculationum liber*, in *Mechanics in Italy*, 231–32.

18. Pascal, *Physical Treatises*, 6. Pascal meant that the force is inversely as the distance, as he states below, or the product between force and distance is constant.

19. Ibid., 8–9.

20. This difference of style has been often noticed. See for instance ibid., 21n1. Shapin and Schaffer, *Leviathan*. Dear, *Discipline*, 208–9.

21. Schott, *Mechanica*, 8–14, 117f.

22. *OCH*, vol. 1, *ad indicem*; 13:384. On Pierre Saporta (1613–85), see Chabbert, "L'Académie de Castres," 283; "Saporta."

23. Castelli, *Traicté de la mesure*. Fermat's letter at 84–87, also in Fermat, *Oeuvres*, 1:362–65. Chabbert, "Fermat à Castres." Rolt, *From Sea to Sea*, 29, 35–83, 177–79. Mukerji, "Cartography," 260–61. Andreossy, *Histoire du Canal du Midi*, 346–63, at 348, 359 (Andreossy visited the areas around Milan and Padua). The introduction of this work outlines a history of hydraulic engineering with special emphasis on locks.

24. Dechales, *Cursus*, 2:177–211, at 203, 208. His *Cursus* deserves a broader study. His extensive *Tractatus* 17 is on navigation (2:237); *Tractatus* 23 deals with the trajectory of projectiles and refers to Galileo and Torricelli (3:83, 85); *Tractatus* 25 (3:765) is on indivisibles. Maffioli, *Out of Galileo*, 186–87.

25. The work by the Académie has been reconstructed on the basis of unpublished documents in Blay, "Recherches," esp. 95. See also Blay, *Naissance*, 332–68, esp. 332–42. For Huygens's contributions, see Hall, *Ballistics*, 111–19.

26. *HOC*, 19:120–21, containing a small table of experimental values. The relationship is not far removed from more modern values, but this is due to compensating errors. The academicians did not take into account the contraction of the vein, as later observed by Newton, thus they overestimated the water's cross-section. On the other hand, the correct equation for the force contains a factor of 2, $F = 2w$, where w is the weight of the cylinder of water mentioned above. As it happens, the two errors approximately offset each other. Blay, "Recherches," 92–93 and passim.

27. *HOC*, 19:122. Dugas, *Mécanique*, 533. See sec. 8.2.

28. *HOC*, 19:127, third experiment. Dugas, *Mécanique*, 534.

29. *HOC*, 19:127–43. Dugas, *Mécanique*, 534–35. Hall, *Ballistics*, 115. The contemporary investigations at the Royal Society focused more on the resistance to motion than the moving force of fluids. See sec. 7.4.

30. Mariotte, *Traité du mouvement des eaux*. English translation by Desagulier, *The Motion of Water and Other Fluids*. Italian translation in the Florence *Raccolta*, 2:1–167; see 101–2, 126, 144–67. I refer to the second French edition, 225–28, 284, 348–90. I discuss the resistance of solids in sec. 8.5.

31. Mariotte, *Traité*, 248, 256; Florence *Raccolta*, 2:110, 113.

32. Mariotte, *Traité*, 282–332, at 284–89; Florence *Raccolta*, 2:125–43, at 126–28. Maffioli, *Out of Galileo*, 190–91 (see also 188–90). The conceptual significance of the efflux problem is that it was linked to the notion of lateral pressure. On Newton, see chap. 10.

33. Salusbury, *Mathematical Collections and Translations*, vol. 1, second part. Salusbury adopts the revised version of p. 82 (see fig. 6.1). On the history of this edition, see Drake's introduction in the 1967 facsimile edition. On the draining of the fens, see Willmoth, *Sir Jonas Moore*. Feingold, "A Friend of Hobbes," 272, reports an unpublished English translation of Castelli's work from 1635.

34. Herivel, *Background*, 183–91. McGuire and Tamny, *Certain Philosophical Questions*, 278n1. Newton's work is discussed in chap. 9. Newton would also have been familiar with Descartes' *Principia*, where Castelli's proposition is used on several occasions, as we saw in sec. 5.6.

35. Wallis, *Mechanica*, 708–46; *A Discourse of Gravity and Gravitation*, 9, 23. On Wallis's familiarity with Torricelli, see De Gandt, "Evolution," at 106–7.

36. Maffioli, *Out of Galileo*, 148–49, 166–70. Cavazza, *Settecento inquieto*, 44–51.

37. Maffioli, *Out of Galileo*, 243–49, esp. 248n39.

38. Guglielmini, *Mensura*, address to the reader, in Florence *Raccolta*, 1:313–420, at 317. In those years at Bologna there raged a dispute between the empirical medical school of Sbaraglia and the rational school of Malpighi. Guglielmini's claim that medicine relies on anatomy and mathematics puts him squarely in Malpighi's camp.

39. Ibid., 1:315–16, 320, 322, prop. 2 at 337, prop. 5 at 340. Maffioli, *Out of Galileo*, 186–87, argues for a much closer similarity between Dechales and Guglielmini than seems warranted.

40. Guglielmini, *Mensura*, in Florence *Raccolta*, 1:333–37. On Pepoli and Donnelli, see Maffioli, *Out of Galileo*, 139, 173.

41. Guglielmini, *Mensura*, in Florence *Raccolta*, 1:366–68, scolio.

42. Maffioli, "Science in Bologna." Maffioli, *Out of Galileo*, 223–36.

43. Maffioli, *Out of Galileo*, 237–42. Florence *Raccolta*, 5:1–174, lists forty-eight reports by Guglielmini on river management.

44. Guglielmini, *Della natura*, 12, 16–17. The treatise was reprinted in the Florence *Raccolta*, vol. 2, with preface and annotations by Eustachio Manfredi. In 1688 Guglielmini also wrote a treatise on the geometric structure of salts. His views on air were inspired by Boyle, *A Defence of the Doctrine Touching the Spring and Weight of the Air*, BWHD, 3:84.

45. Guglielmini, *Della natura*, 52, contains an implicit reference to Papin.

46. Ibid., 69. Maffioli, *Out of Galileo*, 262–65.

47. Daniel Bernoulli argued that in the study of the pierced cistern the sum of three terms proportional to the square of the speed, the pressure, and the height of fall, respectively, is a constant. For our purposes, we can write Bernoulli's equation in modern notation as $p + \rho gh + \rho v^2/_2$ = constant, where p is the pressure, ρ the density per unit of mass, g the gravitation constant, h the height, and v the speed. Blay, "La loi d'écoulement."

48. Ramazzini, *De fontium*, 12–13 and chap. 7, 73–78.

49. A copy of the first edition of Ramazzini's work in the Countway Library, Harvard Medical School, is bound with a medical work by Ramazzini on the epidemics of 1690 in the Mod-

ena countryside. For more information on the reception of Ramazzini's treatise in Italy, see Maffioli, *Out of Galileo*, 434n20. The partial English translation by Robert St. Clair was titled *The Abyssinian Philosophy Confuted*.

Seven • Projected, Oscillating, and Orbiting Bodies

Epigraph: Robert Hooke at the Royal Society, 23 May 1666, in Birch, *History of the Royal Society*, 2:91.

1. Aiton, *Vortex*, chaps. 2−3. Bertoloni Meli, *Equivalence*, chaps. 1−2.

2. Initially Galileo too thought that the Medicean planets moved in a straight line. Dear, *Discipline*, 108.

3. Bertoloni Meli, "Shadows and Deception," 391−95.

4. Borelli, *Theoricae*, 3−5.

5. Ibid., 45ff. Roberval, *Aristarchus*, 1−2. Koyré, *Astronomical Revolution*, 467−513. Auger, *Roberval*, 104−16; "Les idées de Roberval." Aiton, *Vortex*, 90−95. Bertoloni Meli, *Equivalence*, sec. 8.2.

6. Koyré, *Astronomical Revolution*, 487. *GOF*, 8:205−6. For the analogy between the motion of the pendulum and that of the moon, see *GOF*, 7:477.

7. In arguing his case, Borelli used mathematical language stating that the power moving the orbiting body is not indivisible but *"quanta."* Borelli, *Theoricae*, 57, 63. Koyré, *Astronomical Revolution*, 489−90, 498−505.

8. Kepler's law states that the squares of the periods of revolution are proportional to the third power of the major axes of planetary ellipses. In the 1680s Cassini extended the law to satellites. Bertoloni Meli, *Equivalence*, 36. According to Wilson, "Kepler to Newton," 225, Leibniz was the first to refer to Kepler's laws as such.

9. Koyré, *Astronomical Revolution*, 501. Kepler's area law states that the radius from the planet to the sun sweeps out equal areas in equal times. The law implies that the circular component of the orbital speed, not the orbital speed itself, is inversely proportional to the distance from the center. Few mathematicians adopted the correct form of Kepler's area law before Newton.

10. Koyré, *Astronomical Revolution*, 507n20. Descartes, *Principia*, III, 120. Borelli, *Theoricae*, 35−45.

11. Since the excellent Koyré, "Fall," and Galluzzi, "Galileo contro Copernico," analyze the material extensively, in this chapter I focus primarily on the use of objects and the problem of the estimation of orders of magnitudes. Gregory, "An Account." For ease of reference I cite Koyré's account, which provides extensive quotations of the originals, except in a few cases where his text requires corrections.

12. Galluzzi, "Galileo contro Copernico," 131ff. Koyré, "Fall," 391.

13. Galluzzi reconstructed the events of the early phase of the controversy in "Galileo contro Copernico," 105ff. Koyré, "Fall," 358−60.

14. Koyré, "Fall," 363, 362.

15. Ibid., 365.

16. Galluzzi, "Galileo contro Copernico," 116−21, 147−48, relies on unpublished manuscripts that are difficult to interpret, but the experimental nature of Borelli's attempts is beyond question.

17. Koyré, "Fall," 385–86, 371–73. There is considerable confusion on the period of the pendulums in the claims and counterclaims by degli Angeli and his new ally, Geminiano Montanari. Applying conservation of angular momentum to the case in which the radius of the circle described in the cone is reduced to one-quarter, the period would become one-sixteenth.

18. Borelli, *Risposta*, 14. Koyré, "Fall," 373, incorrectly writes eight-twelfths.

19. Koyré, "Fall," 374–76. Borelli calculated the deviation for 1″ of arc rather than 1′ (see 376). The eastward deviation on a rotating Earth results from the difference in the speeds at the top and at the bottom of the tower, multiplied by the time of fall. Calling h the height of the tower, ω the angular rotation of the Earth, and t the time, the deviation is $h\omega t$. Borelli was right in believing that gravity decreases the eastward deviation, though not in the way he thought. The correction due to gravity acting along lines that are not parallel reduces the deviation above by one-third, yielding an eastward deviation equal to $\frac{2}{3}h\omega t$. Bertoloni Meli, "St. Peter," 440.

20. Koyré, "Fall," 385.

21. Blondel, *L'art*, 167. His treatise was reprinted three more times in the Low Countries before the end of the century. Blay, "Balistique," 46.

22. Boyle, *A Continuation of New Experiments* (Oxford, 1669), *BWHD*, 6:137–40.

23. [Magalotti], *Saggi*, chaps. 247–51. Gunther, *Early Science*, 6:373. Hall, *Ballistics*, 119. Borelli, *De vi*, 169–75. Fabri, *Dialogi physici* (Lyons, 1669), challenged Borelli, 221ff. *OCH*, 7:283–85, Wallis to Oldenburg, 15 and 22 November 1670. Hall, "Gunnery," 26–27. *GOF*, 7:181, 194; 8:279. Middleton, *Experimenters*, 240–43, at 242.

24. Birch, *History of the Royal Society*, 1:20, 33, 49, 205; 2:254, 261, 350, 352–53, 461, 465, 467. Hall, *Ballistics*, 111, 119; "Gunnery," 124–25, 127. On the English scene, see also Dugas, *Mécanique*, 537–41. Gunther, *Early Science*, 6:115–16, 200–2, 347. Hooke's experiment was repeated several times.

25. Gregory is not very clear in the text about the meaning of "uniformly retarded." From fig. 6 of his *Tentamina*, however, it is clear that he drew a parabola wi th time on the ordinates *(AK)* and distances on the abscissae *(VK)*. Flamsteed, *Correspondence*, 1:826–29, at 828, Flamsteed to Molyneux, 12 October 1681.

26. Anderson, *Use and Effects of the Gunne*, 26–27, 30–31. Streete's tables amount to sixty-four pages.

27. Gregory, *Tercentenary*, 285–87, Collins to Gregory, London, 25 September 1674. Anderson also circulated manuscript challenges to Gregory's *Tentamina*. For Collins's antipathy for Anderson, see ibid., ad indicem. Hall, *Ballistics*, 125; "Gunnery," 123. Hooke, *Diary*, 11, 17, 23 September 1674. On May 20 1674 Hooke bought a copy of Anderson's work. Willmoth, *Moore*, 170. Willmoth, "Mathematical Sciences," 128, points out that Brouncker was at Blackheath representing the Navy Board rather than the Royal Society, of which he was president. Gunther, *Early Science*, 7:434.

28. Flamsteed, *Correspondence*, 1:557–58, 25 May 1677. Birch, *History of the Royal Society*, 3:400. Hall, *Ballistics*, 119–20. Flamsteed's test with a steel crossbow was probably a response to Anderson's similar tests.

29. Rigaud, *Correspondence*, 2:587–89, Wallis to Collins, 24 August 1674. *NC*, 1:309–12, at 309, Newton to Collins, 20 June 1674. Hall, *Ballistics*, 122. *NMW*, 6:6–8.

30. *HOC*, 19:102–19, 144–57. See also the "Avertissement" on 79–92. Huygens's theory was published in 1690 and is mentioned again in chap. 10.

31. Blay, "Developpement," 42, 45. Blondel, *Bombes*, 336, 416–24. See on 424 a table of ex-

perimental results compared with theoretical predictions. *LSB*, III, 2:119–48, Leibniz to Fabri, 17 May 1677, at 136–37.

32. Yoder, *Time*, 26–32; see 28n26 for Stevin. For a further refinement of this experiment, see the end of this section.

33. *HOC*, 17:27–73. Mahoney, "Christiaan Huygens," 236–38.

34. Yoder, *Time*, 63.

35. Pardies, *Oeuvres*, 330–33. Brouncker, "Demonstratio." Ziggelaar, "Aux origines"; "Les premières." Yoder, *Time*, 62, 158. Gregory, *Tentamina. NMW*, 3:391–401.

36. Accounts of Huygens's private itinerary are in Yoder, *Time*, 48–64, and Mahoney, "Huygens and the Pendulum." On Huygens's use of geometrical diagrams, see Mahoney, "Drawing Mechanics." Others too, such as Galileo in the *Discorsi* and Torricelli in *De motu*, superimposed on the same diagram physical and more abstract spaces.

37. An accurate clock would keep the time of the port whence the ship set sail. By comparing the time marked by the clock with the local time at the place reached by the ship, it was possible to determine the difference in longitude, whereby one hour implied 15 degrees' longitude. Andrewes, *Longitude*.

38. Leopold, "Instrument Makers." Mahoney, "Christiaan Huygens"; "Huygens and the Pendulum." Leopold, "Longitude Timekeepers."

39. Huygens, *Pendulum*, 28, 19–23. Huygens's alternative method for drawing cycloids is on 22–23 and in his fig. 4.

40. Huygens, *Pendulum*, 42–43, props. 5–7. Yoder, *Time*, 59–61. Galileo, *Two New Sciences*, 178. The preliminary version of Galileo's result announced to dal Monte in 1602 is discussed above in fig. 3.2.

41. Huygens, *Pendulum*, 81, 83. Pugliese, "Hooke," 194–96, provides an account of Hooke's 1667 work on the conical pendulum.

42. Huygens, *Pendulum*, 105. *HOC*, 16:353, claims that Huygens did not give serious study to the works by Descartes and Roberval, but he referred to "the recently published letters of Descartes" on the first page of *Horologium oscillatorium*, part 4. Lagrange, *Méchanique*, 171–77ff, where Lagrange extends his account to the eighteenth century. According to Gabbey, "Huygens and Mechanics," 174–75, and "Huygens and Roberval," portions of part 4 of *Horologium oscillatorium* are due to Roberval. Auger, *Roberval*, 98–100.

43. Huygens, *Pendulum*, 108–10. Lagrange, *Méchanique*, 172–73. See also *HOC*, 18:38–39.

44. A tentative reconstruction of how Huygens found this result is in *HOC*, 16:470n6. In definition 13 of part 4, p. 108, Huygens addressed the problem of dimensional homogeneity, claiming that "when we say that weights are multiplied in straight lines, this is to be understood to refer to numbers or lines which represent the quantities of weights and their mutual ratios." *HOC*, 16:416; 414–22. Mahoney, "Sketching Mechanics." Vilain, "Oscillation."

45. Huygens, *Pendulum*, 166–67. I have slightly altered the translation. Huygens tried to overcome the difficulty by constraining the bob to move horizontally. *HOC*, 18:46–47, 346–49, 427–28.

46. Huygens, *Pendulum*, 167–69, prop. 25. *HOC*, 16:354–56. Mahoney, "Christiaan Huygens," 258–59.

47. Huygens, *Pendulum*, 170–72, prop. 26. See the perceptive comments by Bos in the introduction, xvi–xx, xxiii–xxiv. Koyré, *Metaphysics and Measurement*," 110–13.

48. Yoder, *Time*, 18–19.

49. For a critical account of posthumous editions, see ibid., 186n2.

50. Bertoloni Meli, "Relativization"; *Equivalence*, 47–48; "Inherent," sec. 1. Notice also the similarity between the figure in *HOC*, 16:269, and *GOF*, 7:242 (Galileo, *Dialogue*, 216).

51. Dollo, *Filosofia*, 345–46, Borelli to Leopoldo de' Medici, Rome, 9 August 1673.

52. Huygens, *Pendulum*, 33. *HOC*, 21:427–99, at 455. See sec. 10.4.

53. Birch, *History of the Royal Society*, 2:90–92, quotation at 92. Wallis, "Essay." Wilson, "Horrocks," 254; "Origin," 92. Bennett, "Hooke and Wren," 43–49. Westfall, *Force*, 209–10, objects to Hooke's reliance on statics. Hooper, "Tides," 237–41. Gal, *Foundations*, 31–44.

54. Gunther, *Early Science*, 8:209–71. Bennett, "Hooke and Wren," 49–60.

55. Gunther, *Early Science*, 8:27–28. Bennett, "Hooke and Wren," 33–39, at 38, states that Wren carefully annotated a copy of the second 1656 edition of Galileo's *Discorsi* soon after its publication; it is at the Bodleian Library, Oxford. Gal, *Foundations*, 168–71.

56. Bennett, "Hooke and Wren," 39–44, at 42. Hooke's attempts to measure gravity at different heights were partly inspired by the work by Henry Power, who had claimed that a weight was lighter at the bottom of a coal mine. Ibid., 41.

57. Birch, *History of the Royal Society*, 3:257–305. Newton, *Papers and Letters*, 177–235, at 179–80. See also *De aere et aethere*, in Newton, *Scientific Papers*, 221–28, at 227–28.

58. *NC*, 2:297–98, Hooke to Newton, 24 November 1679. For the reference to longitude, see *NC*, 2:309, Hooke to Newton, 6 January 1679/80.

59. *NC*, 2:300–303, Newton to Hooke, 28 November 1679.

60. *NC*, 2:304–6, Hooke to Newton, 9 December 1679.

61. *NC*, 2:307–8, Newton to Hooke, 13 December 1679. Nauenberg, "Early Computational Method." Bertoloni Meli, "Inherent"; "Who Is Afraid?" Kollerstrom, "Halley's Comet." Around the same time, Newton had recourse to centrifugal force in a letter to Thomas Burnet, in which he argued that the Earth is spherical by comparing it to other planets and especially Jupiter, where the *vis centrifuga* at the equator is much greater than on the Earth. *NC*, 2:329, Newton to Burnet, January 1680/81.

62. *NC*, 2:309–10, 6 January 1679/80, and n9 at 312. Hooke supposed that the attractions are as $1/r^2$ and as v^2 or the square of the speeds; therefore v is as $1/r$. Experiments with the three balls were performed outdoors. Hooke performed trials indoors as well. *NC*, 2:312–13, Hooke to Newton, 17 January 1679/80. In fact, for such trials no southward deviation occurs. See Bertoloni Meli, "St. Peter's"; "Public Claims"; "Inherent"; "Who Is Afraid?"

63. Westfall, *Never at Rest*, 387–88n145. Whiteside in Newton, *Preliminary Manuscripts*, xiv and xx–xxi n53. Brackenridge, "Critical Role." Gal, *Foundations*, 200–204.

64. *NC*, 2:340–47, Newton to Crompton for Flamsteed, 28 February 1680/81, quotation at 341 and 342; 348–54, Flamsteed to Crompton for Newton, 7 March 1680/81; 358–62, Newton to [Crompton?], 358–62, quotations at 361; 363–67, Newton to Flamsteed, 16 April 1681, where Newton still defends the two-comet theory. Westfall, *Never at Rest*, 381–401. Wilson, "Kepler's Laws," 147–56. Bertoloni Meli, *Equivalence*, sec. 8.3.

65. Ruffner, "Comets."

Eight • Colliding Bodies, Springs, and Beams

Epigraph: Robert Hooke, *De potentia restitutiva*, in Gunther, *Early Science*, 8:340.

1. Webster, "Boyle's Law," 451–54.

2. *BWHD*, 3:59. Webster, "Boyle's Law," 470–84. Power's work contained tables of experimental data. Ibid., 475; "Henry Power."

3. *BWHD*, 3:62.

4. Hooke, *Micrographia*, 224–28. Breger, "Elastizität." For classic treatments of the history of elasticity, see ibid., 112n1.

5. In addition to Huygens, Marci was known at the Royal Society in 1668 as an author who had published on impact together with Descartes and Borelli. Birch, *History of the Royal Society*, 2:320. Later investigators of impact include the Oratorian Nicholas Malebranche, who engaged in a lengthy dispute with Leibniz. Mouy, *Les lois du choc*. Malebranche, *Des loix de la communication des mouvemens* (Paris, 1692), in Malebranche, *Oeuvres*, XVII.1, 9–236.

6. Jammer, *Mass*, esp. chaps. 5–7.

7. Borelli, *De vi*, prop. 8. The Aristotelian notion had been questioned since medieval times. Borelli provided several other refutations of Aristotle's proposition. Westfall, *Force*, 215–18, 223–30. Westfall's claim, at 215, that Borelli "produced the most heterogeneous and bizarre dynamics" of the century seems unjustified.

8. Borelli, *De vi*, chap. 5.

9. Structure and behavior of matter are discussed in ibid., chap. 26, prop. 16.

10. Ibid., props. 18 and 19. The statement about conservation is in the corollary to prop. 19. Huygens criticized props. 18 and 19, but the differences between him and Borelli are in the initial conditions. Borelli was considering perfect inelastic impacts. *HOC*, 16:203n19. Huygens also criticized props. 64 and 119.

11. Borelli, *De vi*, chap. 11, props. 31ff; 43–45, prop. 18.

12. Ibid., 113–14 and 117–18, props. 61–62. The relevant propositions are 61–64.

13. Ibid., 120, props. 62–63.

14. Ibid., 121–23, prop. 64. In vectorial notation, in an elastic impact between two bodies 1 and 2 with mass m_1 and m_2 moving with speeds v_1 and v_2 where we select v_1 as the positive direction, their speeds u_1 and u_2 after the impact are $u_i = 2V_{cm} - v_i$, where i stands for either 1 or 2 and V_{cm} is the speed of the center of mass. After simple operations we have $u_1(m_1 + m_2) = 2m_2v_2 + v_1(m_1 - m_2)$; therefore m_1 can be greater than m_2 and yet since v_1 and v_2 have opposite signs, u_1 can be negative, meaning that body 1 is reflected.

15. *HOC*, 1:167, 186, 260, 302–3; 2:2, 86–87, 93, 103, 115, 123; 16:3–186. Gabbey, "Huygens and Mechanics," esp. 168–70.

16. *HOC*, 16:9 and 29–91; 98 and 132–36. Vilain, *Mécanique*, emphasized Huygens's empiricism (see, e.g., 69). Westfall, *Force*, 146–58.

17. *HOC*, 16:46–47. This principle, together with that whereby the relative speed between the two bodies before and after the impact is equal, is valid only for elastic impacts.

18. *HOC*, 16:21–25 and 95n10.

19. See sec. 4.3 and Bertoloni Meli, "Axiomatic Tradition."

20. *HOC*, 2:79–80, Huygens to Sluse, 2 November 1657, at 79. Huygens opens his reflection on Descartes' rules by stating, "Mihi primum suspectae esse caeperunt quod experimentis omnibus repugnarent." Ibid., 114–16, Huygens to Sluse, at 115: "Experientias me sectari ne extimes, scio enim lubricas esse."

21. *HOC*, 16:172 and 181. The others present included Moray, Brouncker, Neile, and Goddard. *OCH*, 5:192–94, Wallis to Oldenburg, 19 November 1669.

22. *HOC*, 16:171–78, at 173f. Hall, "Mechanics."

23. *HOC*, 16:182–86.

24. *OCH*, 5:265nn1 and 4.

25. Wallis, "Summary Account."

26. Wren, "Lex naturae." Bennett, *Wren*, 71–73, 118–20. Westfall, *Force*, 203–6. Wren's diagram can be explained algebraically starting from the same vectorial equation employed above. In the case of an elastic impact between two bodies 1 and 2 moving with speeds $\mathbf{v}_1$ and $\mathbf{v}_2$, their speeds $\mathbf{u}_1$ and $\mathbf{u}_2$ after the impact are $\mathbf{u}_i = 2\mathbf{V}_{cm} - \mathbf{v}_i$, where i stands for either 1 or 2 and $\mathbf{V}_{cm}$ is the speed of the center of mass. If $\mathbf{V}_{cm} = 0$, the speeds are simply reversed. If $\mathbf{V}_{cm} \neq 0$, it can be represented in Wren's diagram as *ae*, and the speeds after the impact will differ from the speeds before the impact by the amount 2*ae*.

27. *HOC*, 16:210. The manuscript is tentatively dated 1689.

28. Hall, "Mechanics," 34–36. *HOC*, 16:48–49 and 180, rule 5.

29. This principle is equivalent to the modern notion of conservation of momentum for all types of impacts, whether elastic or not. Huygens's essay is in *HOC*, 16:179–81. He formulated the conservation of the motion of the center of gravity of a system of bodies only for spherical bodies.

30. *HOC*, 16:180. On living force, or *vis viva*, see sec. 10.2.

31. *OCH*, vols. 5 and 6, *ad indicem*. Neile's essay is in 5:519–28. *OCH*, 6:161–65, at 164, Huygens to Oldenburg, 31 July 1669; 289–92, at 292, 20 October 1669. Neile's concerns were shared by others such as John Locke. Stein, "Newton's Metaphysics," 275–77. After a long exchange with Wallis, Neile apparently surrendered (at *OCH*, 6:268–71) but still wondered about the physical causes of collision.

32. Wallis, *Mechanica*, 661: "Ubi enim hoc contingit [namely the deformation of soft bodies], virium pars aliqua in deformando corpore absumitur, nec tota in obstaculum impenditur: cujus itaque seorsum est habenda ratio." Westfall, *Force*, 236–39. On 240 Westfall outlines problems with Wallis's account. Newton, *New Translation*, 424–25, however, listed the work by Wallis and Wren as being equivalent with Huygens's.

33. Wallis, *Mechanica*, 665, 668, 669, 695. See for example ibid., 705, prop. 10.

34. Fabri, *Dialogi physici*, dialogues 2 and 3.

35. I refer to Pardies, *Discours*, by numbered section. His work was promptly translated into English as *A Discourse of Local Motion* (London, 1670) with no reference to the author's name. The dramatic twist occurs in §31. In some respects Pardies' views on impact resemble Neile's. In §26 Pardies criticized one of Riccioli's arguments against the motion of the Earth relying on oblique collision.

36. Pardies, *Discours*, §36. In the final Remarks he refers to the *Météores* and *Physica*. *HOC*, 16:206–7. Costabel, "Mariotte," 76–77. Westfall, *Force*, 194–200.

37. *LSB*, VI, 2:219–76. Garber, "Leibniz: Physics and Philosophy," 271–81.

38. Mariotte, *Traité de la percussion ou chocq des corps, dans lequel les principales regles du mouvement contraires à celle que Mr Des Cartes, et quelques autres modernes ont voulu establir, sont demonstrés*, 8. On Mariotte, see Dugas, *Mécanique*, 293–98. Costabel, "Mariotte."

39. Mariotte, *Traité*, 4. The example he provided, the conical pendulum, surely required some elaboration with regard to direction, as Costabel pointed out, in "Mariotte," 81–82.

40. Mariotte, *Traité*, 74, 87. Dugas, *Mécanique*, 295–96.

41. Mariotte, *Traité*, 239, claims that a body falling in air begins to fall with a finite, not infinitely small, speed. *Traité* (1684^3), 116.

42. *HOC*, 16:206. Dechales, *Traitté*, 136, 376ff.

43. Dechales, *Traitté*, 413–15, prop. 13.

44. Ibid., 433, prop. 19; 464, prop. 28. Props. 23–27 deal with oblique impact.

45. The section on collision was included under *Mechanices* in the second edition of Dechales, *Cursus*, 2:197–232. See 219, prop. 15; 224, prop. 25; and 232, prop. 37. The new edition contains about thirty-five additional folio pages on elastic and nonelastic impact and was completed sometime before Dechales' death in 1678.

46. Fabri, *Physica*, 1:42–43; 44, referring to Boyle's *vas pneumaticum*; 56, prop. 26; 155, prop. 155: "Scopletum pneumaticum totam suam vim compressioni debet." Benvenuto, *Structural Mechanics*, 254–57.

47. Fabri, *Physica*, 1:64, prop. 63. Benvenuto, *Structural Mechanics*, 255.

48. *HOC*, 7:424–25, Huygens to Jean Gallois, February 1675; 18, 479–596. Mahoney, "Sketching Mechanics." On incitation, see Westfall, *Force*, 177–81. I disagree with his inclusion (at 178) of centrifugal endeavor as an example of incitation.

49. Iliffe, "Warehouse."

50. Hooke, *De potentia restitutiva*, in Gunther, *Early Science*, 8:333, 335–36, 345. As for air, Hooke refers to his *Micrographia*, 221–28. Benvenuto, *Structural Mechanics*, 262–65.

51. Hooke, in Gunther, *Early Science*, 8:337–38. Webster, "Boyle's Law," 462 and n70.

52. Hooke, in Gunther, *Early Science*, 8:339–41. The reference to capillarity is on 338. Henry, "Hooke."

53. Hooke, in Gunther, *Early Science*, 8:347–48.

54. Ibid., 349.

55. Hooke, *Lampas*, in Gunther, *Early Science*, 8:186–87. Westfall, *Force*, 207–8; "Robert Hooke," 100–101. In modern terms, the integral of force over a given distance would correspond to a work, with the dimension of energy. On changing attitudes about indivisibles and infinitesimals, see Andersen, "Method"; Malet, "Barrow."

56. Birch, *History of the Royal Society*, 2:338–39 and 126–27. Westfall, *Force*, 206–13, esp. 207–8. On 212 Westfall is rather harsh on Hooke. See also Westfall, "Robert Hooke," esp. 103. Westfall erroneously believes that Hooke derived the relation $v^2 \propto s$ from Galileo's law for uniformly accelerated motion. However, it was well known at the time that the same relation holds also for the motion of a pendulum where s measures the vertical height by which the pendulum has risen. Therefore the same relation holds also when acceleration is not uniform, as in the case of the pendulum, where it is proportional to the displacement. Despite its inaccuracies, Hooke's analysis is better than Westfall believes. Hooke, in Gunther, *Early Science*, 8:350, 352. Hooke's diagram includes other curves that are not discussed here. A correct way to prove isochronism would be to use the average speed, which is proportional to AC; therefore time is indeed constant. Newton, *New Translation*, 460, corollary 2 to prop. I, 10.

57. Hooke, in Gunther, *Early Science*, 8:355–56, refers to water fountains and fire engines as described in the twelfth book of Vitruvius, edited by Claude Perrault.

58. The first publication is a reply by Blondel to Wurtz, *Epistola*. In "Second discours, ou lettre au Sieur B.," 526–30, Blondel criticized Galileo's claim that if the spaces are as the speeds, the times are the same, arguing that the proposition is true for uniform motions but cannot be applied to accelerated motions. Benvenuto, *Structural Mechanics*, 235–36, 239–40.

59. Marchetti, *De resistentia*, preface (pages not numbered). Marchetti may have made his

dubious inference from *GOF*, 8:177–78. For the term *periculum*, see Schmitt, "Zabarella and Galileo," 115–23.

60. Marchetti, *De resistentia*, 124. Benvenuto, *Structural Mechanics*, 235–46, details the subsequent controversy between Guido Grandi and Marchetti over Blondel's publication, as well as Viviani's work on the subject.

61. Fabri, *Physica*, vol. 1, *Tractatus 2*, book 5, *De resistentia solidorum*, 514–648, at 514.

62. Fabri, *Physica*, 1:521–22. Fabri acknowledged that in bending a bow there is some compression, but he argued that physically the situation would be the same if there were only tension. Benvenuto, *Structural Mechanics*, 256.

63. Fabri, *Physica*, 1:589–92. Somewhat incongruously, he considered a weightless beam.

64. Pardies, *Statique*, props. 70, 73–74, 77. Truesdell, *Flexible or Elastic Bodies*, 50–53. The problem of the weights attached at equal horizontal intervals is called the suspension bridge. Huygens had already solved the problem in the 1640s. Ibid., 44–46. At 52–53 Truesdell makes much more of Pardies' work on elasticity than seems warranted. The resistance of materials is discussed in props. 88–114. At the end of his treatise Pardies treated the problem of the motion of a sailboat. For an earlier similar proof, see *SWD*, 208–11. Benvenuto, *Structural Mechanics*, 257–61.

65. Birch, *History of the Royal Society*, 1:436–37, dated 1664.

66. *HOC*, 19:18 and 69–75. Huygens's researches remained unpublished.

67. Mariotte, *Mouvement des eaux*, 370–400; Florence *Raccolta*, 2:150–64. Benvenuto, *Structural Mechanics*, 265–68. Szabó, *Geschichte*, 360–62.

68. *LSB*, III, 2:404–8, at 405–6, Mariotte to Leibniz, 28 April 1678; *LSB*, III, 3:669–72, 675–81, 705–7, 712–13, 772–76, 793–98, 829–32, all dating from mid-1682 to mid-1683. Truesdell, *Flexible or Elastic Bodies*, 59–60.

69. *LMG*, 6:106–12. *LSB*, III, 3:793–98, at 795, Leibniz to Mariotte, spring 1683.

Nine • *A New World-System*

Epigraph: Newton, *New Translation*, 699.

1. This is the perspective adopted by Cohen and Smith in the *Guide to Newton's "Principia,"* in Newton, *New Translation*.

2. *De gravitatione* was initially published in Newton, *Unpublished Scientific Papers*, 89–156. Dobbs, *Janus*, 138–46, dates it to the mid-1680s. Stein, "Metaphysics," 263ff. Domski, "Constructible." Guicciardini, "Geometry."

3. Since we know the details of Halley's visit from a later recollection, we are not sure about the exact question he posed. Newton's answer would have been more accurate had he said a conic section. *NMW*, 6:16ff. Cohen, *Introduction*, 47–142, discusses at length how long it took Newton to compose the *Principia*. Wilson, "Kepler's Laws," 156–70.

4. *NC*, 2:441–43, Halley to Newton, 29 June 1686. I have expanded Halley's contractions.

5. Pugliese, "Hooke." Nauenberg, "Computational Method"; "Hooke."

6. Prominent interpretations are by Westfall, *Force*, chaps. 7 and 8; *Never at Rest*. Dobbs, *Janus, ad indicem* (principles). Henry, "Occult Qualities." Dugas, *Mécanique*, 313–14.

7. Dobbs, *Janus*, 133, for example, argues that the regularity of planetary motions precluded aethereal mechanisms. Stein's perceptive comments in "Metaphysics" are relevant to the pres-

ent discussion, in particular the remark that, unlike Descartes, Newton developed a tentative metaphysics "*after* the special sciences themselves have been established" (261).

8. Newton, *New Translation*, 926. Kubrin, "Cyclical Cosmos." Schechner Genuth, "Comets." Cohen and Smith, "Introduction," 28, state rather categorically, "It is a fact of record that in his writing in mathematics, in the *Principia*, and in his writings about optics proper, there was no trace of his concern for these esoteric subjects."

9. Newton, *New Translation*, 795, note bb.

10. Stein, "Metaphysics," 256.

11. Newton, *New Translation*, 793. *NC*, 3:155–56. Westfall, *Never at Rest*, 459.

12. Newton, *New Translation*, 381–83, at 382.

13. Cohen pointed out the primacy of density in "Force and Mass," 59n9. Ghetaldi, *Archimedes*. Cohen in *New Translation, Guide*, 86–87. Newton, *New Translation*, III, 6, p. 807; general scholium to II, 40, pp. 722–73 (in later editions Newton moved his account of the experiment to the end of sec. 6).

14. Newton, *New Translation*, 404. Cohen, "Force and Mass," 60–62. Bertoloni Meli, "Inherent."

15. Herivel, *Background*, 28, calls *vis insita* "a species of *potential* force." See also ibid., 306, from *De motu corporum in medijs regulariter cedentibus:* "Def 12. Corporis vis insita innata et essentialis est potentia qua id perseverat in statu suo quiescendi vel movendi uniformiter in linea recta, estque corporis quantitati proportionalis, exercetur vero proportionaliter mutationi status et quatenus exercetur dici potest corporis vis exercita, conatus et reluctatio. Huius una specie est vis centrifuga gyrantium." I have corrected Herivel's transcription on the basis of the manuscript Cambridge University Library, Add. 3965.5ᵃ, fol. 26r, as reproduced in Newton, *Preliminary Manuscripts*, 31. See also the deleted "Def. 14, Corporis vis exercita," Add. 3965.5ᵃ, fol. 25v, in ibid., 30, and Herivel, *Background*, 317. Westfall, *Force*, 451, argues for a link between *vis exercita* and the third law of motion.

16. Edleston, *Correspondence*, 311. Referring to prop. 11 in Leibniz's "Tentamen de motuum coelestium causis" (published in the *Acta eruditorum* for 1689), Newton stated, "Sed ubi fit in Orbe excentrico Propositio vera non est. Conatus centrigufus semper aequalis est vi gravitatis & in contraris partes dirigitur per tertiam motus Legem in Principiis Mathematicis Newtoni." Bertoloni Meli, "Inherent."

17. Newton, *New Translation*, 405–8, 810, corollary 5 (4 in *Principia, editio princeps*).

18. A philosophical classic is Stein, "Newtonian Space-Time." Rynasiewicz, "Properties."

19. Newton, *New Translation*, 416. Cohen, "'Quantum in se est.'" Gabbey, "Force," 276ff.

20. Mach, *Mechanik*, chap. 2, sec. 3, par. 1.

21. Mersenne, *Cogitata, Hydraulica*, 51. Newton referred to *Cogitata, Ballistica*, in *Principia, editio princeps* 370. Smith, "Newtonian Style," 267–69, 302–4.

22. The relevant proposition is translated in the appendix to Smith, "Newtonian Style," 302–4. Cohen, *Introduction*, 182, reports Fatio's struggle to convince Newton. In reading Newton's *Principia*, Huygens criticized prop. II, 37. *HOC*, 9:331–32. *NC*, 3:37–40, at 39n2; 5:67–68, nn3–5. MacPike, *Halley*, 147–50, 222–23.

23. See corollary 22 to props. I, 66, and III, 39, and the three preceding lemmas. Wilson, "Achievement," 269–70; "Equinoxes," 238–42. Dobson, "Rigid Body." Bertoloni Meli, "Emergence," 327–33. Gabbey, "The Case of Mechanics," 515.

24. Mahoney, "Diagrams and Dynamics"; "Algebraic Techniques." Pourciau, "Preliminary." Guicciardini, *Reading,* offers new and thoughtful comments on this topic.

25. Newton, *New Translation,* 437–38. *NMW,* 6:114–17. Blay, "Force," 236–37, 243.

26. Newton, *Preliminary Manuscripts,* 13–27; *Unpublished Scientific Papers,* 243–92, at 286. Compare also the passage from *De gravitatione,* ibid., 147. At 287 Newton referred to Hooke's experiment on falling bodies deviating toward the east and, erroneously, the south. *NMW,* 6:74–81.

27. Newton, *New Translation,* 412–13. Descartes, *Principia,* II, 25. Stein, "Newtonian Space-Time," 268–70. Rynasiewicz, "Properties."

28. Newton, *New Translation,* 428.

29. Newton, *Principia, editio princeps,* 302–3; *New Translation,* 698–99. Newton used the properties of light too in a similar way in sec. 14 of book I, where he discussed the inflection of light rays when they pass near the edge of sharp bodies, as observed by Grimaldi and as he had himself "diligently observed." *New Translation,* 625.

30. Newton, *New Translation,* 424–27, at 426.

31. Ibid., 711–12. Smith "Newtonian Style," 259. Smith's work is the best study of book II.

32. They are, respectively, at Newton, *New Translation,* 713–18, 719–20, 721–22, and 722–23. The third edition presents slight variations from the first. In later editions the general scholium followed prop. 31, rather than prop. 40 as in the first edition. In the first edition Newton briefly discussed the decay of the motion of a conical pendulum at 351–52, but he later deleted that passage.

33. Smith, "Newtonian Style," 257–64. Newton, *New Translation,* 718.

34. Newton, *Principia, editio princeps,* 339, translation in Smith, "Newtonian Style," 311. This material was excised from the first edition. See also *Principia, editio princeps,* II, 38, corollary 2, p. 336: "Hallucinantur igitur qui credunt resistentiam projectilium per infinitam divisionem partium Fluidi in infinitum diminui."

35. Newton, *New Translation,* 718. This passage made perfect sense in the first edition, in conjunction with the paragraph referring to the corollaries to II, 40, but in later editions it must have seemed cryptic. Newton, *Preliminary Manuscripts,* 13–27, at 23; *Unpublished Scientific Papers,* 286. Bertoloni Meli, *Equivalence,* 98.

36. Newton, *New Translation,* 721.

37. Smith, "Newtonian Style," 263–64.

38. Newton, *New Translation,* 403–4.

39. Ibid., 722–3. Westfall, *Force,* 374–77, dates Newton's experiment to 1679, but I see no reason to date it before the mid-1680s. Dobbs, *Janus,* 130–46, esp. 136–37. At 139–46 Dobbs presents her arguments for redating *De aere et aethere* and *De gravitatione.*

40. Newton, *New Translation,* 806–7, 700–701, 217–18. The first reference to this experiment is in the set of definitions in *De motu corporum in mediis regulariter cedentibus,* in Newton, *Preliminary Manuscripts,* 30. For the link connecting the proportionality of mass and weight, the third law of motion, and universal gravity, see Wilson, "Eötvös," 24–25.

41. Newton, *New Translation,* 810 and 201–4 for pertinent comments on corollary 2. The corollary went through important changes in later editions. The text from the first edition is translated in the footnote (ibid., 776–77). Bertoloni Meli, *Equivalence,* 98 and 223, line 53.

42. Descartes, *Principia philosophiae,* IV, 25 and also 23–24.

43. Westfall, *Never at Rest*, 455–56; *Force*, 496–98. Westfall, however, is entirely right in noticing that II, 41, aims at undermining Descartes' view of light as pressure, by arguing that it cannot be reconciled with its rectilinear propagation. See esp. the figure accompanying II, 41, and my main text.

44. *BWHD*, 1:229–32, experiment 27; 228–29, experiment 26. McGuire and Tamny, *Certain Philosophical Questions*, 390. Newton, *New Translation*, 776–77. The passage from the first edition is translated in the footnote (ibid.). In later editions Newton clarified that he was neglecting friction (ibid.).

45. Newton, *New Translation*, 776–77, note aa. At 762–63 Newton mentioned the experiment of the spherical propagation of waves after they pass through a narrow aperture. See for example Descartes, *Principia*, III, 64. McGuire and Tamny, *Certain Philosophical Questions*, 380–83. Mersenne, *Cogitata, Ballistica*, 138–40. Dostrovski, "Vibration," 209–18.

46. Newton, *New Translation*, 697. For an excellent account closer to the text, see Cohen's *Guide* in Newton, *New Translation*. Cohen, *Newtonian Revolution*, with the review by Shapiro in *Science*, 1982.

47. Newton, *New Translation*, 697–99, at 699.

48. *NMW*, 6:48–49. Newton, *New Translation*, 444–47, 624–25. Bertoloni Meli, *Equivalence*, 231, lines 329–35.

49. This was Leibniz's opinion. Bertoloni Meli, *Equivalence*, 113–25.

50. Newton, *Principia, editio princeps*, 44–45, 48–49.

51. The matter is still controversial. Guicciardini, *Reading*, 54–58, 217–33. *NMW*, 6:46–49. See also the profound Smith, "Phenomenon," showing that if Kepler's first two laws hold very nearly, it does not follow that the inverse-square law also holds very nearly.

52. Newton, *New Translation*, 136–40.

53. Ibid., 535–45, 147–48. Smith, "Methodology," 144. Valluri et al., "Precession Theorem."

54. Newton, *New Translation*, 561, quotation at 588–89. Bertoloni Meli, "Excerpts," 482.

55. Newton, *New Translation*, 789–90. A criticism of Newton's application of Castelli's theorem is in Neményi, "Fluid Dynamics," 74. Newton proposed experiments to test Descartes' explanation of tides, which, as we saw in chap. 5, relied on the same ideas formalized by Castelli. McGuire and Tamny, *Certain Philosophical Questions*, 400–401, 175–80. Truesdell, *Mechanics*, 144–49.

56. *NC*, 2:408–12, Flamsteed to Newton, 5 January 1684/85, 408 and n1. In addition to the exchange on comets discussed in sec. 8.6, Newton's surviving correspondence with Flamsteed occurred between December 1684 and September 1686 and is in *NC*, vol. 2. Westfall, *Never at Rest*, 409, 433–34. *NC*, 2:450–51n8. For Flamsteed's belief in the solar vortex, see also *NC*, 2:336–40, Flamsteed to Halley, 17 February 1680/81. Johns, *Nature*, chap. 8.

57. Herivel, *Background*, 65–76. Newton, *System of the World*, 551.

58. Lunar motion has generated several controversies. Compare for example the historically sensitive account by Wilson, "Moon," and Nauenberg, "Perturbation Methods."

59. A classic study is Koyré, *Newtonian Studies*, 261–72. A table charting the fate of the nine hypotheses is in Newton, *New Translation*, 794.

60. Newton, *Principia, editio princeps*, 403–4.

61. Newton, *New Translation*, 810.

62. Subsequent editions provided a table of degrees of longitude, length of the seconds pendulum, and length of one degree of meridian. Ibid., 828. Greenberg, *Earth's Shape*, 1–14.

Ten • *Causes, Conservation, and the New Mathematics*

Epigraph: De L'Hôpital, "Solution d'un problème," *MASP,* 1700 (1703), 13.

1. Ferrone, *Scienza,* 289. Guicciardini, *Reading,* 179–84.

2. Cohen, *Introduction,* 145–57; "Halley"; "Review."

3. *SL, Sonderheft 14,* is devoted to Leibniz's calculus. Hofmann, *Paris.*

4. Blay, *Naissance,* 18–27. Aiton, *Vortex,* chap. 7. Mazzone and Roero, *Hermann.*

5. The classic paper is Bos, "Differentials." Bertoloni Meli, *Equivalence,* 56–73.

6. Bertoloni Meli, *Equivalence,* 151–55, 202, 212–13.

7. Truesdell, *Flexible or Elastic Bodies,* 64–75. Bos, "Huygens and Mathematics," 142. Benvenuto, *Structural Mechanics,* 271–74. Goldstine, *Variations,* 30–66. Blay, *Naissance,* 63–109. Robinet, *Iter,* 238–54, 369–74.

8. Bertoloni Meli, *Equivalence,* 24. Blay, *Naissance,* 101.

9. Johann Bernoulli, *Briefwechsel,* 1:262–63, L'Hôpital to Johann Bernoulli, 19 February 1695, at 263. Bertoloni Meli, "Centrifugal Force," 36–37. Blay, *Naissance,* 98–109.

10. *LMG,* 2:10–3, Jakob Bernoulli to Leibniz, 15 December 1687. Leibniz's reply was delayed by his Italian journey. *LMG,* 2:13–20, 24 September 1690. Truesdell, *Flexible or Elastic Bodies,* 63–64.

11. Jakob Bernoulli, *Opera,* 2:976–89. Truesdell, *Flexible or Elastic Bodies,* 105–9. Benvenuto, *Structural Mechanics,* 274–80.

12. Jakob Bernoulli, "Narratio controversiae," *AE,* July 1686, and "Demonstratio centri oscillationis ex natura vectis," *AE,* July 1691, *Opera,* 1:277–81 and 460–65 (see also 454–60); "Démonstration générale du centre de balancement ou d'oscillation, tirée de la nature du levier," *MASP* 1703, and "Demonstration du principe de Mr. Huygens touchant le centre de balancement, et de l'identité de ce centre avec celui de percussion," *MASP,* 1704, *Opera,* 2:930–36 and 947–53. Lagrange, *Mécanique,* 174–77. Truesdell, *Essays,* 248–52.

13. Quoted and translated in Benvenuto, *Structural Mechanics,* 86. A copy of the edition of Descartes' treatise published at Kiel (Reumann, 1672) is at the Niedersächsische Landesbibliothek and in all probability was owned by Leibniz. Contrary to its title page, that was not the first edition. Descartes' text had been first published in Paris in 1668.

14. Leibniz, "Brevis demostratio," in *LMG,* 6:117–19. The term *vis viva* first appeared in print in Leibniz's "Tentamen de motuum coelestium causis," published in the *Acta eruditorum* for 1689. Bertoloni Meli, *Equivalence,* 86–87. The relevant passage from the "Tentamen" is translated at 133. On Leibniz's discovery of the conservation of *vis viva,* see Fichant, "Dynamique"; Concepts"; "Einblicke." Garber, "Physics and Philosophy," 309–21. The most recent account of the *vis viva* controversy is Terrall, "*Vis viva.*"

15. *LMG,* 6:234–46, 246–54. Buchdahl, *Metaphysics,* chap. 7. Garber, "Physics and Philosophy," 289–93. Bertoloni Meli, *Equivalence,* 198.

16. *LPG,* 2:153–63, after November 1698, at 154, 156. Westfall, *Force,* 298–301, quotation at 298, argued that instead of integrating solicitation over a distance, Leibniz integrated over time, which would lead to simple velocity rather than velocity squared. In order to move from solicitation to velocity, however, Leibniz did not have to integrate over time, since solicitation is an incomparably small velocity. Bertoloni Meli, *Equivalence,* 88–91, 119–20.

17. *LMG*, 6:215–31, at 226–31. Breger, "Elastizität," 120–21. Bertoloni Meli, *Equivalence*, 54–55.

18. Newton, *New Translation*, 420; *Opticks*, query 31, 397–401, at 398. Dobbs, *Janus*.

19. Vailati, *Leibniz and Clarke*. Bertoloni Meli, "Caroline."

20. Halley, "Synopsis," 1886. Part 4 of Thrower, *Giants*, 231–411, is devoted to comets. *HOC*, 21:474. Derham, "Experimenta," 3, 13, 28. Smith, "Newtonian Style," 264–82, extends to experiments on falling bodies in the third (1726) edition. Dostrovsky, "Vibration," 201–9.

21. *HOC*, 21:413–26, 478–88. Koyré, *Newtonian Studies*, 115–24. Guicciardini, *Reading*, 121–35. Schliesser and Smith, "Report." *HOC*, 21:472. Dugas, *Mécanique*, 446–58.

22. Bertoloni Meli, *Equivalence*.

23. The best account is Aiton, "Resisting Medium," which does not link Leibniz's framework to Descartes. Aiton, *Vortex*, 271–72. *NMW*, 6:68–71n109.

24. *NC*, 4:266–67. On reading Newton's *Principia*, largely in England, see Iliffe, "Butter."

25. The biographical note by Pierre Costabel in Johann Bernoulli, *Briefwechsel*, 2:5–14, at 9, mentions Varignon's didactic activities. Aiton, *Vortex*, 126. Blay, *Naissance*, 193–221, 277–321.

26. Varignon, *Project*, preface. Descartes, *Epistolae* (Amsterdam, 1668), 2:107. Lagrange, *Mécanique*, 6–7. Duhem, *Statics*, 395–99, 420–23. Dugas, *Mechanics*, 172–73, 224–26. Wallis, *Mechanica*, chaps. 1–3. Westfall, *Force*, 231–43.

27. Varignon's works are listed in Johann Bernoulli, *Briefwechsel*, 2:387–408; O.24, "Manière générale," at 23. Blay, *Naissance*, 180–93. Ibid., 152–79, takes into account Varignon's unpublished memoirs presented to the Paris Académie.

28. *LMG*, 6:276–80. Bertoloni Meli, *Equivalence*, 80–83. Blay, *Naissance*, 185–86; "Force."

29. Guicciardini, *Reading*, 217–47. *NMW*, 8:312–424. Blay, *Naissance*, 216–21.

30. This is the main thesis of Guicciardini, *Calculus*.

31. Varignon, "Du mouvement en général"; "Des forces centrales"; "Autre règle général," listed in Johann Bernoulli, *Briefwechsel*, 2:O.25–O.27. Bertoloni Meli, *Equivalence*, 203–4. The style of the Paris Académie emerges beautifully from Terrall, *Maupertuis*.

Conclusion • *Mapping the Transformations of Mechanics*

1. Other types of objects inviting reflection include Galileo's water tank on a moving barge to explain tides, the hanging chain, and Newton's rotating bucket to detect absolute rotation.

2. Pappus, *Book VII of the Collection*, 66–70, 82–84.

3. Edgerton, "Scientific Illustration." Mahoney, "Diagrams." Topper, "Epistemology," 241–45. Mahoney added a new dimension in "Huygens and the Pendulum" and "Drawing Mechanics." Drake, *Galileo at Work*, 395. For works on visualization, conceptualization, and computation, see Netz, *Deduction;* Cat, "Understanding"; Kaiser, *Dispersion*.

4. I found the following works especially helpful: Leigh Star and Griesemer, "Ecology"; Kohler, *Lords;* Klein, "Modelling."

5. *HOC*, 16:32–33. Huygens does not talk of reference frames but of bodies or barges with respect to which motion is observed.

6. Gabbey, "The Case of Mechanics," 496–99, quotation at 496.

7. On this *topos* in the history of seventeenth-century science, see Cassirer, *Erkenntnisproblem.* Friedman, "Kuhn," 29–35.

REFERENCES

Abbreviations

AE	*Acta eruditorum*
AHES	*Archive for History of Exact Sciences*
AIHS	*Archives internationales d'histoire des sciences*
AIMSSF	*Annali dell'Istituto e Museo di Storia della Scienza di Firenze*
BJHS	*British Journal for the History of Science*
BSSM	*Bollettino di storia delle scienze matematiche*
BWHD	Boyle, Robert. *Works.* Edited by Michael Hunter and Edward B. Davis. 14 vols. London and Brookfield, Vt.: Pickering and Chatto, 1999−2000.
DOAT	Descartes, René. *Oeuvres.* Edited by Charles Adam and Paul Tannery. 12 vols. Paris: Léopold Cerf, 1897−1913. Rev. ed. Paris: Vrin, 1964−76.
DSB	*Dictionary of Scientific Biography.* Edited by Charles C. Gillispie. 16 vols. New York: Scribner, 1970−90.
ESM	*Early Science and Medicine*
GOF	Galilei, Galileo. *Opere.* Edited by Antonio Favaro. 20 vols. Florence: G. Barbèra, 1890−1909. Reprinted 1968.
HOC	Huygens, Christiaan. *Oeuvres complètes.* 22 vols. The Hague: Martinus Nijhoff, 1888−1950.
HS	*History of Science*
JHA	*Journal for the History of Astronomy*
JHI	*Journal of the History of Ideas*
LMG	Leibniz, Gottfried Wilhelm. *Mathematische Schriften.* Edited by Karl Immanuel Gerhardt. 7 vols. Berlin and Halle, 1849−63. Reprinted, Hildesheim: Olms, 1962.
LPG	Leibniz, Gottfried Wilhelm. *Die philosophischen Schriften.* Edited by Karl Immanuel Gerhardt. 7 vols. Berlin, 1875−90. Reprinted, Hildesheim: Olms, 1960−61.

LSB	Leibniz, Gottfried Wilhelm. *Sämtliche Schriften und Briefe.* Darmstadt, Leipzig, and Berlin, 1923–.
MASP	*Mémoires de l'Académie Royale des Sciences de Paris*
MCA	Malpighi, Marcello. *Correspondence.* Edited by Howard B. Adelmann. 5 vols. Ithaca: Cornell University Press, 1975.
MMC	Mersenne, Marin. *Correspondance.* 17 vols. Paris: G. Beauchesne, 1932–88.
NC	Newton, Isaac. *The Correspondence.* Edited by Herbet W. Turnbull, Joseph F. Scott, A. Rupert Hall, and Laura Tilling. 7 vols. Cambridge: Cambridge University Press, 1959–77.
NMW	Newton, Isaac. *The Mathematical Papers.* Edited by Derek T. Whiteside. 6 vols. Cambridge: Cambridge University Press, 1967–81.
NRRSL	*Notes and Records of the Royal Society of London*
OCH	Oldenburg, Henry. *Correspondence.* Edited by A. Rupert Hall and Marie Boas Hall. 1965–86. 13 vols. Vols. 1–9, Madison: University of Wisconsin Press. Vols. 10–11, London: Mansell. Vols. 12–13, London: Taylor and Francis.
PS	*Perspectives on Science*
PT	*Philosophical Transactions of the Royal Society of London*
RHS	*Revue d'histoire des sciences*
SHPS	*Studies in History and Philosophy of Science/Modern Physics*
SIC	*Science in Context*
SL	*Studia Leibnitiana*
SR	*Studies in the Renaissance*
SWD	Stevin, Simon. *The Principal Works.* 5 vols. Amsterdam: Swets & Zeitlinger, 1955–66. Vol. 1 ed. Eduard Jan Dijksterhuis.

Primary Sources

Anderson, Robert. *The Genuine Use and Effects of the Gunne, with Tables of Projections by Thomas Streete.* London: J. Dorby for William Berry, 1674.

Archimedes. *Archimedis de iis quae vehuntur in aqua.* Edited by Federico Commandino. Bologna: Ex officina A. Benacii, 1565.

[pseudo] Aristotle. *Mechanical Problems.* In *Minor Works,* trans. Walter Stanley Hett, 327–411. Loeb Classical Library. Cambridge, Mass.: Harvard University Press; London, Heinemann, 1936.

Arriaga, Roderigo de. *Cursus philosophicus.* Antwerp: Plantin, 1632; Paris: Apud Franciscvm Piot, 1647[4].

Baldi, Bernardino. *In mechanica Aristotelis problemata exercitationes.* Mainz: Vidua Ioannis Albini, 1621.

Baliani, Giovanni Battista. *De motu naturali gravium solidorum et liquidorum.* Genoa: Farroni, 1646[2]. Originally published without the portion on liquids (Genoa: Farroni, Pesagni, & Barberi, 1638). Both are reprinted in the volume edited by Giovanna Baroncelli (Florence: Giunti, 1988).

Barattieri, Giovanni Battista. *Architettura d'acque*. 2 vols. Piacenza: Giovanni Bazachi, 1656, 1663.

Beeckman, Isaac. *Journal*. Edited by Cornelis de Waard. 4 vols. The Hague: M. Nijhoff, 1939–53.

Benedetti, Giovanni Battista. *Demonstratio propotionum motuum localium contra Aristotilem et omnes philosophos*. Venice: [Bartolomeo Cesano], 1554 = **Demonstratio;* ides of February 1554 [*more veneto* 1555] = ***Demonstratio*.

———. *Diversarum speculationum liber*. Turin: Apud haeredem Nicolai Bevilaquae, 1585.

———. *Resolutio omnium Euclidis problematum*. Venice: [Bartolomeo Cesano], 1553.

Bernoulli, Jakob. *Opera*. 2 vols. Geneva, 1744. Facsimile reproduction, Bruxelles: Culture et Civilisation, 1967.

Bernoulli, Johann. *Briefwechsel*. Basel: Birkhäuser, 1955–.

———. *Opera omnia*. 4 vols. Lausanne & Geneva: Sumptibus Marci-Michaelis Bousquet & Sociorum, 1742.

Biringuccio, Vannoccio. *De la pirotechnia*. Venice: Curzio Navò et fratelli, 1540.

Blondel, Nicholas François. *L'art de jetter les bombes*. Paris: Chez l'auteur et Nicolas Langlois, 1683.

———. *Epistola ad P.W. in qua famosa Galilaei propositio discutitur*. Paris: Franciscus Clousier, 1661. Reprinted in *Histoire de l'Académie Royale des Sciences* 5 (1729), 477–94.

———. "Second discours, ou lettre au Sieur B." *Histoire de l'Académie Royale des Sciences* 5 (1729), 494–530.

Borelli, Giovanni Battista. *De motionibus naturalibus a gravitate pendentibus*. Regio Iulio: D. Ferri, 1670.

———. *De vi percussionis*. Bologna: Ex typographia Iacobi Montii, 1667.

———. *Risposta alle considerazioni*. Messina, 1668.

———. *Theoricae Mediceorum planetarum*. Florence: Ex typographia S[erenissimi] M[agni] D[ucis], 1666.

[Brouncker, Viscount William]. "Demonstratio synchronismi vibrationum peractarum in cycloide." *PT* 8 (1673), 6032.

Cabeo, Niccolò. *In quatuor libros metereologicorum Aristotelis commentaria*. Rome: Typis hæredum Francisci Corbelletti, 1646.

Cardano, Gerolamo. *Opera omnia*. 10 vols. Lyon, 1663. Reprinted, Stuttgart-Bad Cannstatt: F. Frommann, 1966.

Castelli, Benedetto. *Carteggio*. Edited by Massimo Bucciantini. Florence: Olschki, 1988.

———. *Della misura dell'acque correnti*. Bologna: Per gli hh. del Dozza, 1660³. Originally published in Rome, 1628.

———. *Traicté de la mesure*. Translated by Pierre Saporta. Castres: Bernard Barcovda, Imprimeur du Roy, de la Chambre de l'Edict, 1664.

Cavalieri, Bonaventura. *Carteggio*. Edited by Giovanna Baroncelli. Florence: Olschki, 1987.

dal Monte, Guidobaldo. *De cochlea*. Venice: Evangelista Deuchino, 1615.

———. *In duos Archimedis aequiponderantium libros paraphrasis scholijs illustrata*. Pesaro: Gerolamo Concordia, 1588.

———. *Mechanicorum liber*. Pesaro: Gerolamo Concordia, 1577; Venice: Evangelista Deuchino,

1615[2]. Italian translation by Filippo Pigafetta, *Le mechaniche* (Venice: Francesco di Franceschi, 1581; Venice: Evangelista Deuchino, 1615).

Dechales, Claude-François Milliet. *Cursus seu mundus mathematicus.* 3 vols. Lyon: Ex officina Anissoniana, 1674, 1690[2].

———. *Traitté du mouvement local, et du ressort.* Lyon: Anisson et Posuel, 1682.

Derham, William. "Experimenta & observationes de soni motu, aliisque ad id attinentibus." *PT* 26 (1708–9), 2–35.

Descartes, René. *Le monde: Ou traité de la lumière.* Translated and with an introduction by Michael Sean Mahoney. New York: Abaris Books, 1979.

———. *Principles of Philosophy.* Translated by Valentine Rodger Miller and Reese P. Miller. Dordrecht: Reidel, 1983.

Discepoli di Galileo Galilei, Opere. Edited by Paolo Galluzzi and Maurizio Torrini. 2 vols. Florence: Giunti-Barbèra, 1975–.

Eden, Richard. *A Very Necessarie and Profitable Booke Concerning Navigation.* London: Richard Jugge, 1579.

Edleston, J. *Correspondence of Isaac Newton and Professor Cotes.* London, 1850; London: F. Cass, 1969[2].

Fabri, Honoré. *Dialogi physici.* Lyon: Antonius Molin, 1669.

———. *Physica, id est scientia rerum corporearum.* 4 vols. Lyon: Anisson, 1669–71.

Fermat, Pierre de. *Oeuvres.* 5 vols. Paris: Gauthier-Villars et fils, 1891–1922.

Flamsteed, John. *Correspondence.* Edited by Eric G. Forbes, Lesley Murdin, and Frances Willmoth. 3 vols. Bristol, U.K., and Philadelphia: Institute of Physics 1995–.

Galilei, Galileo. *Dialogo sopra i due massimi sistemi del mondo, Tolemaico e Copernicano: Edizione critica.* Edited by Ottavio Besomi and Mario Helbing. Padua: Antenore, 1998.

———. *Dialogue Concerning the Two Chief World Systems.* Translated by Stillman Drake. Berkeley: University of California Press, 1967[2].

———. *Discorsi e dimostrazioni matematiche intorno a due nuove scienze.* Edited by Enrico Giusti. Turin: Einuadi, 1990.

———. *Le mecaniche.* Edited by Romano Gatto. Florence: Olschki, 2002.

———. *On Motion and On Mechanics.* Translated and with an introduction by Stillman Drake and Israel E. Drabkin. Madison: University of Wisconsin Press, 1960.

———. *Two New Sciences, Including Centers of Gravity and Force of Percussion.* Translated by Stillman Drake. Madison: University of Wisconsin Press, 1974.

Gassendi, Pierre. *Opera omnia.* 6 vols. Lyon, 1658. Reprinted, with an introduction by Tullio Gregory, Stuttgart-Bad Cannstatt: F. Frommann, 1964.

Ghetaldi, Marino. *Promotus Archimed[e]s.* Rome: Apud Aloysium Zannettum, 1603.

Gregory, James. "An Account of a Controversy betwixt Stephano de Angelis, Professor of the Mathematicks in Padua, and Joh. Baptista Riccioli Jesuite." *PT* 3 (1668), 693–98.

———. *Tentamina quaedam geometrica de motu penduli et projectorum.* In Patrick Gregory, *The Great and New Art of Weighing Vanity.* Glasgow: Robert Sanders, 1672.

———. *Tercentenary Memorial Volume.* Edited by Herbert W. Turnbull. London: Bell, 1939.

Guglielmini, Domenico. *Aquarum fluentium mensura.* 2 vols. Bologna: Ex typographia Pisariana, 1690–91.

———. *Della natura de' fiumi trattato fisico-matematico.* Bologna: Eredi di Antonio Pisarri, 1697.

Gunther, Robert Theodore, ed. *Early Science in Oxford.* 14 vols. Vols. 6−8. Oxford: Printed for the author, 1930−31.

Halley, Edmond. "Astronomiae cometicae synopsis." *PT* (24), 1705, 1882−99.

———. Review of Newton, *Principia. PT* 16 (1687) 291−97.

Hermann, Jakob. *Phoronomia, sive de viribus et motibus corporum solidorum libri duo.* Amsterdam: Apud R. & G. Wetstenios, 1716.

Hooke, Robert. *The Diary.* London: Taylor and Francis, 1935.

———. *Micrographia.* London: J. Martyn and J. Allestry, 1665. Facsimile reprint, New York: Dover, 1962.

Huygens, Christiaan. *The Pendulum Clock.* Translated by Richard J. Blackwell. Ames: Iowa State University Press, 1986.

Lagrange, Joseph Louis. *Méchanique analitique.* Paris: Desaint, 1788.

Mach, Ernst. *Die Mechanik in ihrer Entwicklung: historisch-kritisch dargestellt.* Darmstadt: Wissenschaftliche Buchgesellschaft, 1988. Originally published 1883, 1933[9].

[Magalotti, Lorenzo]. *Saggi di naturali esperienze fatte nell'Accademia del Cimento.* Florence: Giuseppe Cocchini, 1667.

Magiotti, Raffaello. *Renitenza certissima dell'acqua alla compressione.* Rome: Per F. Moneta, 1648.

Malebranche, Nicholas. *Oeuvres complètes.* Paris: Vrin, 1958−.

Marchetti, Alessandro. *De resistentia solidorum.* Florence: Typis Vincentii Vangelisti, & Petri Matini, Typographi S.M.D., 1669.

Marci, Johannes Marcus. *De proportione motus.* Prague: Typis Ioannis Bilinæ, 1639. Facsimile reproduction in *Acta historiae naturalium necnon technicarum,* special issue, 3 (1967), 131−258.

Mariotte, Edme. *Traité du mouvement des eaux et des autres corps fluides.* Paris: E. Michallet, 1686; Paris: J. Jombert, 1700[2]. Translated by John Theophilus Desagulier as *The Motion of Water and Other Fluids* (London: Printed for J. Senex and W. Taylor, 1718).

———. *Traitté de la percussion ou chocq des corps.* Paris: É. Michallet, 1673, 1684[3].

Mazzoni, Jacopo. *In universam Platonis, et Aristotelis philosophiam praeludia, siue de comparatione Platonis, & Aristotelis.* Venice: Apud Ioannem Guerilium, 1597.

Mechanics in Sixteenth-Century Italy: Selections from Tartaglia, Benedetti, Guido Ubaldo, and Galileo. Translated and annotated by Stillman Drake and Israel E. Drabkin. Madison: University of Wisconsin Press, 1969.

The Medieval Science of Weights. Edited and translated by Ernst A. Moody and Marshall Clagett. Madison: University of Wisconsin Press, 1960.

Mersenne, Marin. *Cogitata physico-mathematica.* Paris: Antonius Bertier, 1644.

———. *Harmonie universelle.* Paris, 1636−37. Reprinted, with an introduction by François Lesure, Paris: CNRS, 1963.

———. *Les méchaniqves de Galilée, mathématicien & ingénieur du Duc de Florence.* Paris: H. Guenon, 1634. Critical edition, ed. Bernard Rochot, Paris: PUF, 1966.

———. *Les nouvelles pensées de Galilée, mathématicien et ingénieur de Duc de Florence.* Paris: Henry Guenon, Pierre Rocolet, 1639. Critical edition, ed. Pierre Costabel and Michel-Pierre Lerner, Paris: Vrin, 1973.

———. *Novarum observationum physico-mathematicarum tomus III.* Paris: Antonius Bertier, 1647.

———. *Quaestiones celeberrimae in Genesim.* Paris: Sebastian Cramoisy, 1623.

———. *Traité des mouvemens, et de la cheute des corps pesans.* Paris: Iacques Villery, 1634. Reprinted in *Corpus: Revue de philosophie* 2 (1986), 25–58.

Michelini, Famiano. *Trattato della direzione de' fiumi.* Florence: Nella Stamperia della Stella, 1664.

Mousnerius, Petrus [from lectures by Honorè Fabri]. *Tractatus physicus de motu locali.* Lyon: Champion, 1646.

Newton, Isaac. *Opticks.* London, 1730^4. Reprinted, New York: Dover, 1952.

———. *Papers and Letters on Natural Philosophy.* Edited by I. Bernard Cohen. Cambridge, Mass.: Harvard University Press, 1958.

———. *Philosophia naturalis principia mathematica.* London: Joseph Streater, 1687. Facsimile reprint, London: Dawson, 1954.

———. *Philosophia naturalis principia mathematica.* 3rd ed. (1726) with variant readings. Edited by Alexandre Koyré and I. Bernard Cohen, with the assistance of Anne Whitman. Cambridge: Cambridge University Press, 1972.

———. *The Preliminary Manuscripts for Isaac Newton's "Principia," 1684–1686.* Edited by Derek T. Whiteside. Cambridge: Cambridge University Press, 1989.

———. *The Principia: Mathematical Principles of Natural Philosophy. A New Translation by I. Bernard Cohen and Anne Whitman, Assisted by Julia Budenz.* Berkeley: University of California Press, 1999.

———. *The System of the World.* London, 1728. Reprinted in *Principia,* trans. Andrew Motte and Florian Cajori, 549–626 (Berkeley: University of California Press, 1932).

———. *Unpublished Scientific Papers.* Edited by by A. Rupert Hall and Marie Boas Hall. Cambridge: Cambridge University Press, 1962.

Pappus of Alexandria. *Book VII of the Collection.* Edited and translated by Alexander Jones. New York: Springer, 1986.

———. *La collection mathématique.* Translated by Paul ver Eecke. Paris-Bruges: Desclée de Brouwer, 1933.

———. *Mathematicae collectiones a Federico Commandino Urbinate in latinum conversæ, et commentariis illustratæ.* Pesaro: Gerolamo Concordia, 1588.

Pardies, Ignace Gaston. *Oeuvres.* Lyon: Freres Bruyset, 1725.

———. *Statique.* Paris: S. Mabre-Cramoisy, 1673.

Pascal, Blaise. *Oeuvres.* 14 vols. Paris: Hachette, 1914–25.

———. *The Physical Treatises.* Translated by Isidore H. B. Spiers and Alexander G. H. Spiers. New York: Columbia University Press, 1937.

———. *Traitez de l'equilibre des liqueurs et de la pesanteur de la masse de l'air.* Paris: Chez Guillaume Desprez, 1663.

[Pfautz, Christoph]. Review of Newton, *Principia. AE* 7 (1688), 303–15.

Plato. *Timaeus.* Translated by Donald J. Zeyl. Indianapolis: Hackett, 2000.

Raccolta d'autori che trattano del moto dell'acque. 9 vols. Florence: Stamperia di Sua Altezza reale, G. Cambiagi, $1765–74^2$.

Raccolta d'autori italiani che trattano del moto delle acque. 10 vols. Bologna: Jacopo Marsigli, Cardinali e Frulli, $1821–26^4$.

Raccolta di varie scritture e notizie concernenti l'interesse della remotione del Reno dalle valli. Bologna: G. Monti, 1682.

Ramazzini, Bernardino. *De fontium Mutinensium admiranda scaturigine.* Modena: Typis haer. Suliani, 1691. Translated by in part by Robert St. Clair as *The Abyssinian Philosophy Confuted: or, Telluris Theoria Neither Sacred, Nor Agreeable to Reason: Being, for the Most Part, a Translation of Petrus Ramazzini, of the Wonderful Springs of Modena* (London: Printed for the author and sold by W. Newton, 1697).

Riccioli, Giovanni Battista. *Almagestum novum.* 2 vols. Bologna: Haeredis Victorij Benatij, 1651.

———. *Astronomia reformata.* Bologna: Ex Typographia Hæredis Victorij Benatij, 1665.

———. *Geographiae et hydrographiae reformatae nuper recognitae, & auctae.* Venice: Typis Ioannis La Noù, 1672[2].

Rigaud, Stephen J. *Correspondence of Scientific Men of the Seventeenth Century.* 2 vols. Oxford, 1841. Reprinted, Hildesheim: G. Olms, 1965.

[Roberval, Gilles Personne de]. *Aristarchi Samii De mvndi systemate.* In Mersenne, *Novarum observationum tomus III.* Paris: Antonius Bertier, 1647[2]. Originally published separately by Bertier, 1644.

Salusbury, Thomas, ed. *Mathematical Collections and Translations.* 2 vols. London: William Leybourne, 1661, 1665. Reprinted, with an introduction by Stillman Drake (London: Dawsons; Los Angeles: Zeitlin and Ver Brugge, 1967).

Sarpi, Paolo. *Pensieri naturali, metafisici e matematici.* Edited by Luisa Cozzi and Libero Sosio. Milan: Ricciardi, 1996.

Schott, Gaspar. *Mechanica hydraulico-pneumatica.* Frankfurt/M: Sumptu heredum J. G. Schönwetteri, 1657.

Taisnier, Jean. *Opusculum perpetua memoria dignissimum, de natura magnetis, et eius effectibus.* Cologne: Ioann Birckmann, 1562.

Tartaglia, Niccolò. *Nova scientia.* Venice: Stephano da Sabio, 1537.

———. *Quesiti et inventioni diverse.* Venice: Venturino Ruffinelli, 1546.

Torricelli, Evangelista. *Opere.* Edited by Gino Loria and Giuseppe Vassura. 4 vols. Faenza: G. Montanavi, 1919–44.

Varignon, Pierre. "Autre règle général." *MASP* 1701 (1704), 20–38.

———. "Des forces centrales." *MASP* 1700 (1703), 218–37.

———. "Du mouvement en général." *MASP* 1700 (1703), 83–101.

———. "Manière générale." *MASP* 1700 (1703), 22–27.

———. *Project d'une nouvelle mechanique.* Paris: La veuve d'Edme Martin, Jean Boudot & Estienne Martin, 1687.

Wallis, John. *A Discourse of Gravity and Gravitation.* London: John Martyn, 1675.

———. "An Essay Exhibiting His Hypothesis about the Flux and Reflux of the Sea" and the following "Appendix." *PT* 1 (1666), 263–89.

———. *Mechanica, sive de motu tractatus geometrico.* London: William Godbid, 1670–71.

———. "A Summary Account of the General Laws of Motion." *PT* 3 (1669), 864–66.

Wren, Christopher. "Lex naturae de collisione corporum." *PT* 3 (1669), 867–68.

Secondary Sources

Aiton, Eric J. "The Contribution of Isaac Newton, Johann Bernoulli and Jakob Hermann to the Inverse Problem of Central Forces." In Heinz-Jürgen Hess and Fritz Nagel, eds., *SL,* Sonderheft 17 (1989).

———. "Descartes's Theory of Tides." *AS* 11 (1955), 337–48.

———. "Ioannes Marcus Marci (1595–1667)." *AS* 26 (1970), 153–64.

———. "Leibniz on Motion in a Resisting Medium." *AHES* 9 (1972), 257–74.

———. *The Vortex Theory of Planetary Motion.* London: Macdonald; New York: American Elsevier, 1972.

Andersen, Kirsti. "Cavalieri's Method of Indivisibles." *AHES* 31 (1985), 291–367.

———. "The Method of Indivisibles: Changing Understanding." *SL*, Sonderheft 14 (1986), 14–25.

Andreossy, Antoine-François. *Histoire du Canal du Midi.* Paris, XIII: Crapelet, 1804.

Andrewes, William J. H., ed. *The Quest for Longitude.* Cambridge, Mass.: Collection of Historical Scientific Instruments, Harvard University, 1996.

Applebaum, Wilbur, ed. *Encyclopedia of the Scientific Revolution from Copernicus to Newton.* New York and London: Garland, 2000.

Ariotti, P. E. "Aspects of the Conception and Development of the Pendulum in the Seventeenth Century." *AHES* 8 (1972), 329–410.

Auger, Léon. "Les idées de Roberval sur le système du monde." *RHS* 10 (1957), 226–34.

———. *Un savant méconnu: Gilles Personne de Roberval.* Paris: Blanchard, 1962.

Baldini, Ugo. *Legem impone subactis. Studi su filosofia e scienza dei Gesuiti in Italia, 1540–1632.* Rome: Bulzoni, 1992.

Beaulieu, Armand. *Mersenne: Le grand Minim.* Brussels: Fondation Nicolas-Claude Fabri de Peiresc, 1995.

Bedini, Silvio A. "Galileo and Scientific Instrumentation." In *Reinterpreting Galileo*, ed. Wallace, 127–53.

Bennett, James A. "The Challenge of Practical Mathematics." In *Science, Culture and Popular Belief in Renaissance Europe*, ed. Stephen Pumfrey, Paolo Rossi, and Maurice Slawinski, 176–90. Manchester: Manchester University Press, 1991.

———. "Hooke and Wren and the System of the World: Some Points towards an Historical Account." *BJHS* 8 (1975), 32–61.

———. *The Mathematical Science of Christopher Wren.* Cambridge: Cambridge University Press, 1982.

———. "The Mechanics' Philosophy and the Mechanical Philosophy." *HS* 24 (1986), 1–28.

———. "Practical Geometry and Operative Knowledge." *Configurations* 6 (1998), 195–222.

———. "Shopping for Instruments in Paris and London." In *Merchants and Marvels*, ed. Smith and Findlen, 370–95.

Benvenuto, Edoardo. *An Introduction to the History of Structural Mechanics.* 2 vols. Vol. 1, *Statics and Resistance of Solids.* New York: Springer, 1991.

Berkel, Klaas van; Albert Van Helden; and L. C. Palm. "Descartes' Debt to Beeckman: Inspiration, Cooperation, Conflict." In *Descartes' Natural Philosophy*, ed. Gaukroger et al., 46–59.

———. *A History of Science in the Netherlands.* Leiden: Brill, 1999.

Bernstein, Howard R. "*Conatus*, Hobbes, and the Young Leibniz." *SHPS* 11 (1980), 25–38.

Bertoloni Meli, Domenico. "The Axiomatic Tradition in Seventeenth-Century Mechanics." In *Synthesis and the Growth of Knowledge*, ed. Michael Dickson and Mary Domski, forthcoming.

———. "Caroline, Leibniz, and Clarke." *JHI* 60 (1999), 469–86.

———. "Dynamics." In *Encyclopedia*, ed. Applebaum, 196–97.

————. *Equivalence and Priority.* Oxford: Oxford University Press, 1993.

————. "Guidobaldo dal Monte and the Archimedean Revival." *Nuncius* 7, no. 1 (1992), 3–34.

————. "Inertia." In *Encyclopedia*, ed. Applebaum, 326–28.

————. "Inherent and Centrifugal Forces in Newton." *AHES* 60 (2006), 319–35.

————. "Kinematics." In *Encyclopedia*, ed. Applebaum, 348–49.

————. "Leibniz's Excerpts from the *Principia Mathematica.*" *AS* 45 (1988), 477–505.

————. "Mechanics." In *The Cambridge History of Early Modern Science*, ed. L. Daston and K. Park, Cambridge: Cambridge University Press, 2006.

————. "Public Claims, Private Worries: Newton's Principia and Leibniz's Theory of Planetary Motion." *SHPS* 22 (1991), 415–49.

————. "The Relativization of Centrifugal Force." *Isis* 81 (1990), 23–43.

————. "The Role of Numerical Tables in Galileo and Mersenne." *PS* 12 (2004), 164–90.

————. "St. Peter and the Rotation of the Earth: The Problem of Fall around 1800." In *The Investigation of Difficult Things*, ed. Shapiro and Harman, 421–47.

————. "Shadows and Deceptions: From Borelli's *Theoricae* to the *Saggi* of the Cimento." *BJHS* 31 (1998), 383–402.

————. "Who Is Afraid of Centrifugal Force?" *ESM* 10 (2005), 535–41.

Biagioli, Mario. "Etiquette, Interdependence, and Sociability in Seventeenth-Century Science." *Critical Inquiry* 22 (1996), 193–238.

————. *Galileo Courtier.* Chicago: University of Chicago Press, 1993.

————. "The Instability of Authorship: Credit and Responsibility in Scientific Authorship." *FASEB Journal* 12 (1998), 3–16.

————. "The Social Status of Italian Mathematicians, 1450–1600." *HS* 27 (1989), 41–95.

Biener, Zvi. "Galileo's First New Science: The Science of Matter." *PS* 12 (2004), 262–87.

Birch, Thomas. *The History of the Royal Society of London.* 4 vols. London: A. Miller, 1756, 1757. Reprinted, Hildesheim: G. Olms, 1968.

Blay, Michel. "Le developpement de la balistique et la practique du jet des bombes en France a la mort de Colbert." In *De la mort de Colbert a la revocation de l'edit de Nantes: un monde nouveau?* ed. Louise Godard de Donville, 33–51. Marseille: C.M.R., 1984.

————. "Force, Continuity, and the Mathematization of Motion at the End of the Seventeenth Century." In *Natural Philosophy*, ed. Buchwald and Cohen, 225–48.

————. "La loi d'écoulement de Torricelli et sa réception au 17ᵉ siècle." In *Torricelli*, ed. De Gandt, 79–110.

————. *La naissaince de la mécanique analitique.* Paris: PUF, 1992.

————. *Reasoning with the Infinite.* Chicago: University of Chicago Press, 1998.

————. "Recherches sur les forces exercées par les fluides en mouvement a l'Académie Royale des Sciences: 1668–1669." In *Mariotte, savant et philosophe*, 91–124.

Blay, Michel, and Egidio Festa. "Mouvement, continu et composition des vitesses au XVIIe siècle." *AIHS* 48 (1998), 65–118.

Boas, Marie. "L'élasticité de l'air avant Mariotte et en Angleterre." In *Mariotte, savant et philosophe*, 55–63

————. "Hero's *Pneumatica*: A Study of Its Transmission and Influence." *Isis* 40 (1949), 38–48.

Bordiga, Giovanni. *Giovanni Battista Benedetti. Filosofo e matematico veneziano del secolo XVI.* Venice: Palazzo Loredan, 1985. Originally published in 1926.

Borgato, Maria Teresa. "Niccolò Cabeo tra teoria ed esperimenti: Le leggi del moto." In *Gesuiti,* ed. Brizzi and Greci, 361–85.

———. "Riccioli e la caduta dei gravi." In *Giambattista Riccioli e il merito scientifico dei Gesuiti nell'età barocca,* ed. Maria Teresa Borgato, 79–118. Florence: Olschki, 2002.

Bos, Henk J. M. "Differentials, Higher-Order Differentials and the Derivative in the Leibnizian Calculus." *AHES* 14 (1974), 1–81.

———. "Huygens and Mathematics." In *Studies,* ed. Bos et al., 126–46.

———. "On the Representation of Curves in Descartes's *Géométrie.*" *AHES* 24 (1981), 295–338.

Bos, Henk J. M., et al. *Studies on Christiaan Huygens.* Lisse: Swets & Zeitlinger, 1980.

Brackenridge, J. Bruce. "The Critical Role of Curvature in Newton's Developing Dynamics." In *Investigation,* ed. Shapiro and Harman, 231–60.

———. *The Key to Newton's Dynamics.* Berkeley: University of California Press, 1995.

Breger, Herbert. "Elastizität als Strukturprinzip der Materie bei Leibniz." *SL,* Sonderheft 13 (1984), 112–21.

Brizzi, Gian Paolo, and Robero Greci, eds. *Gesuiti e università in Europa.* Bologna: CLUEB, 2002.

Brockliss, Lawrence W. B. *French Higher Education in the Seventeenth and Eighteenth Centuries: A Cultural History.* Oxford: Oxford University Press, 1987.

Bucciantini, Massimo. "Atomi goemetria e teologia nella filosofia galileiana di Benedetto Castelli." In *Geometria,* ed. Bucciantini and Torrini, 171–91.

———. "Il trattato *Della misura dell'acque correnti* di Benedetto Castelli. Una discussione sulle acque all'interno della scuola galileiana." *AIMSSF* 8, no. 2 (1983), 103–40.

Bucciantini, Massimo, and Maurizio Torrini, eds. *Geometria e atomismo nella scuola Galileiana.* Florence: Olschki, 1992.

Buchdahl, Gerd. *Metaphysics and the Philosophy of Science: The Classical Origins: Descartes to Kant.* Oxford: Blackwell, 1969.

Buchwald, Jed Z., and I. Bernard Cohen, eds. *Isaac Newton's Natural Philosophy.* Cambridge, Mass.: MIT Press, 2001.

Burke, John G., ed. *The Uses of Science in the Age of Newton.* Berkeley: University of California Press, 1983.

Büttner, Jochen; Peter Damerow; Jürgen Renn; and Matthias Schemmel. "The Challenging Images of Artillery." In *The Power of Images in Early Modern Science,* ed. Wolfgang Lefèvre, Jürgen Renn, and Urs Schoepflin, 3–27. Basel: Birkhäuser, 2003.

Camerota, Michele. *Galileo Galilei e la cultura scientifica nell'età della Controriforma.* Rome: Salerno Editrice, 2004.

———. *Gli scritti "De motu antiquiora" di Galileo Galilei: Il Ms Gal 71.* Cagliari: CUEC, 1992.

Camerota, Michele, and Mario O. Helbing. "Galileo and Pisan Aristotelianism: Galileo's *De motu antiquiora* and the *Quaestiones de motu elementorum* of the Pisan Professors." *ESM* 5 (2000), 319–65.

Carli, Alarico, and Antonio Favaro. *Bibliografia Galileiana.* Rome, 1896. Reprinted, Bologna: Libreria Antiquaria Brighenti, 1972.

Cassirer, Ernst. *Das Erkenntnisproblem in der Philosophie und Wissenschaft der neueren Zeit.* Vol. 1. Berlin: Bruno Cassirer, 1920. Reprinted, Darmstadt: Wissenschaftliche Buchgesellschaft, 1994.

Cat, Jordi. "On Understanding: Maxwell on the Methods of Illustration and Scientific Metaphor." *SHPS* 32 (2001), 395–41.

Cavazza, Marta. *Settecento inquieto.* Bologna: Il Mulino, 1990.

Caverni, Raffaello. *Storia del metodo sperimentale in Italia.* 6 vols. New York: Johnson Reprint Corp., 1972. Originally published 1891–1900.

Chabbert, Pierre. "L'Académie de Castres." In *Castres et Pays Tarnais,* 271–87. [Albi]: Éditions de la "Revue du Tarn," 1972.

———. "Fermat à Castres." *RHS* 20 (1967), 237–48.

———. "P. Saporta." *Bulletin de la société de sciences, arts et belles-lettres du Tarn,* n.s., 32, 1976 (1974), 77–91.

Clagett, Marshall. *Archimedes in the Middle Ages.* 5 vols. Madison: University of Wisconsin Press; Philadelphia: American Philosophical Society, 1964–84.

———. *The Science of Mechanics in the Middle Ages.* Madison: University of Wisconsin Press, 1959.

Clavelin, Maurice. *La philosophie naturelle de Galilée.* Paris: Armand Colin, 1968. Translated by by A. J. Pomerans as *The Natural Philosophy of Galileo* (Cambridge, Mass.: MIT Press, 1974).

Cohen, H. Floris. *Quantifying Music: The Science of Music at the First Stage of the Scientific Revolution, 1580–1650.* Dordrecht: Kluwer, 1984.

Cohen, I. Bernard. "Halley's Two Essays on Newton's *Principia.*" In *Standing on the Shoulders of Giants,* ed. Norman J. W. Thrower, 91–108. Berkeley: University of California Press, 1990.

———. *Introduction to Newton's "Principia."* Cambridge: Cambridge University Press; Cambridge, Mass.: Harvard University Press, 1971.

———. *The Newtonian Revolution.* Cambridge: Cambridge University Press, 1980.

———. "Newton's Concepts of Force and Mass, with Notes on the Laws of Motion." In *Cambridge Companion,* ed. Cohen and Smith, 57–84.

———. "'Quantum in se est': Newton's Concept of Inertia in Relation to Descartes and Lucretius." *NRRSL* 19 (1964), 131–55.

———. "The Review of the First Edition of Newton's *Principia* in the *Acta Eruditorum,* with Notes on the Other Reviews." In *Investigation,* ed. Harman and Shapiro, 323–53.

Cohen, I. Bernard, and George Smith, eds. *The Cambridge Companion to Newton.* Cambridge: Cambridge University Press, 2002.

———. "Introduction." In *Cambridge Companion,* ed. Cohen and Smith, 1–32.

Concina, Ennio. *L'Arsenale della Repubblica di Venezia.* Milan: Electa, 1984.

Cook, Harold J. "The New Philosophy in the Low Countries." In *The Scientific Revolution in National Context,* ed. Roy Porter and Mikulás Teich, 115–49. Cambridge: Cambridge University Press, 1992.

Costabel, Pierre. "La controverse Descartes-Roberval au sujet du centre d'oscillation." *Revue des sciences humaine,* n.s., 61 (1951), 74–86.

———. "Mariotte et les règles du mouvement." In *Mariotte, savant et philosophe,* 75–89.

Cottingham, John, ed. *The Cambridge Companion to Descartes.* Cambridge: Cambridge University Press, 1992.

Crombie, Alistair C. "Mathematics, Music and Medical Science." In *Actes du XIIe Congrès International d'Histoire des Sciences,* 295–310. Paris: Blanchard, 1971.

———. "Le proprietá primarie e le qualitá secondarie nella filosofia naturale di Galileo." In *Galileo*, ed. Adriano Carugo and Paul Tannery, 207–37. Milan: ISEDI, 1978.

Cultura, scienze e tecniche nella Venezia del cinquecento: Atti del Convegno Internazionale di Studio Giovan Battista Benedetti e il suo tempo. Venice: Istituto veneto di scienze, lettere ed arti, 1987.

Cuomo, Serafina. *Pappus of Alexandria and the Mathematics of Late Antiquity.* Cambridge: Cambridge University Press, 2000.

———. "Shooting by the Book: Notes on Niccolò Tartaglia's *Nova Scientia.*" *HS* 35 (1997), 155–88.

Damerow, Peter; Gideon Freudenthal; Peter McLaughlin; and Jürgen Renn. *Exploring the Limits of Preclassical Mechanics.* New York: Springer, 2004[2].

Daston, Lorraine. "Baconian Facts, Academic Civility, and the Prehistory of Objectivity." *Annals of Scholarship* 8 (1991), 337–63.

Dear, Peter. *Discipline and Experience: The Mathematical Way in the Scientific Revolution.* Chicago: University of Chicago Press, 1995.

———. *Mersenne and the Learning of the Schools.* Ithaca: Cornell University Press, 1988.

De Gandt, François. *Force and Geometry in Newton's "Principia."* Princeton: Princeton University Press, 1995.

———. "L'evolution de la theorie des indivisibles et l'apport de Torricelli." In *Geometria e atomismo*, ed. Bucciantini and Torrini, 103–18.

———., ed. *L'Oeuvre de Torricelli.* Nice: Diffusion, 1987.

De Pace, Anna. "Archimede nella discussione su aristotelismo e platonismo di Jacopo Mazzoni." In *Archimede*, ed. Dollo, 165–97.

Des Chene, Dennis. *Physiologia: Natural Philosophy in Late Aristotelian and Cartesian Thought.* Ithaca: Cornell University Press, 1996.

Dijksterhuis, Eduard Jan. *Archimedes.* Translated by C. Dickshoorn, with a new bibliographic essay by Wilbur R. Knorr. Princeton: Princeton University Press, 1987.

———. *The Mechanization of the World Picture: Pythagoras to Newton.* Amsterdam, 1950. Translated by C. Dikshoorn and reprinted, Princeton: Princeton University Press, 1986.

———. *Simon Stevin: Science in the Netherlands around 1600.* The Hague: Martinus Nijhoff, 1970.

Dobbs, Betty Jo Teeter. *The Janus Faces of Genius.* Cambridge: Cambridge University Press, 1991.

Dobson, Geoffrey J. "Newton's Problems with Rigid Body Dynamics in Light of His Treatment of the Precession of the Equinoxes." *AEHS* 53 (1998), 125–45.

Dollo, Corrado, ed. *Archimede. Mito, tradizione, scienza.* Florence: Biblioteca di *Nuncius*, 4, 1992.

———. *Filosofia e scienze in Sicilia.* Padova: CEDAM, 1979.

———. "Galilei e la fisica del Collegio Romano." *Giornale Critico della Filosofia Italiana* 71 (1992), 161–201.

Domski, Mary. "The Constructible and the Intelligible in Newton's Philosophy of Geometry." *Philosophy of Science* 70 (2003), 1114–24.

Dostrovsky, Sigalia. "Early Vibration Theory: Physics and Music in the Seventeenth Century." *AHES* 14 (1975), 169–218.

Drabkin, Israel E. "G. B. Benedetti and Galileo's *De motu.*" In *Proceedings of the Tenth International Congress of the History of Science*, 627–30. Paris: Hermann, 1964.

Drake, Stillman. "Free Fall from Albert of Saxony to Honoré Fabri." *SHPS* 5 (1975), 347–66.

———. *Galileo at Work: His Scientific Biography.* Chicago: University of Chicago Press, 1978.

———. "Galileo's 1604 Fragment on Falling Bodies." *BJHS* 4 (1969), 340–58.

———. "Renaissance Music and Experimental Science." *JHI* 31 (1970), 483–500.

Dubarle, Dominique. "La théorie du centre d'oscillation et le principe de la conservation des forces vives." *Revue des Questions Scientifiques,* 5th ser., 16 (1955), 352–78.

Dugas, René. *A History of Mechanics.* Translated by J. R. Maddox. Neuchâtel: Editions du Griffon, 1955.

———. *La mécanique au XVIIe siècle.* Neuchâtel: Editions du Griffon, 1954.

Duhem, Pierre. *The Origins of Statics.* Dordrecht: Kluwer, 1991. Originally published in 1905–6, trans. Grant F. Leneaux, Victor N. Vagliente, and Guy H. Wagener.

Edgerton, Samuel Y., Jr. "The Renaissance Development of the Scientific Illustration." In *Science,* ed. Shirley and Hoeniger, 168–97.

Favaro, Antonio. "Due lettere inedite di Guidobaldo del Monte a Giacomo Contarini." *Atti del reale istituto veneto di scienze, lettere, e arti* 59 (1899–1900), II, 303–12.

Feingold, Mordechai. "A Friend of Hobbes and an Early Translator of Galileo: Robert Payne of Oxford." In *The Light of Nature,* ed. John D. North and John J. Roche, 265–80. Dordrecht: M. Nijhoff, 1985.

Ferrone, Vincenzo. *Scienza, natura, religione.* Napoli: Jovene, 1982.

Festa, Egidio. "Gassendi lecteur de Cavalieri." In *Quadricentenaire de la naissance de Piere Gassendi, 1592–1992,* 2 vols., 2:355–64. Digne-Les Bains: Société scientifique et littéraire des Alpes de Haute-Provence, 1994.

Fichant, Michel. "Les concepts fondamentaux de la mécanique selon Leibniz, en 1676." *SL Supplementa* 17 (1978), 219–232.

———. "Neue Einblicke in Leibniz' Reform seiner Dynamik." *SL* 22 (1990), 48–68.

———. "La 'réforme' leibnizienne de la dynamique, d'après des textes inédits." *SL Supplementa* 13 (1974), 195–214.

Field, Judith V., and Frank A. J. L. James, eds. *Renaissance and Revolution: Humanists, Scholars, Craftsmen, and Natural Philosophers in Early Modern Europe.* Cambridge: Cambridge University Press, 1993.

Fiocca, Alessandra. "Ferrara e i gesuiti periti in materia d'acque." In *Gesuiti,* ed. Brizzi and Greci, 339–59.

Fredette, Raymond. "Galileo's *De motu antiquiora.*" *Physis* 14 (1972), 321–48.

Friedman, Michael. "Kuhn and Logical Empiricism." In *Thomas Kuhn,* ed. Thomas Nickles, 19–44. Cambridge: Cambridge University Press, 2003.

Gabbey, Alan. "Between *Ars* and *Philosophia Naturalis:* Reflections on the Historiography of Early Modern Mechanics." In *Renaissance and Revolution,* ed. Field and James, 133–45.

———. "The Case of Mechanics: One Revolution or Many?" In *Reappraisals,* ed. Lindberg and Westman, 493–528.

———. "Descartes's Physics and Descartes's Mechanics: Chicken and Egg?" In *Essays on Descartes,* ed. Voss, 311–23.

———. "Force and Inertia in the Seventeenth Century: Descartes and Newton." In *Descartes,* ed. Gaukroger, 230–320.

———. "Huygens and Mechanics." In *Studies on Huygens,* ed. Bos et al., 166–99.

———. "Huygens et Roberval." In *Huygens et la France,* 69–83. Paris: Vrin, 1982.

————. "Newton's *Mathematical Principles of Natural Philosophy:* A Treatise on 'Mechanics'?" In *The Investigation,* ed. Shapiro and Harman, 305–22.

————. "What Was 'Mechanical' about the 'Mechanical Philosophy'"? In *Reception,* ed. Palmerino and Thijssen, 11–23.

Gal, Ofer. *Meanest Foundations and Nobler Superstructures: Hooke, Newton and the "Compounding of the Celestiall Motions of the Planet."* Dordrecht: Kluwer, 2002.

Galison, Peter. *Image and Logic.* Chicago: University of Chicago Press, 1997.

Galluzzi, Paolo. "Evangelista Torricelli: Concezione della matematica e segreto degli occhiali." *AIMSSF* 1, no. 1 (1976), 71–95.

————. "Galileo contro Copernico." *AIMSSF* 2, no. 2 (1977), 87–148.

————. "Gassendi and *l'Affaire Galilée* of the Laws of Motion." In *Galileo in Context,* ed. Renn, 239–75.

————. *Mechanical Marvels: Invention in the Age of Leonardo.* Florence: Giunti, 1997.

————. *Momento. Studi Galileiani.* Rome: Ateneo e Bizzarri, 1979.

————., ed. *Novità celesti e crisi del sapere.* Florence: Giunti Barbèra, 1984.

————. "Il Platonismo nel tardo Cinquecento e la filosofia di Galileo." In *Ricerche sulla cultura dell'Italia moderna,* ed. Paola Zambelli, 37–79. Bari: Laterza, 1973.

Gamba, Enrico. "Documenti di Muzio Oddi per la storia del compasso di riduzione e di proporzione." *Physis,* n.s., 31 (1994), 799–815.

————. "Guidobaldo dal Monte matematico e ingegnere." In *Giambattista Aleotti e gli ingegneri del Rinascimento,* ed. Alessandra Fiocca, 341–51. Florence: Olschki, 1998.

Gamba, Enrico, and Vico Montebelli. *Le scienze a Urbino nel tardo Rinascimento.* Urbino: QuattroVenti, 1988.

Garber, Daniel. "Descartes and Experiment in the *Discourse* and the *Essays.* In *Essays on Descartes,* ed. Voss, 288–310.

————. "Descartes, Mechanics, and the Mechanical Philosophy." *Midwest Studies in Philosophy* 26 (2002), 185–204.

————. *Descartes' Metaphysical Physics.* Chicago: University of Chicago Press, 1992.

————. "Descartes' Physics." In *Cambridge Companion to Descartes,* ed. Cottingham, 286–334.

————. "A Different Descartes: Descartes and the Programme for a Mathematical Physics in His Correspondence." In *Natural Philosophy,* ed. Gaukroger et al., 113–30.

————. "Leibniz: Physics and Philosophy." In *The Cambridge Companion to Leibniz,* ed. Nicholas Jolley, 270–352. Cambridge: Cambridge University Press, 1995.

————. "On the Frontlines of the Scientific Revolution: How Mersenne Learned to Love Galileo." *PS* 12 (2004), 135–63.

Gascoigne, John. "A Reappraisal of the Role of the Universities in the Scientific Revolution." In *Reappraisals,* ed. Lindberg and Westman, 207–60.

Gaukroger, Stephen. *Descartes: An Intellectual Biography.* Oxford: Oxford University Press, 1995.

————, ed. *Descartes: Philosophy, Mathematics and Physics.* Brighton: Harvester Press, 1981.

————. "Descartes' Project for a Mathematical Physics." In *Descartes,* ed. Gaukroger, 97–140.

Gaukroger, Stephen, and John Schuster. "The Hydrostatic Paradox and the Origins of Cartesian Dynamics." *SHPS* 33 (2002), 535–72.

Gaukroger, Stephen; John Schuster; and John Sutton, eds. *Descartes' Natural Philosophy*. London and New York: Routledge, 2000.

Giusti, Enrico. "Aspetti matematici della cinematica galileiana." *BSSM* 1, no. 2 (1981), 3–42.

———. "Elements for the Relative Chronology of Galileo's *De motu antiquiora*." *Nuncius* 13, no. 2 (1998), 427–60.

———. *Euclides reformatus: La teoria delle proporzioni nella scuola galileiana*. Turin: Bollati Boringhieri, 1993.

———. "Galilei e le leggi del moto." In Galileo, *Discorsi*, ix–lx.

———. "A Master and His Pupils: Theories of Motion in the Galilean School." In *Reception*, ed. Palmerino and Thijssen, 119–35.

———. "Ricerche galileiane: Il trattato *De motu aequabili* come modello della teoria delle proporzioni." *BSSM* 6, no. 2 (1986), 89–108.

Goldstine, Hermann H. *A History of the Calculus of Variations*. New York: Springer, 1980.

Greenberg, John Leonard. *The Problem of the Earth's Shape from Newton to Clairaut*. Cambridge: Cambridge University Press, 1995.

Gregory, Tullio. *Scetticismo ed empirismo. Studio su Gassendi*. Bari: Laterza, 1961.

Guicciardini, Niccolò. *The Development of Newtonian Calculus in Britain*. Cambridge: Cambridge University Press, 1989.

———. "Geometry and Mechanics in the Preface to Newton's *Principia*: A Criticism of Descartes' *Géométrie*." *Graduate Faculty Philosophy Journal* 25 (2004), 119–59.

———. *Reading the "Principia": The Debates on Newton's Mathematical Methods for Natural Philosophy from 1687 to 1736*. Cambridge: Cambridge University Press, 1999.

Hall, A. Rupert. *Ballistics in the Seventeenth Century*. Cambridge: Cambridge University Press, 1952.

———. "Gunnery, Science, and the Royal Society." In *Uses of Science*, ed. Burke, 111–41.

———. "Mechanics and the Royal Society." *BJHS* 3 (1966), 24–38.

———. *Philosophers at War: The Quarrel between Newton and Leibniz*. Cambridge: Cambridge University Press, 1980.

Hattab, Helen. "From Mechanics to Mechanism: The *Quaestiones mechanicae* and Descartes' Physics." Forthcoming.

Heinekamp, Albert, ed. *300 Jahre "Nova Methodus" von G. W. Leibniz, 1684–1984. SL*, Sonderheft 14, 1986.

Helbing, Mario O. *La filosofia di Francesco Buonamici, professore di Galileo a Pisa*. Pisa: Nistri-Lischi, 1989.

Henninger-Voss, Mary. "Working Machines and Noble Mechanics: Guidobaldo del Monte and the Translation of Knowledge." *Isis* 91 (2000), 233–59.

Henry, John. "Occult Qualities and the Experimental Philosophy: Active Principles in Pre Newtonian Matter Theory." *HS* 24 (1986), 335–81.

———. "Robert Hooke, the Incongruous Mechanist." In *Robert Hooke*, ed. Hunter and Schaffer, 149–80.

Herivel, John. *The Background to Newton's "Principia."* Oxford: Oxford University Press, 1965.

Hill, David K. "Pendulums and Planes: What Galileo Didn't Publish." *Nuncius* 9, no. 2 (1994), 499–515.

Hofmann, Joseph E. *Leibniz in Paris, 1672–1676*. Cambridge: Cambridge University Press, 1974.

Hon, Giora. "Putting Error to Historical Work: Error as a Tell-Tale in the Studies of Kepler and Galileo." *Centaurus* 46 (2004), 58–81.

Hooper, Wallace. "Seventeenth-Century Theories of the Tides as a Gauge of Scientific Change." In *Reception,* ed. Palmerino and Thijssen, 199–242.

Howard, Nicole. "Christiaan Huygens: The Construction of Texts and Audiences." Ph.D. diss., Indiana University, Bloomington, 2003.

Hunter, Michael, and Simon Schaffer. *Robert Hooke: New Studies.* Woodbridge: Boydell, 1989.

Iliffe, Rob. "Butter for Parsnips: Authorship, Audience, and the Incomprehensibility of the *Principia.*" In *Scientific Authorship: Credit and Intellectual Property in Science,* ed. Mario M. Biagioli and Peter Galison, 33–65. New York: Routledge, 2003.

––––––. 'In the Warehouse': Privacy, Property and Priority in the Early Royal Society." *HS* 30 (1992), 29–68.

Jammer, Max. *Concepts of Mass in Classical and Modern Physics.* Cambridge, Mass.: Harvard University Press, 1961.

Jardine, Nicholas. "Demonstration, Dialectic, and Rhetoric in Galileo's *Dialogue.*" In *Shapes of Knowledge in Early Modern Europe,* ed. Donald R. Kelley and Richard H. Popkin, 101–21. Dordrecht: Kluwer, 1991.

––––––. "Epistemology and the Sciences." In *The Cambridge History of Renaissance Philosophy,* ed. Charles B. Schmitt and Quentin Skinner, 685–711. Cambridge: Cambridge University Press, 1988.

Jesseph, Douglas M. "Galileo, Hobbes, and the Book of Nature." *PS* 12 (2004), 191–211.

––––––. *Squaring the Circle.* Chicago: University of Chicago Press, 1999.

Johns, Adrian. *The Nature of the Book: Print and Knowledge in the Making.* Chicago: University of Chicago Press, 1998.

Kaiser, David. *Drawing Theories Apart: The Dispersion of Feynman Diagrams in Postwar Physics.* Chicago: University of Chicago Press, 2005.

Keller, A. G. "Mathematicians, Mechanics, and Experimental Machines in Northern Italy in the Sixteenth Century." In *The Emergence of Science in Western Europe,* ed. Maurice Crosland, 15–34. New York: Science History Publications, 1976.

Klein, Ursula. "Techniques of Modelling and Paper-Tools in Classical Chemistry." In *Models as Mediators: Perspectives on Natural and Social Science,* ed. Mary S. Morgan and Margaret Morrison, 146–67. Cambridge: Cambridge University Press, 1999.

Koertge, Noretta. "Galileo and the Problem of Accidents." *JHI* 38 (1977), 277–318.

Kohler, Robert E. *Lords of the Fly.* Chicago: University of Chicago Press, 1994.

Kollerstrom, Nicholas. "The Path of Halley's Comet." *AS* 56 (1999), 31–56.

Koyré, Alexandre. "A Documentary History of the Problem of Fall from Kepler to Newton." *Transactions of the American Philosophical Society,* n.s., 45 (1955), 329–95.

––––––. *From the Closed World to the Infinite Universe.* Baltimore: Johns Hopkins Press, 1957.

––––––. *Galileo Studies.* Translated by John Mepham. Trowbridge: Harvester Press, 1978.

––––––. *Metaphysics and Measurement.* Cambridge, Mass.: Harvard University Press, 1968.

––––––. *Newtonian Studies.* Cambridge, Mass.: Harvard University Press; London: Chapman & Hall, 1965.

Krafft, Fritz. "Die Anfänge einer theoretischen Mechanik und die wandlung ihrer Stellung zur Wissenschaft der Natur." In *Beiträge zur Methodik der Wissenschaftsgeschichte,* ed. Walter Baron, 12–33. Wiesbaden: Steiner, 1967.

Kubrin, David Charles. "Newton and the Cyclical Cosmos: Providence and the Mechanical Philosophy." *JHI* 28 (1967), 325–46.

Kuhn, Thomas S. *The Essential Tension: Selected Studies in Scientific Tradition and Change.* Chicago: University of Chicago Press, 1977.

———. "The Function of Measurement in Modern Physical Science." In Kuhn, *Essential Tension,* 178–224.

———. "Mathematical Versus Experimental Traditions in the Development of Physical Science." In Kuhn, *Essential Tension,* 31–65.

———. *The Structure of Scientific Revolutions.* Chicago: University of Chicago Press, 1970^2, 1996^3.

Laird, William R. "Archimedes among the Humanists." *Isis* 82 (1991), 628–38.

———. "Patronage of Mechanics and Theories of Impact in Sixteenth-Century Italy." In *Patronage,* ed. Moran, 51–66.

———. "The Scope of Renaissance Mechanics." *Osiris,* n.s., 2 (1986), 43–68.

———. *The Unfinished "Mechanics" of Giuseppe Moletti.* Toronto: University of Toronto Press, 2000.

Lefèvre, Wolfgang, ed. *Picturing Machines, 1400–1700.* Cambridge, Mass.: MIT Press, 2004.

Leigh Star, Susan, and James R. Griesemer. "Institutional Ecology, 'Translation,' and Boundary Objects: Amateurs and Professionals in Berkeley's Museum of Vertebrate Zoology, 1907–1939." *Social Studies of Science* 19 (1989), 387–420.

Lenoble, Robert. *Mersenne, ou la naissance du mécanisme.* Paris: Vrin, 1943.

Leopold, John H. "Christiaan Huygens and His Instrument Makers." In *Studies on Christiaan Huygens,* ed. Bos et al., 221–33.

———. "The Longitude Timekeepers of Christiaan Huygens." In *Quest for Longitude,* ed. Andrewes, 102–14.

Libri, Guglielmo. *Histoire des sciences mathématiques en Italie.* 4 vols. Paris: J. Renouardi, 1838–41.

Lindberg, David C., and Robert S. Westman, eds. *Reappraisals of the Scientific Revolution.* Cambridge: Cambridge University Press, 1990.

Long, Pamela O. *Openness, Secrecy, Authorship: Technical Arts and the Culture of Knowledge from Antiquity to the Renaissance.* Baltimore: Johns Hopkins University Press, 2001.

———. "Picturing the Machine: Francesco di Giorgio and Leonardo da Vinci in the 1490s." In *Picturing Machines,* ed. Lefèvre, 117–41.

Maccagni, Carlo. "Mechanics and Hydrostatics in the Late Renaissance: Relations between Italy and the Low Countries." In *Italian Scientists in the Low Countries in the Seventeenth and Eighteenth Centuries,* ed. Cesare S. Maffioli and L. C. Palm, 79–99. Amsterdam: Rodopi, 1989.

Machamer, Peter. "Galileo's Machines, His Mathematics, and His Experiments." In *The Cambridge Companion to Galileo,* ed. Peter Machamer, 53–89. Cambridge: Cambridge University Press, 1998.

MacLachlan, James. "Mersenne's Solution for Galileo's Problem of the Rotating Earth." *Historia Mathematica* 4 (1977), 173–82.

MacPike, Eugene Fairfield, ed., *Correspondence and Papers of Edmond Halley.* Oxford: Oxford University Press, 1932.

Maffioli, Cesare S. "Guglielmini vs. Papin, 1691–1697: Science in Bologna at the End of the Seventeenth Century through a Debate on Hydraulics." *Janus* 71 (1984), 63–104.

———. *Out of Galileo: The Science of Waters, 1628–1718*. Rotterdam: Erasmus, 1994.

Mahoney, Michael S. "Algebraic vs. Geometric Techniques in Newton's Determination of Planetary Orbits." In *Action and Reaction*, ed. Paul Theerman and Adele F. Seeff, 183–205. Newark: University of Delaware Press, 1993.

———. "Christiaan Huygens: The Measurement of Time and the Longitude at Sea." In *Studies on Christiaan Huygens*, ed. Bos et al., 234–70.

———. "Diagrams and Dynamics: Mathematical Perspectives on Edgerton's Thesis." In *Science*, ed. Shirley and Hoeniger, 198–220.

———. "Drawing Mechanics." In *Picturing Machines*, ed. Lefèvre, 281–306.

———. "Huygens and the Pendulum: From Device to Mathematical Relation." In *The Growth of Mathematical Knowledge*, ed. Herbert Breger and Emily Grosholz, 17–39. Dordrecht: Kluwer, 2000.

———. "Infinitesimals and Transcendent Relations: The Mathematics of Motion in the Late Seventeenth Century." In *Reappraisals*, ed. Lindberg and Westman, 461–91.

———. *The Mathematical Career of Pierre de Fermat, 1601–1665*. Princeton: Princeton University Press, 1973; rev. ed. 1994.

———. "The Mathematical Realm of Nature." In *The Cambridge History of Seventeenth-Century Philosophy*, ed. Daniel E. Garber and Michael Ayers, 2 vols., 1:702–55. Cambridge: Cambridge University Press, 1998.

Malet, Antoni. "Barrow, Wallis, and the Remaking of Seventeenth-Century Indivisibles." *Centaurus* 39 (1997), 67–92.

Mancosu, Paolo. *Philosophy of Mathematics and Mathematical Practice in the Seventeenth Century*. Oxford: Oxford University Press, 1996.

Manno, Antonio. "Giulio Savorgnan: *Machinatio* e *ars* fortifictoria a Venezia." In *Cultura, scienze e tecniche*, 227–45.

Marcucci, Ettore. *Lettere edite e inedite di Filippo Sassetti*. Florence: Le Monnier, 1855.

Mariotte, savant et philosophe. Paris: Vrin, 1986.

Mazzone, Silvia, and Clara Silvia Roero. *Jacob Hermann and the Diffusion of the Leibnizian Calculus in Italy*. Florence: Olschki, 1997.

McGuire, James E, and Martin Tamny. *Certain Philosophical Questions: Newton's Trinity Notebook*. Cambridge: Cambridge University Press, 1983.

McKirahan, Richard D., Jr. "Aristotle's Subordinate Sciences." *BJHS* 11 (1978), 197–220.

Micheli, Gianni. "Guidobaldo del Monte e la meccanica." In *La matematizzazione dell'universo*, ed. Lino Conti, 87–104. Assisi: Porziuncola, 1992.

———. *Le origini del concetto di macchina*. Florence: Biblioteca di *Physis*, 4, 1995.

Middleton, W. E. Knowles. *The Experimenters: A Study of the Accademia del Cimento*. Baltimore: Johns Hopkins Press, 1971.

———. *The History of the Barometer*. Baltimore: Johns Hopkins Press, 1964.

Molland, A. Geroge. "The Atomisation of Motion: A Facet of the Scientific Revolution." *SHPS* 13 (1982), 31–54.

Montucla, Jean Etienne. *Histoire des mathématiques*. 4 vols. Paris: H. Agasse, [1799]–1802.

Moody, Ernst A. "Galileo and Avempace: The Dynamics of the Leaning Tower Experiment." *JHI* 12 (1951), 163–93, 375–422.

Moody, Ernst A., and Marshall Clagett. *The Medieval Science of Weights*. Madison: University of Wisconsin Press, 1960.

Moran, Bruce, ed. *Patronage and Institutions.* Rochester: Boydell, 1991.

Moscovici, Serge. *L'expérience du mouvement: Jean-Baptiste Baliani disciple et critique de Galilée.* Paris: Hermann, 1967.

Mouy, Paul. *Le développement de la physique cartésienne, 1646–1712.* Paris: Vrin, 1934.

———. *Les lois du choc des corps d'après Malebranche.* Paris: Vrin, 1927.

Mukerji, Chandra. "Cartography, Entrepreneurialism, and Power in the Reign of Louis XIV: The Case of the Canal du Midi." In *Merchants and Marvels,* ed. Smith and Findlen, 248–76.

Murdoch, John E., and Edith D. Sylla. "The Science of Motion." In *Science in the Middle Ages,* ed. David C. Lindberg, 206–64. Chicago: University of Chicago Press, 1978.

Napolitani, Pier Daniele. "La geometrizzazione della realta' fisica: Il peso specifico in Ghetaldi e in Galileo." *BSSM* 8 (1988), 139–247.

Nauenberg, Michael. "Hooke, Orbital Motion, and Newton's *Principia.*" *American Journal of Physics* 62 (1994), 331–350.

———. "Newton's Early Computational Method for Dynamics." *AHES* 46 (1994), 221–52.

———. "Newton's Perturbation Methods for the Three-Body Problem and Their Application to Lunar Motion." In *Natural Philosophy,* ed. Buchwald and Cohen, 189–224.

Naylor, Ronald H. "Galileo and the Problem of Free Fall." *BJHS* 7 (1974), 105–34.

———. "Galileo's Theory of Motion: Processes of Conceptual Change in the Period 1604–6." *AS* 34 (1977), 365–92.

———. "Galileo's Theory of Projectile Motion." *Isis* 71 (1980), 550–70.

Neményi, P. F. "The Main Concepts and Ideas of Fluid Dynamics in Their Historical Development." *AHES* 2 (1962–5), 52–86.

Netz, Reviel. *The Shaping of Deduction in Greek Mathematics.* Cambridge: Cambridge University Press, 1999.

Palisca, Claude V. *Humanism in Italian Renaissance Musical Thought.* New Haven: Yale University Press, 1985.

Palmerino, Carla Rita. "Galileo's Theories of Free Fall and Projectile Motion as Interpreted by Pierre Gassendi." In *Reception,* ed. Palmerino and Thijssen, 137–64.

———. "Gassendi's Reinterpretation of the Galilean Theory of Tides." *PS* 12 (2004), 212–37.

———. "Infinite Degrees of Speed: Marin Mersenne and the Debate over Galileo's Law of Free Fall." *ESM* 4 (1999), 269–328.

———. "Una nuova scienza della materia per la *Scienza nova* del moto." In *Atomismo e continuo nel XVII secolo,* ed. Egidio Festa, Romano Gatto, and Michel Bitbol, 275–319. Naples: Vivarium, 2000.

———. "Two Jesuit Responses to Galileo's Science of Motion: Honoré Fabri and Pierre Le Cazre." In *The New Science and Jesuit Science: Seventeenth-Century Perspectives,* ed. Mordechai Feingold, 187–227. Dordrecht: Kluwer, 2003.

Palmerino, Carla Rita, and J. M. M. Hans Thijssen, eds. *The Reception of the Galilean Science of Motion in Seventeenth-Century Europe.* Dordrecht: Kluwer, 2004.

Palmieri, Paolo. "Re-examining Galileo's Theory of Tides." *AHES* 53 (1998), 223–375.

Passalacqua, Lorena. "Le *Collezioni* di Pappo." *BSSM* 14 (1994), 91–156.

Picolet, Guy. "Sur la biographie de Mariotte: Etat des recherches récentes." In *Mariotte, savant et philosophe,* 245–76.

Popplow, Marcus. "Why Draw Pictures of Machines? The Social Contexts of Early Modern Machine Drawing." In *Picturing Machines,* ed. Lefèvre, 17–48.

Pourciau, Bruce. "The Preliminary Mathematical Lemmas of Newton's *Principia.*" *AHES* 52 (1998), 279–95.

Pugliese, Patri J. "Robert Hooke and the Dynamics of Motion in a Curved Path." In *Robert Hooke,* ed. Hunter and Schaffer, 181–205.

Purnell, Frederick, Jr. "Jacopo Mazzoni and Galileo." *Physis* 14 (1972), 273–94.

Renn, Jürgen, ed., *Galileo in Context.* Cambridge: Cambridge University Press, 2001. Originally published in *SIC* 13 (2000).

Renn, Jürgen; Peter Damerow; and Simone Rieger; with an appendix by Domenico Giulini. "Hunting the White Elephant: When and How Did Galileo Discover the Law of Fall?" In *Galileo,* ed. Renn, 29–149.

Renn, Jürgen, and Matteo Valleriani. "Galileo and the Challenge of the Arsenal." *Nuncius* 16, no. 2 (2001), 481–503.

Robinet, André. *G. W. Leibniz. Iter Italicum.* Florence: Olschki, 1988.

Rodis-Lewis, Geneviève. "Descartes' Life and the Development of His Philosophy." In *Cambridge Companion,* ed. Cottingham, 21–57.

Rolt, L. T. C. *From Sea to Sea.* [Athens]: Ohio University Press, 1974.

Rose, Paul Lawrence. *The Italian Renaissance of Mathematics.* Geneva: Droz, 1975.

———. "Jacomo Contarini, 1536–1595: A Venetian patron and Collector of Mathematical Instruments and Books." *Physis* 18 (1976), 117–30.

———. "The Origin of the Proportional Compass from Mordente to Galileo." *Physis* 10 (1968), 53–69

Rose, Paul Lawrence, and Stillman Drake. "The Pseudo-Aristotelian 'Questions of Mechanics' in Renaissance Culture." *SR* 18 (1971), 65–104.

Rosen, Edward. "The Invention of the Reduction Compass." *Physis* 10 (1968), 306–8.

Rossi, Paolo. *I filosofi e le macchine.* Milan: Feltrinelli, 1962, 1984.

Ruffner, James A. "Newton's Propositions on Comets: Steps in Transition, 1681–84." *AHES* 54 (2000), 259–77.

Rynasiewicz, Robert. "By Their Properties, Causes and Effects: Newton's Scholium on Time, Space, Place and Motion." *SHPS* 26 (1995), 133–53, 295–321.

Schechner Genuth, Sarah. "Comets, Teleology, and the Relationship of Chemistry to Cosmology in Newton's Thought." *AIMSSF* 10, no. 1 (1985), 31–65.

Schliesser, Eric, and George Smith. "Huygens's 1688 report to the Directors of the Dutch East India Company." Forthcoming in *AHES.*

Schmitt, Charles B. *Aristotle and the Renaissance.* Cambridge, Mass.: Harvard University Press, 1983.

———. "Experience and Experiment: A Comparison of Zabarella's Views with Galileo's in *De motu.*" *SR* 16 (1969), 80–138.

———. "Experimental Evidence for and against a Void: The Sixteenth-Century Arguments." *Isis* 58 (1967), 352–66.

———. "The Faculty of Arts at Pisa at the Time of Galileo." *Physis* 14 (1972), 243–72.

———. "A Fresh Look at Mechanics in Sxiteenth-Century Italy." *SHPS* 1 (1970), 161–75.

———. "Galileo and the Seventeenth-Century Text-book Tradition." In *Novità celesti,* ed. Galluzzi, 217–28.

Schneider, Ivo. *Archimedes: Ingenieur, Nuturwissenschaftler, Mathematiker.* Darmstadt: Wissenschaftliche Buchgesellschaft, 1979.

Segre, Michael. *In the Wake of Galileo.* New Brunswick, N.J.: Rutgers University Press, 1991.

————. "Torricelli's Correspondence on Ballistics." *AS* 40 (1983), 489–99.

Settle, Thomas B. "An Experiment in the History of Science." *Science* 133 (1961), 19–23.

————. "Galileo and Early Experimentation." In *Springs of Scientific Creativity: Essays of Founders of Modern Science,* ed. Rutherford Aris, H. Ted Davis, and Roger H. Stuewer, 3–20. Minneapolis: University of Minnesota Press, 1983.

————. "Galileo's Use of Experiment as a Tool of investigation." In *Galileo, Man of Science,* ed. Ernan McMullin, 315–38. New York: Basic Books, 1967.

Shapin, Steven. *A Social History of Truth: Civility and Science in Seventeenth-century England.* Chicago: University of Chicago Press, 1994.

Shapin, Steven, and Simon Schaffer. *Leviathan and the Air Pump: Hobbes, Boyle, and the Experimental Life.* Princeton: Princeton University Press, 1985.

Shapiro, Alan E. Review of Cohen, *Newtonian Revolution. Science* 215 (1982), 960–61.

————. "Newton's Experiments on Diffractions and the Delayed Publication of the *Opticks.*" In *Natural Philosophy,* ed. Buchwald and Cohen, 47–76.

Shapiro, Alan E., and Peter M. Harman, eds. *The Investigation of Difficult Things.* Cambridge: Cambridge University Press, 1992.

Shea, William R. *Galileo's Intellectual Revolution.* New York: Science History Publications; London: Macmillan, 1972.

————. *The Magic of Numbers and Motion: The Scientific Career of René Descartes.* Nantucket, Mass.: Science History Publications, 1991.

Shiefsky, Mark J. "Art and Nature in Ancient Mechanics." Forthcoming.

Shirley, John W., and F. David Hoeniger, eds. *Science and the Arts in the Renaissance.* Washington, D.C.: Folger Shakespeare Library: London: Associated University Presses, 1985.

Smith, George. "From the Phenomenon of the Ellipse to an Inverse-Square Force: Why Not?" In *Reading Natural Philosophy,* ed. David Malament, 31–70. Chicago: Open Court, 2002.

————. "The Newtonian Style in Book II of the *Principia.*" In *Natural Philosophy,* ed. Buchwald and Cohen, 249–98.

Smith, Pamela H., and Paula Findlen, eds. *Merchants and Marvels.* New York: Routledge, 2002.

Sørensen, Knud Erik. "A Study of the *De proportione motus* by Marcus Marci de Kronland." *Centaurus* 20 (1976), 50–76; 21 (1977), 246–77.

Souffrin, Pierre. "Du mouvement uniforme au mouvement uniformément accéléré." *BSSM* 6, no. 1 (1986), 135–44.

Stein, Howard. "Newtonian Space-Time." In *The "Annus Mirabilis" of Sir Isaac Newton, 1666–1966,* ed. Robert Palter, 258–84. Cambridge, Mass.: MIT press, 1967.

————. "Newton's Metaphysics." In *Cambridge Companion,* ed. Cohen and Smith, 256–307.

Steinle, Friedrich. "From Principles to Regularities: Tracing 'Laws of Nature' in Early Modern France and England." In *Law, Nature, Laws of Nature: Historical Perspectives,* ed. Lorraine Daston and Michael Stolleis, forthcoming.

Sylla, Edith D. "Compounding Ratios: Bradwardine, Oresme, and the First Edition of Newton's *Principia.*" In *Transformation and Tradition in the Sciences,* ed. E. Mendelsohn, 11–43. Cambridge: Cambridge University Press, 1984.

————. "Galileo and the Oxford *Calculatores:* Analytical Languages and the Mean-Speed Theorem for Accelerated Motion." In *Reinterpreting Galileo,* ed. Wallace, 53–108.

Szabó, István. *Geschichte der mechanischen Prinzipien und ihrer wichtigsten Anwendungen.* Basel: Birkhäuser, 1977, 1987³.

Terrall, Mary. *The Man Who Flattened the Earth: Maupertuis and the Sciences in the Enlightenment.* Chicago: University of Chicago Press, 2002.

———. "*Vis viva* Revisited." *HS* 42 (2004), 189–209.

Topper, David. "Towards an Epistemology of Scientific Illustrations." In *Picturing Knowledge,* ed. Brian S. Baigrie, 215–49. Toronto: University of Toronto Press, 1996.

Torrini, Maurizio. *Dopo Galileo: Una polemica scientifica 1684–1711.* Florence: Olschki, 1979.

Truesdell, Clifford A. *Essays in the History of Mechanics.* Berlin: Springer, 1968.

———. *The Rational Mechanics of Flexible or Elastic Bodies, 1638–1788.* In Leonhard Euler, *Opera omnia,* ser. 2, vol. 11. Zurich: Orell Füssli, 1960.

Vailati, Ezio. *Leibniz and Clarke: A Study of Their Correspondence.* New York: Oxford University Press, 1997.

Valleriani, Matteo. "A View on Galileo's *Ricordi autografi.* Galileo Practitioner in Padua." In *Largo campo di filosofare,* ed. José Montesinos and Carlo Solís, 281–91. La Orotava: Fundación Canaria, 2001.

Valluri, See Ram; Curtis Wilson; and William Harper. "Newton's Apsidal Precession Theorem and Eccentric Orbits." *JHA* 28 (1997), 13–27.

van Helden, Albert. "The Birth of the Modern Scientific Instrument, 1550–1700." In *The Uses of Science,* ed. Burke, 49–84.

Verbeek, Theo. *Descartes and the Dutch: Early Reactions to Cartesian Philosophy, 1637–1650.* Carbondale: Southern Illinois University Press, 1992.

Vilain, Christiane. *La mécanique de Christian Huygens.* Paris: A. Blanchard, 1996.

———. "La question du 'centre d'oscillation' de 1660 à 1690." *Physis* 37 (2000), 21–51.

Voss, Stephen, ed. *Essays on the Philosophy and Science of René Descartes.* Oxford: Oxford University Press, 1993.

Wallace, William A. *Galileo and His Sources: The Heritage of the Collegio Romano in Galileo's Science.* Princeton: Princeton University Press, 1984.

———. *Prelude to Galileo.* Dordrecht: Reidel, 1981.

———, ed. *Reinterpreting Galileo.* Washington, D.C.: Catholic University of America Press, 1986.

Warner, Deborah J. "What Is a Scientific Instrument, When Did It Become One, and Why?" *BJHS* 23 (1990), 83–94.

Warwick, Andrew, *Masters of Theory: Cambridge and the Rise of Mathematical Physics.* Chicago: University of Chicago Press, 2003.

Webster, Charles. "The Discovery of Boyle's Law, and the Concept of the Elasticity of Air in the Seventeenth Century." *AHES* 2 (1965), 441–502.

Westfall, Richard S. "Floods along the Bisenzio: Science and Technology in the Age of Galileo." *Technology and Culture* 30 (1989), 879–907.

———. *Force in Newton's Physics: The Science of Dynamics in the Seventeenth Century.* London: MacDonald; New York: American Elsevier, 1971.

———. *Never at Rest: A Biography of Isaac Newton.* Cambridge: Cambridge University Press, 1980.

———. "Newton and the Fudge Factor." *Science* 179 (1973), 751–58.

———. "Robert Hooke, Mechanical Technology, and Scientific Investigation." In *The Uses of Science*, ed. Burke, 85–110

———. "Scientific Patronage: Galileo and the Telescope." *Isis* 76 (1985), 11–30.

Westman, Robert S. "The Astronomer's Role in the Sixteenth Century: A Preliminary Study." *HS* 18 (1980), 105–47.

———. "The Reception of Galileo's *Dialogue:* A Partial World Census of Extant Copies." In *Novitá celesti*, ed. Galluzzi, 329–71.

Whiteside, Derek T. "The Prehistory of the 'Principia' from 1664 to 1686." *NRRSL* 45 (1991), 11–61.

Willmoth, Frances. "Mathematical Sciences and Military Technology: The Ordnance Office in the Reign of Charles II." In *Renaissance and Revolution*, ed. Field and James, 117–31.

———. *Sir Jonas Moore: Practical Mathematics and Restoration Science*. Woodbridge and Rochester: Boydell Press, 1993.

Wilson, Curtis. *Astronomy from Kepler to Newton*. London: Variorum, 1989.

———. "D'Alembert versus Euler on the Precession of the Equinoxes and the Mechanics of Rigid Bodies." *AHES* 37 (1987), 233–73.

———. "From Kepler's Laws, So Called, to Universal Gravitation: Empirical Factors." *AEHS* 6 (1970), 89–170.

———. "From Kepler to Newton: Telling the Tale." In *The Foundations of Newtonian Scholarship*, ed. Richard H. Dalitz and Michael Nauenberg, 223–42. Singapore: World Scientific Publishing, 2000.

———. "Horrocks, Harmonies, and the Exactitude of Kepler's Third Law." *Studia Copernicana* 16 (1978), 235–59.

———. "Newton and the Eötvös Experiment." Annapolis: St. John's College Press, 1976. Reprinted in Wilson, *Astronomy*, 9:1–29.

———. "The Newtonian Achievement in Astronomy." In *The General History of Astronomy*, ed. René Taton and Curtis Wilson, 2:233–74. Cambridge: Cambridge University Press, 1989.

———. "Newton on the Moon's Variation and Apsidal Motion: The Need for a Newer 'New Analysis.'" In *Natural Philosophy*, ed. Buchwald and Cohen, 139–88.

———. "On the Origin of Horrocks's Lunar Theory." *JHA* 18 (1987), 77–94.

Wisan, Winifred L. "Galileo's Scientific Method: A reexamination." In *New Perspectives on Galileo*, ed. Robert E. Butts and Joseph C. Pitt, 1–57. Dordrecht: Reidel, 1978.

———. "The New Science of Motion: A Study of Galileo's *De motu locali.*" *AHES* 13 (1974), 103–306.

Wise, Norton. "Mediating Machines." *SIC* 2 (1988), 77–113.

Yoder, Joella G. *Unrolling Time*. Cambridge: Cambridge University Press, 1988.

Ziggelaar, August. "Aux origines de la théoriedes vibrations harmoniques: Le Père Ignace Gaston Pardies (1636–1673)." *Centaurus* 11 (1965), 145–51.

———. "Les premières démonstrations du tautochronisme de la cycloïde, et une conséquence pour la théorie de la vibration harmonique." *Centaurus* 12 (1967), 21–37.

action 296

no Bradwardine
would I want to write a book in this style?
   does this book make more unnecessary?
does he really prove the connection via object
   p 312, 314   map on p. 309 ??

315 emphasis on objects seems artificial or strained
   what if you look directly at the publications?

more detail in captions   what is his theory
                                               of scientific change

   descriptive   synthetic
      non argumentative

   objects - Exposition or Explanation?
      math - scale invariance as expt

   139 Descartes redrew the boundaries of mechanics

142 £25   rotation of rigid body vs. acceleration
              constrained   unconstrained

R   144   Descartes vs Galileo
             inexact status of mechanics

    156   importance of laws of impact

    187   use of proportions

*   214   theory incarnate vs thinking with objects

    220   capsule biog Newton
    222   centrifugal force dep on ref frame
    223. needs better description of conical pendulum
    288   impact of Galileo vs Newton
    277   math vs. physics
    304   Varignon
*   314   of Koyré - review
    315   vs Gabbey re limits